145 Isoperimetric Inequalities

This introduction treats the classical isoperimetric inequality in Euclidean space and contrasting rough inequalities in noncompact Riemannian manifolds. In Euclidean space the emphasis is on quantitative precision for very general domains, and in Riemannian manifolds the emphasis is on qualitative features of the inequality that provide insight into the coarse geometry at infinity of Riemannian manifolds.

The treatment in Euclidean space features a number of proofs of the classical inequality in increasing generality, providing in the process a transition from the methods of classical differential geometry to those of modern geometric measure theory; and the treatment in Riemannian manifolds features discretization techniques and applications to upper bounds of large time heat diffusion in Riemannian manifolds.

The result is an introduction to the rich tapestry of ideas and techniques of isoperimetric inequalities, a subject that has beginnings in classical antiquity and that continues to inspire fresh ideas in geometry and analysis to this very day – and beyond.

Isaac Chavel is Professor of Mathematics at the City College of The City University of New York. He received his Ph.D. in mathematics from Yeshiva University under the direction of Professor Harry E. Rauch. He has published in international journals in the areas of differential geometry and partial differential equations, especially the Laplace and heat operators on Riemannian manifolds. His other books include *Eigenvalues in Riemannian Geometry* and *Riemannian Geometry: A Modern Introduction.*

He has been teaching at the City College of The City University of New York since 1970, and he has been a member of the doctoral program of The City University of New York since 1976. He is a member of the American Mathematical Society.

145 Isoperimetric Inequalities

Isaac Chavel
The City University of New York

Isoperimetric Inequalities

Differential Geometric and Analytic Perspectives

CAMBRIDGE
UNIVERSITY PRESS

CAMBRIDGE UNIVERSITY PRESS
Cambridge, New York, Melbourne, Madrid, Cape Town,
Singapore, São Paulo, Delhi, Tokyo, Mexico City

Cambridge University Press
The Edinburgh Building, Cambridge CB2 8RU, UK

Published in the United States of America by Cambridge University Press, New York

www.cambridge.org
Information on this title: www.cambridge.org/9781107402270

First published 2001
First paperback edition 2011

A catalogue record for this publication is available from the British Library

Library of Congress Cataloguing in Publication data

Chavel, Isaac.

Isoperimetric inequalities : differential geometric and analytic perspectives / Isaac Chavel.

p. cm. – (Cambridge tracts in mathematics ; 145)

Includes bibliographical references and index.

ISBN 0-521-80267-9

1. Geometry, Differential. 2. Isoperimetric inequalities. 3. Riemannian manifolds.
I. Title. II. Series.

QA641 .C45 2001

516.3′6 – dc21 2001016177

ISBN 978-0-521-80267-3 Hardback
ISBN 978-1-107-40227-0 Paperback

Contents

Preface

This book discusses two venues of the isoperimetric inequality: (i) the sharp inequality in Euclidean space, with characterization of equality, and (ii) isoperimetric inequalities in Riemannian manifolds, where precise inequalities are unavailable but rough inequalities nevertheless yield qualitative global geometric information about the manifolds.

In Euclidean space, a variety of proofs are presented, each slightly more ambitious in its application to domains with irregular boundaries. One could easily go directly to the final definitive theorem and proof with little ado, but then one would miss the extraordinary wealth of approaches that exist to study the isoperimetric problem. An idea of the overwhelming variety of attack on this problem can be quickly gleaned from the fundamental treatise of Burago and Zalgaller (1988); and I have attempted on the one hand to capture some of that variety, and on the other hand to find a more leisurely studied approach that covers less material but with more detail.

In Riemannian manifolds, the treatment is guided by two motifs: (a) the dichotomy between the local Euclidean character of all Riemannian manifolds and the global geometric properties of Riemannian manifolds, this dichotomy pervading the study of nearly all differential geometry, and (b) the discretization of Riemannian manifolds possessing bounded geometry (some version of local uniformity). The dichotomy between local and global is expressed, in our context here, as the study of properties of Riemannian manifolds that remain invariant under the replacement of a compact subset of the manifold with another of different geometry and topology, as long as the new one fits smoothly in the manifold across the boundary of the deletion of the original compact subset. Thus, we do not seek fine results, in that we study coarse robust invariants that highlight the "geometry at infinity" of the manifold. Our choice of isoperimetric constants will even be invariant with respect to the discretization of the Riemannian manifold. The robust character of these new isoperimetric constants will then allow us to use this discretization to show how the geometry at infinity influences large time heat diffusion on Riemannian manifolds.

Regrettably, there is hardly any discussion of isoperimetric inequalities on compact Riemannian manifolds. That would fill a book – quite different from this one – all by itself.

* * *

A summary of the chapters goes as follows:

Chapter I starts with posing the isoperimetric problem in Euclidean space and gives some elementary arguments toward its solution in the Euclidean plane. These arguments are essentially a warm-up. They are followed by a summary of background definitions and results to be used later in the book. Thus the discussion of the isoperimetric problem, proper, begins in Chapter II.

Chapter II starts with uniqueness theory, under the assumption that the boundary of the solution domain is C^2. We first show that, if a domain Ω with C^2 boundary is a solution to the isoperimetric problem for domains with C^2 boundaries, Ω must be an open disk. Then we strengthen the result a bit – we show that if a domain is but an extremal for isoperimetric problems, then it must be a disk. Then we consider the existence of a solution to the isoperimetric problem for domains with C^1 boundaries. We give M. Gromov's argument that for such domains the disk constitutes a solution to the isoperimetric problem. But only if one restricts oneself to convex domains with C^1 boundaries does his argument imply that the disk is the unique solution.

Chapter III is the heart of the first half of the book. It expands the isoperimetric problem in that it considers all compacta and assigns the Minkowski area to each compact subset of Euclidean space to describe the size of the boundary. In this setting, using the Blaschke selection theorem and Steiner symmetrization, one shows that the closed disk constitutes a solution to the isoperimetric problem. Since the Minkowski area of a compact domain with C^1 boundary is the same as the differential geometric area of the boundary, the result extends the solution of the isoperimetric problem from the C^1 category to compacta. Moreover, one can use the traditional calculations to show that the disk is the unique solution to the isoperimetric problem in the C^1 category. But uniqueness in the more general collection of compacta is too difficult for such elementary arguments.

Then, in Chapter III, we recapture Steiner's original intuition that successive symmetrizations could be applied to any compact set to ultimately have it converge to a closed disk – in the topology of the Hausdorff metric on compact sets. We use this last argument to prove the isoperimetric inequality for compacta with finite perimeter. The perimeter, as a measure of the area of the boundary, seems to be an optimal general setting, since one can not only prove the isoperimetric inequality for compacta with finite perimeter, but can also characterize the case of equality.

In Chapter IV we introduce Hausdorff measure for subsets of Euclidean space and develop the story sufficiently far to prove that the perimeter of a Lipschitz domain in n-dimensional Euclidean space equals the $(n-1)$-dimensional Hausdorff measure of its boundary. The proof involves the *area formula,* for which we include a proof.

Chapter V begins a new view of isoperimetric inequalities, namely, rough inequalities in a Riemannian manifold. The goal of Chapters V–VIII is to show how these geometric isoperimetric inequalities influence the qualitative rate of decay, with respect to time, of heat diffusion in Riemannian manifolds.

In Chapter V we summarize the basic notions and results concerning isoperimetric inequalities in Riemannian manifolds, and in Chapter VI we give their implications for analytic Sobolev inequalities on Riemannian manifolds. Chapter V consists, almost entirely, of a summary of results from my *Riemannian Geometry: A Modern Introduction,* and I have included just those proofs that seemed to be important to the discussion here. The discussion of Sobolev inequalities in Chapter VI has received extensive treatment in other books, but our interest is restricted to those inequalities required for subsequent applications. Moreover, we have also treated the relation of Sobolev inequalities on Riemannian manifolds and their discretizations, one to the other. To my knowledge, this has yet to be treated systematically in book form.

Chapter VII introduces the Laplacian and the heat operator on Riemannian manifolds and is devoted to setting the stage for the "main problem" in large time heat diffusion; its formulation and solution are presented in Chapter VIII. The book ends with an introduction to the new arguments of A.A. Grigor'yan, the full possibilities of which have only begun to be realized.

* * *

I have attempted to strike the right balance between merely summarizing background material (of which there is quite a bit) and developing preparatory arguments in the text. Also, although I have summarized the necessary basic definitions and results from Riemannian geometry at the beginning of Chapter V, I occasionally require some of that material in earlier chapters, and I use it as though the reader already knows it. This seems the lesser of two evils, the other evil being to disrupt the flow of the arguments in the first half of the book for an excursus that would have to be repeated in its proper context later. Most of that material is quite elementary and standard, so it should not cause any major problems.

In order to clarify somewhat the relation between material quoted and material presented with proofs, I have referred to every result that either is an exercise or that relies on a treatment outside this book as a proposition, and every result proven in the book as a theorem. This is admittedly quite artificial and obviously gives rise to some strange effects, in that the titles *proposition* and *theorem* are often (if not usually) used to indicate the relative significance of the results discussed. That is not the case here.

There are bibliographic notes at the end of each chapter. They are intended to give the reader some guidance to the background material, and to give but an introduction to a definitive study of the literature.

It is a pleasure to thank the many people with whom I have been associated in the study of geometry since I first came to the City College of CUNY in 1970: first and foremost, Edgar A. Feldman and the other geometers of the City University of New York – J. Dodziuk, L. Karp, B. Randol, R. Sacksteder, and J. Velling. Also, I have benefited through the years from the friendship and mathematics of I. Benjamini, M. van den Berg, P. Buser, E. B. Davies, J. Eels, D. Elworthy, A. A. Grigor'yan, E. Hsu, W. S. Kendall, F. Morgan, R. Osserman, M. Pinsky and D. Sullivan. But, as is well known, any mistakes herein are all mine.

The isoperimetric problem has been a source of mathematical ideas and techniques since its formulation in classical antiquity, and it is still alive and well in its ability to both capture and nourish the mathematical imagination. This book only covers a small portion of the subject; nonetheless, I hope the presentation gives expression to some of its beauty and inspiration.

I
Introduction

In this chapter we introduce the subject. We describe the classical isoperimetric problem in Euclidean space of all dimensions, and give some elementary arguments that work in the plane. Only one approach will carry over to higher dimensions, namely, the necessary condition established by classical calculus of variations, that a domain with C^2 boundary provides a solution to the isoperimetric problem only if it is a disk. Then we give a recent proof of the isoperimetric inequality in the plane by P. Topping (which does not include a characterization of equality), and the classical argument of A. Hurwitz to prove the isoperimetric inequality using Fourier series. This is followed by a symmetry and convexity argument in the plane for very general boundaries that proves the isoperimetric inequality, *if* one assumes in advance that the *isoperimetric functional* $D \mapsto L^2(\partial D)/A(D)$ has a minimizer. (So this is a weak version – if the isoperimetric problem has a solution, then the disk is also a solution.) Finally, we present the background necessary for what follows later in our general discussion, valid for all dimensions. The subsections of §I.3 include a proof of H. Rademacher's theorem on the almost everywhere differentiability of Lipschitz functions, and a proof of the general co-area formula for C^1 mappings of Riemannian manifolds. We obtain the usual co-area formula, as well as an easy consequence: Cauchy's formula for the area of the boundary of a convex subset of $\mathbb{R}^n$ with C^1 boundary.

I.1 The Isoperimetric Problem

Given any bounded domain on the real line (that is, an open interval), the discrete measure of its boundary (the endpoints of the interval) is 2. And given any bounded open subset of the line, the discrete measure of its boundary is greater than or equal to 2, with equality if and only if the open set consists of one open interval. This is the statement of the isoperimetric inequality on the line.

In the plane, one has three common formulations of the *isoperimetric problem*:

1. Consider all bounded domains in $\mathbb{R}^2$ with fixed given perimeter, length of the boundary (that is, all domains under consideration are *isoperimetric*). Find the domain that contains the greatest area. The answer, of course, will be the disk. Note that the specific value of the perimeter in question is of no interest, because all domains of perimeter L_1 are mapped by a similarity of $\mathbb{R}^2$ to all domains with perimeter L_2 for any given values of L_1, L_2, and the image under the similarity of an area maximizer for L_1 is an area maximizer for L_2.
2. One insists on a common area of all bounded domains under consideration, and asks how to minimize the perimeter.
3. Lastly, one expresses the problem as an analytic inequality, namely, since we know exactly the values of the area of the disk and the length of its boundary, the isoperimetric problem is then expressed as proving the *isoperimetric inequality*

   $$L^2 \geq 4\pi A, \tag{I.1.1}$$

   where A denotes the area of the domain under consideration, and L denotes the length of its boundary. The inequality is extremely convenient, in that it remains invariant under similarities of $\mathbb{R}^2$, and one has equality if the domain is a disk. One wishes to show that the inequality is always true, with equality if and only if the domain is a disk.

One can consider the above for any $\mathbb{R}^n$, $n \geq 2$. The proposed analytic isoperimetric inequality then becomes

$$\frac{A(\partial\Omega)}{V(\Omega)^{1-1/n}} \geq \frac{A(\mathbb{S}^{n-1})}{V(\mathbb{B}^n)^{1-1/n}}, \tag{I.1.2}$$

where Ω is any bounded domain in $\mathbb{R}^n$ and $\partial\Omega$ its boundary, V denotes n-measure and A denotes $(n-1)$-measure, $\mathbb{B}^n$ is the unit disk in $\mathbb{R}^n$, and $\mathbb{S}^{n-1}$ the unit sphere in $\mathbb{R}^n$. We let $\boldsymbol{\omega}_\mathbf{n}$ denote the n-dimensional volume of $\mathbb{B}^n$ and $\mathbf{c}_{\mathbf{n}-1}$ the $(n-1)$-dimensional surface area of $\mathbb{S}^{n-1}$. It is standard that

$$\mathbf{c}_{\mathbf{n}-1} = \frac{2\pi^{n/2}}{\Gamma(n/2)}, \qquad \boldsymbol{\omega}_\mathbf{n} = \frac{\mathbf{c}_{\mathbf{n}-1}}{n}, \tag{I.1.3}$$

where $\Gamma(x)$ denotes the classical gamma function; and (I.1.2) now reads as

$$\frac{A(\partial\Omega)}{V(\Omega)^{1-1/n}} \geq n{\boldsymbol{\omega}_\mathbf{n}}^{1/n}. \tag{I.1.4}$$

One wants to prove the inequality and to show that equality is achieved if and only if Ω is an n-disk. Note that for $n = 2$ we took in (I.1.4) the square root of (I.1.1).

Remark I.1.1 Throughout the book, *domain* will refer to a connected open set. In general, we consider the isoperimetric problem for relatively compact domains when we are working in the differential geometric setting (Chapters I, II, V–VIII). Therefore, the disks that realize the solution in $\mathbb{R}^n$ are open. In Chapters III and IV, where we work in a more general setting, the isoperimetric problem is considered for compacta. In that setting the disks that realize the solution in $\mathbb{R}^n$ are closed.

Remark I.1.2 We have restricted the isoperimetric problem to domains in $\mathbb{R}^n$; but if we could solve this problem, then the isoperimetric problem for open sets consisting of finitely many bounded domains would easily follow from the solution for single domains. Indeed, assume one has the inequality (I.1.2) for domains in $\mathbb{R}^n$. If

$$\Omega = \Omega_1 \cup \Omega_2 \cup \cdots,$$

where each Ω_j is a relatively compact domain in $\mathbb{R}^n$ such that

$$\operatorname{cl} \Omega_j \cap \operatorname{cl} \Omega_k = \emptyset \qquad \forall\, j \neq k$$

(cl denotes the closure), then Minkowski's inequality implies

$$\begin{aligned} V(\Omega)^{1-1/n} &\leq \sum_j V(\Omega_j)^{1-1/n} \leq \frac{1}{n{\omega_{\mathbf{n}}}^{1/n}} \sum_j A(\partial\Omega_j) \\ &= \frac{1}{n{\omega_{\mathbf{n}}}^{1/n}} A(\partial\Omega). \end{aligned} \tag{I.1.5}$$

So the inequality extends to the union of domains. Note that equality implies that Ω is a domain.

Remark I.1.3 Note that for any domain Ω in $\mathbb{R}^n$, its volume is the n-dimensional Lebesgue measure, and if $\partial\Omega$ is C^1 then the area of $\partial\Omega$ is given by the standard differential geometric surface area of a smooth hypersurface in $\mathbb{R}^n$. However, if $\partial\Omega$ is not smooth, then one must propose an area functional defined on a collection of domains such that the area functional will give a working definition of the area of the boundaries of the domains. Besides a number of natural properties [see the discussions in Burago and Zalgaller (1988)], one requires that the new definition agree with the differential geometric one when applied to a domain with smooth boundary. Then, with this new collection of domains and definition of the area of their boundaries, one wishes to prove the isoperimeric inequality. Also, one wishes to characterize the case of equality in each of these settings.

Remark I.1.4 As soon as one expands the problem to the model spaces of constant sectional curvature, that is, to spheres and hyperbolic spaces, one has no self-similarities of the Riemannian spaces in question. And if the disks on the right hand side of (I.1.2) are to have radius r, then the right hand side of the inequality in (I.1.2) is no longer independent of the value of r. Nonetheless, one still has the isoperimetric inequality in the sense that all domains in question with the same n-volume have the $(n-1)$-area of their boundaries minimized by disks. For $n = 2$, the analytic formulation reads as follows: If $M = \mathbb{M}^2_\kappa$, the model space with constant curvature κ, then the isoperimetric inequality becomes

$$L^2 \geq 4\pi A - \kappa A^2, \tag{I.1.6}$$

with equality if and only if the domain in question is a disk. Of course, one can still consider the isoperimetric problem, whether or not it is to be expressed as an inequality, in the first or second formulation above.

Similarly, one can extend the isoperimetric problem and associated inequalities to surfaces, or, more generally, to Riemannian manifolds. We shall consider such inequalities in Chapter V.

Remark I.1.5 Finally, one can consider a *Bonnesen inequality*. In $\mathbb{R}^2$, such an inequality is of the form

$$L^2 - 4\pi A \geq B \geq 0,$$

where B is a nonnegative geometric quantity associated with the domain that vanishes if and only if the domain is a disk.

I.2 The Isoperimetric Inequality in the Plane

For any C^2 path $\omega : (\alpha, \beta) \to \mathbb{R}^2$ in the plane, the velocity vector field of ω is given by its derivative ω', and acceleration vector field by ω''. We assume that ω is an immersion, that is, ω' never vanishes. The infinitesimal element of arc length ds is given by

$$ds = |\omega'(t)|\, dt.$$

Given any $t_0 \in (\alpha, \beta)$, the *arc length function* of ω based at t_0 is given by

$$s(t) = \int_{t_0}^{t} |\omega'(\tau)|\, d\tau.$$

Let

$$\mathbf{T}(t) = \frac{\omega'(t)}{|\omega'(t)|}$$

denote the unit tangent vector field along ω,

$$\iota : \mathbb{R}^2 \to \mathbb{R}^2$$

the rotation of $\mathbb{R}^2$ by $\pi/2$ radians, and

$$\mathbf{N} = \iota\,\mathbf{T}$$

the oriented unit normal vector field along ω. Then one defines the *curvature* κ of ω by

$$\frac{d\mathbf{T}}{ds} = \kappa \mathbf{N} \tag{I.2.1}$$

(indeed, since $\mathbf{T}$ is a unit vector field, its derivative must be perpendicular to itself). Then the formula for the curvature, relative to the original path, is given by

$$\kappa = \frac{d\mathbf{T}}{ds} \cdot \mathbf{N} = \frac{\omega'' \cdot \iota\omega'}{|\omega'|^3}.$$

One can easily show that

$$\frac{d\mathbf{N}}{ds} = -\kappa \mathbf{T}. \tag{I.2.2}$$

The equations (I.2.1) and (I.2.2) are referred to as the *Frenet formulae.*

One can prove, from (I.2.1), that if the curvature κ is constant, then ω is an arc on a circle (if not the complete circle).

I.2.1 Uniqueness for Smooth Boundaries

As a warm-up, we give the argument from classical calculus of variations. Given the area A, let D vary over relatively compact domains in the plane of area A, with C^1 boundary, and suppose the domain Ω, $\partial\Omega \in C^2$, realizes the minimal boundary length among all such domains D. We claim that Ω is a disk.

Proof Since Ω is relatively compact in $\mathbb{R}^2$, there exists a simply connected domain Ω_0 such that

$$\Omega = \Omega_0 \setminus \{\text{finite disjoint union of closed topological disks}\}.$$

We claim that since Ω is a minimizer, then $\Omega_0 = \Omega$. If not, we may add the topological disks to Ω, which will increase the area of the domain and decrease the length of the boundary, and therefore Ω will not be a minimizer. Thus $\Omega_0 = \Omega$, and is bounded by an imbedded circle.

Let $\Gamma : \mathbb{S}^1 \to \mathbb{R}^2 \in C^2$ be the imbedding of the boundary of Ω. We always assume that the path Γ is oriented so that $\boldsymbol{\nu} = -\mathbf{N}$ at all points of Γ, where $\boldsymbol{\nu}$ is the unit normal exterior vector field along $\partial\Omega$.

One then considers a 1-parameter family $\Gamma_\epsilon : \mathbb{S}^1 \to \mathbb{R}^2$ of imbeddings

$$v : (-\epsilon_0, \epsilon_0) \times \mathbb{S}^1 \to \mathbb{R}^2,$$

such that the variation function $v(\epsilon, t)$ given by

$$v(\epsilon, t) = \Gamma_\epsilon(t) = \Gamma(t) + \Psi(\epsilon, t)\boldsymbol{\nu}(t), \qquad \Psi(0, t) = 0,$$

is C^1. First,

$$\frac{\partial v}{\partial \epsilon} = \frac{\partial \Psi}{\partial \epsilon}\boldsymbol{\nu}.$$

Also

$$\frac{\partial v}{\partial t} = \Gamma' + \left\{\frac{\partial \Psi}{\partial t}\boldsymbol{\nu} + \Psi\boldsymbol{\nu}'\right\} = \{1 + \kappa\Psi\}\,\Gamma' + \frac{\partial \Psi}{\partial t}\boldsymbol{\nu},$$

which implies

$$\left|\frac{\partial v}{\partial t}\right| = \left\{(1 + \kappa\Psi)^2 + \frac{1}{|\Gamma'|^2}\left(\frac{\partial \Psi}{\partial t}\right)^2\right\}^{1/2} |\Gamma'|.$$

Taylor's theorem implies, for

$$\phi(t) := \left.\frac{\partial \Psi}{\partial \epsilon}\right|_{\epsilon=0},$$

the expansion

$$\Psi(\epsilon, t) = \epsilon\phi(t) + o(\epsilon), \qquad \frac{\partial \Psi}{\partial \epsilon} = \phi(t) + o(1), \qquad \frac{\partial \Psi}{\partial t} = O(\epsilon),$$

which implies

$$\left|\frac{\partial v}{\partial t}\right| = |\Gamma'|\,\{1 + \epsilon\kappa\phi + o(\epsilon)\}.$$

Therefore, the area element dA in the curvilinear coordinates (t, ϵ) is given by

$$dA = \left|\frac{\partial v}{\partial \epsilon} \times \frac{\partial v}{\partial t}\right| d\epsilon\, dt = \phi|\Gamma'|\,\{1 + o(1)\}\, d\epsilon\, dt = \{\phi + o(1)\}\, d\epsilon\, ds.$$

For the domain Ω_ϵ determined by Γ_ϵ we have, for sufficiently small ϵ,

$$A(\Omega_\epsilon) - A(\Omega) = \int_0^\epsilon d\sigma \int_\Gamma \{\phi + o(1)\}\, ds.$$

Therefore, if $A(\Omega_\epsilon) = A(\Omega)$ for all ϵ, then

$$\int_\Gamma \phi\, ds = 0.$$

Now let $L(\epsilon)$ denote the length of Γ_ϵ. Since Γ has the shortest length, we have $L'(0) = 0$. Therefore, because

$$L(\epsilon) = \int_{\mathbb{S}^1} \left|\frac{\partial v}{\partial t}\right| dt = \int_{\mathbb{S}^1} |\Gamma'|\, \{1 + \epsilon\kappa\phi + o(\epsilon)\}\, dt = \int_\Gamma \{1 + \epsilon\kappa\phi + o(\epsilon)\}\, ds,$$

we have

$$0 = L'(0) = \int_\Gamma \kappa\phi\, ds, \qquad \int_\Gamma \phi\, ds = 0$$

for any such variation of Γ.

Similarly, given any $\phi \in C^1$ such that $\int_\Gamma \phi\, ds = 0$, there exists a variation v of Γ such that $A(\Omega_\epsilon) = A(\Omega)$ for all ϵ, and $L'(0) = \int_\Gamma \kappa\phi\, ds$. Then, by assumption, we have

$$\int_\Gamma \kappa\phi\, ds = 0 \qquad \forall \phi \in C^1 : \int_\Gamma \phi\, ds = 0.$$

To show that this implies that κ is constant, we argue as follows: Given *any* $\psi : \mathbb{S}^1 \to \mathbb{R}$ in C^1, set

$$\phi = \psi - \int_\Gamma \psi\, ds \Big/ \int_\Gamma ds\ .$$

Then $\int_\Gamma \phi\, ds = 0$, which implies

$$0 = \int_\Gamma \kappa \left(\psi - \frac{1}{L}\int_\Gamma \psi\, ds\right) ds = \int_\Gamma \left(\kappa - \frac{1}{L}\int_\Gamma \kappa\, ds\right) \psi\, ds,$$

where L denotes the length of Γ. Since ψ is arbitrary C^1, a standard argument then implies that

$$\kappa - \frac{1}{L}\int_\Gamma \kappa\, ds = 0, \tag{I.2.3}$$

that is, the curvature κ is constant. Then, as mentioned, (I.2.1) implies that Γ is a circle.

I.2.2 Quick Proof Using Complex Variables

Theorem I.2.1 (Isoperimetric Inequality in $\mathbb{R}^2$) *Let Ω be a relatively compact domain, with boundary $\partial\Omega \in C^1$ consisting of one component. Then*

$$L^2(\partial\Omega) \geq 4\pi A(\Omega).$$

Proof We denote any element of the plane as the complex number

$$z = x + iy,$$

and the area measure as an oriented volume element; so

$$dA = dx \wedge dy = \frac{i}{2} dz \wedge d\bar{z}.$$

Then

$$\begin{aligned} 4\pi A(\Omega) &= \iint_\Omega 2\pi i \, dz \wedge d\bar{z} \\ &= \iint_\Omega dz \wedge d\bar{z} \int_{\partial\Omega} \frac{d\zeta}{\zeta - z} \\ &= \int_{\partial\Omega} d\zeta \iint_\Omega \frac{dz \wedge d\bar{z}}{\zeta - z} \\ &= \int_{\partial\Omega} d\zeta \int_{\partial\Omega} \frac{\bar{\zeta} - \bar{z}}{\zeta - z} dz \\ &\leq L^2(\partial\Omega), \end{aligned}$$

– the second equality follows from the fact that the winding number of $\partial\Omega$ about any point $z \in \Omega$ is 1; the last equality follows from Green's theorem – which implies the claim. ■

I.2.3 The Method of Fourier Series

Lemma I.2.1 (Wirtinger's Inequality) *If f is a C^1, L-periodic function on $\mathbb{R}$, and*

$$\int_0^L f(t)\, dt = 0,$$

then

$$\int_0^L |f'|^2(t)\, dt \geq \frac{4\pi^2}{L^2} \int_0^L |f|^2(t)\, dt,$$

with equality if and only if there exist constants a_{-1} and a_1 such that

$$f(t) = a_{-1} e^{-2\pi i t/L} + a_1 e^{2\pi i t/L}.$$

Proof This is an exercise from Fourier series. The function $f(t)$ admits a Fourier expansion

$$f(t) \leftrightarrow \sum_{k=-\infty}^{\infty} a_k e^{2\pi ikt/L} \quad \text{with} \quad a_k = \frac{1}{L}\int_0^L f(t)e^{-2\pi ikt/L}\,dt.$$

Similarly, we have

$$f'(t) \leftrightarrow \sum_{k=-\infty}^{\infty} b_k e^{2\pi ikt/L} \quad \text{with} \quad b_k = \frac{1}{L}\int_0^L f'(t)e^{-2\pi ikt/L}\,dt.$$

The continuity of f implies $b_0 = 0$, and the hypothesis implies $a_0 = 0$. Integration by parts implies

$$b_k = \frac{2\pi ik}{L} a_k \quad \forall\ |k| \geq 1.$$

Parseval's inequality then implies

$$\begin{aligned}
\int_0^L |f'|^2\,dt = L\sum_{k\neq 0} |b_k|^2 &= L\frac{4\pi^2}{L^2}\sum_{k\neq 0} k^2|a_k|^2 \\
&\geq L\frac{4\pi^2}{L^2}\sum_{k\neq 0} |a_k|^2 \\
&= \frac{4\pi^2}{L^2}\int_0^L |f|^2\,dt,
\end{aligned}$$

which implies the inequality. One has equality if and only if $a_k = 0$ for all $|k| > 1$. ■

Theorem I.2.2 (Isoperimetric Inequality in $\mathbb{R}^2$) *If Ω is a relatively compact domain in $\mathbb{R}^2$, with C^1 boundary consisting of one component, then*

$$L^2(\partial\Omega) \geq 4\pi A(\Omega),$$

with equality if and only if Ω is a disk.

Proof If necessary, we translate Ω to guarantee

$$\int_{\partial\Omega} x\,ds - 0, \qquad x = (x^1, x^2).$$

Let $\mathbf{x} = x^1\mathbf{e}_1 + x^2\mathbf{e}_2$ be the vector field on $\mathbb{R}^2$ with base point $x = (x^1, x^2)$.

One now uses the 2-dimensional divergence theorem, namely, for any vector field $x \mapsto \boldsymbol{\xi}(x) \in \mathbb{R}^2$ with support containing cl Ω, one has

$$\iint_{\Omega} \operatorname{div} \boldsymbol{\xi}\,dA = \int_{\partial\Omega} \boldsymbol{\xi}\cdot\boldsymbol{\nu}\,ds, \tag{I.2.4}$$

where $\boldsymbol{\nu}$ denotes the outward unit normal vector field along $\partial\Omega$. One can obtain the formula (I.2.4) by converting the traditional Green's theorem

$$\iint_\Omega \left\{ \frac{\partial Q}{\partial x} - \frac{\partial P}{\partial y} \right\} dA = \int_{\partial\Omega} P\,dx + Q\,dy, \tag{I.2.5}$$

by choosing

$$P = -\xi^2, \quad Q = \xi^1, \qquad \boldsymbol{\xi} = \xi^1 \mathbf{e}_1 + \xi^2 \mathbf{e}_2.$$

For the left hand side of (I.2.5) one has

$$\frac{\partial Q}{\partial x} - \frac{\partial P}{\partial y} = \frac{\partial \xi^1}{\partial x^1} + \frac{\partial \xi^2}{\partial x^2} = \operatorname{div} \boldsymbol{\xi}.$$

For the right hand side of (I.2.5) one has

$$\begin{aligned} P\,dx + Q\,dy &= -\xi^2\,dx^1 + \xi^1\,dx^2 = \boldsymbol{\xi}\cdot\{dx^2\mathbf{e}_1 - dx^1\mathbf{e}_2\} \\ &= \boldsymbol{\xi}\cdot\{-\iota(d\mathbf{x})\} = \boldsymbol{\xi}\cdot\boldsymbol{\nu}\,ds. \end{aligned}$$

For our vector field $\mathbf{x}$, we have $\operatorname{div}\mathbf{x} = 2$ on all Ω. Then the divergence theorem implies

$$2A(\Omega) = \int_\Omega \operatorname{div}\mathbf{x}\,dA = \int_{\partial\Omega} \mathbf{x}\cdot\boldsymbol{\nu}\,ds,$$

which implies

$$\begin{aligned} 2A(\Omega) &= \int_{\partial\Omega} \mathbf{x}\cdot\boldsymbol{\nu}\,ds \\ &\leq \int_{\partial\Omega} |\mathbf{x}|\,ds \\ &\leq \left\{ \int_{\partial\Omega} |\mathbf{x}|^2\,ds \right\}^{1/2} \left\{ \int_{\partial\Omega} 1^2\,ds \right\}^{1/2} \\ &= L^{1/2}(\partial\Omega) \left\{ \int_{\partial\Omega} |\mathbf{x}|^2\,ds \right\}^{1/2} \end{aligned}$$

– the first inequality is the vector Cauchy–Schwarz inequality, and the second inequality is the integral Cauchy–Schwarz inequality.

Parametrize $\partial\Omega$ with respect to arc length. Note that

$$|\mathbf{x}|^2 = (x^1)^2 + (x^2)^2, \qquad \left|\frac{d\mathbf{x}}{ds}\right|^2 = \left(\frac{dx^1}{ds}\right)^2 + \left(\frac{dx^2}{ds}\right)^2$$

along $\partial\Omega$; so Wirtinger's inequality, applied to each coordinate function $x^1(s)$, $x^2(s)$, implies

$$\begin{aligned} 2A(\Omega) &\le L^{1/2}(\partial\Omega)\left\{\int_{\partial\Omega} |\mathbf{x}|^2\, ds\right\}^{1/2} \\ &\le L^{1/2}(\partial\Omega)\left\{\frac{L^2(\partial\Omega)}{4\pi^2}\int_{\partial\Omega} |\mathbf{x}'|^2\, ds\right\}^{1/2} \\ &= \frac{L^2(\partial\Omega)}{2\pi}, \end{aligned}$$

which is the desired inequality. The case of equality follows easily. ■

I.2.4 A Symmetry-and-Convexity Argument

Theorem I.2.3 *Consider the isoperimetric problem for all relatively compact domains in $\mathbb{R}^2$ with piecewise C^1 boundaries, and assume the isoperimetric problem in $\mathbb{R}^2$ has at least one solution, that is, there is a domain Ω realizing the minimum of the isoperimetric functional*

$$D \mapsto \frac{L^2(\partial D)}{A(D)}.$$

Then the open 2-disk also minimizes the isoperimetric functional.

Remark I.2.1 Thus, we are only proving a weak result, namely, the analytic isoperimetric inequality (I.1.1) is valid if the isoperimetric functional has a minimizer.

Although the arguments from complex variables and from Fourier series given above are also valid when the boundaries are only piecewise C^1, here the proof *requires* that the variational problem be defined over the extended class of domains.

Proof Let Ω be a minimizer. Pick a line Π that divides Ω into two open sets Ω_1, Ω_2 of equal area. Then

$$L(\partial\Omega \cap \operatorname{cl}\Omega_1) = L(\partial\Omega \cap \operatorname{cl}\Omega_2).$$

If not, and $L(\partial\Omega \cap \operatorname{cl}\Omega_1) < L(\partial\Omega \cap \operatorname{cl}\Omega_2)$, then one could replace Ω by Ω', consisting of Ω_1 union its reflection in Π, which would imply that Ω is not a minimizer, which is a contradiction. So we may replace Ω by a domain Ω^* that is symmetric with respect to the line Π.

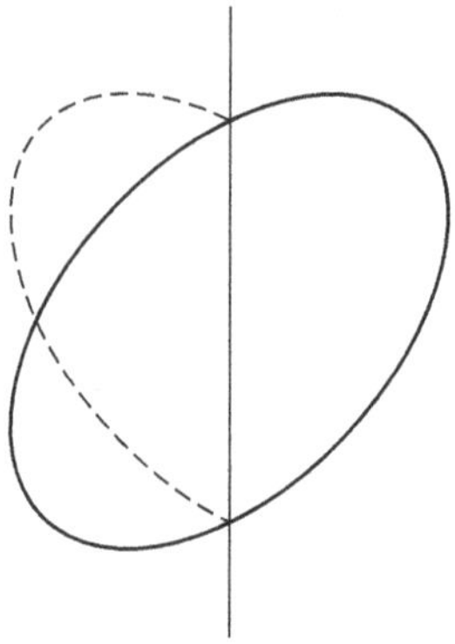

Figure I.2.1: The line through the origin intersecting $\partial\Omega$.

By carrying out this argument successively, one may assume that there exists a coordinate system in $\mathbb{R}^2$ such that Ω is symmetric with respect to the two coordinate axes. This implies that every line through the origin divides Ω into two open sets of equal area and, therefore (by the above argument), equal bounding length.

Next, we may assume $\mathrm{cl}\,\Omega$ is convex. If not, we may replace it by its convex hull, which increases the area and decreases the bounding length. Since the origin is in Ω, we may think of $\partial\Omega$ as the image of a piecwise C^1 map defined on the unit circle $\mathbb{S}^1$ in $\mathbb{R}^2$.

Let $w_0 \in \partial\Omega$ have a tangent line, not equal to the line from the origin through w_0. (See Figure I.2.1.) We claim that the line from the origin to w_0 must intersect $\partial\Omega$ perpendicular to the tangent line at w_0. If not, it will be possible to carry out the construction $\Omega \mapsto \Omega^*$ described in the first paragraph above so that $\mathrm{cl}\,\Omega^*$ is nonconvex, which is a contradiction.

Therefore, Ω is convex with every point of differentiablity of $\partial\Omega$ having its tangent line orthogonal to the position vector. Let θ denote the local coordinate on $\mathbb{S}^1$, $\mathbf{x}$ the position vector on $\partial\Omega$, and $\mathbf{n}$ the position vector on $\mathbb{S}^1$. Then

$$\mathbf{x} = \tau\mathbf{n}$$

at all points of differentiability of $\partial\Omega$, which implies

$$\frac{\partial\mathbf{x}}{\partial\theta} = \frac{\partial\tau}{\partial\theta}\mathbf{n} + \tau\frac{\partial\mathbf{n}}{\partial\theta}.$$

Because $\mathbf{n}$ is perpendicular to the tangent space, we conclude that τ is differentiable, with $\partial\tau/\partial\theta = 0$, at all points of differentiability of $\partial\Omega$, which implies τ is locally constant on the C^1 arcs of $\partial\Omega$. Thus $\partial\Omega$ consists of a finite number of circular arcs connected by segments of lines that (the lines, not the segments) pass through the origin. The convexity of Ω implies $\partial\Omega$ is a circle. ■

Remark I.2.2 In higher dimensions one has to be careful about the convexity. For example, it is not true that passing to the convex hull of a domain decreases its isoperimetric quotient. Pick a standard imbedded solid torus in $\mathbb{R}^3$, where the circle in the xy-plane has radius 1, and the rotated circle perpendicular to the xy-plane has radius $\epsilon \ll 1$. Then its isoperimetric quotient is asymptotic to const.$\epsilon^{-1/3}$ as $\epsilon \downarrow 0$. On the other hand, the convex hull has isoperimetric quotient asymptotic to const.$\epsilon^{-2/3}$ as $\epsilon \downarrow 0$.

I.3 Preliminaries

We describe here most of the background required for the rest of these notes. The Riemannian geometric background is discussed in Chapter V.

Notation Let X be any space, A a subset of X. We let $\mathcal{I}_A$ denote the *indicator* (or *characteristic*) *function* of A, that is, $\mathcal{I}_A(x) = 1$ for $x \in A$, and $\mathcal{I}_A(x) = 0$ for $x \in X \setminus A$. The cardinality of A is denoted by card A.

Notation In any topological space X we let supp f denote the support of a function f on X. We denote continuous functions of compact support by $C_c(X)$. When the target is some vector space V, we write $C_c(X; V)$.

Notation Whenever discussing a measure space $(X, \mathfrak{M}, d\mu)$, we let $\| \ \|_p$ denote the associated L^p norm. (In general, the σ-algebra $\mathfrak{M}$ is fixed, so we rarely mention it.) If there is doubt as to which space or measure is intended, then we also indicate our choice in the subscript: $\| \ \|_{p,X}$ or $\| \ \|_{p,d\mu}$, as required. In general we write $\| \ \| = \| \ \|_2$. We also indicate the L^2 inner product by $(\ , \ )$. Unless otherwise indicated, these are spaces of real-valued functions.

Proposition I.3.1 *If the sequence (f_j) converges to f in $L^p(\mu)$ for some $p \in [1, +\infty]$, then (f_j) has a subsequence that converges pointwise to f a.e.-$[d\mu]$.*

Proposition I.3.2 *Let X be a complete, separable metric space, Y a Hausdorff space, $f : X \to Y$ a continuous map, and μ a measure on Y such that every closed subset of Y is μ-measurable. Then $f(B)$ is μ-measurable for every Borel subset B of X.*

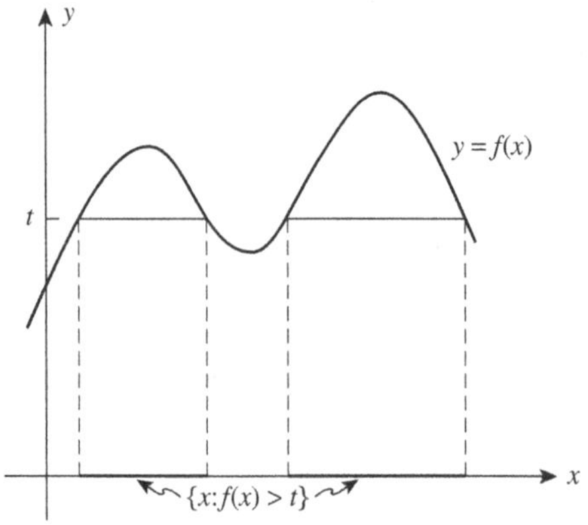

Figure I.3.1: Equation (I.3.1) when $p = 1$, $d\mu(x) = dx$.

Proposition I.3.3 (Cavalieri's Principle) *Let ν be a measure on Borel sets in $[0, \infty)$, ϕ its indefinite integral, given by*

$$\phi(t) = \nu([0, t)) < +\infty \qquad \forall\, t > 0,$$

(Ω, Σ, μ) a measure space, and f a nonnegative Σ-measurable function on Ω. Then

$$\int_\Omega \phi(f(x))\, d\mu(x) = \int_0^\infty \mu(\{f > t\})\, d\nu(t),$$

or equivalently,

$$\int_\Omega d\mu(x) \int_0^{f(x)} d\nu(t) = \int_0^\infty d\nu(t) \int_\Omega \mathcal{I}_{\{f>t\}}\, d\mu.$$

In particular, if $d\nu(t) = pt^{p-1}\, dt$, $p > 0$, then we have (see Figure I.3.1)

$$\int_\Omega f^p\, d\mu = p \int_0^\infty t^{p-1}\, dt \int_\Omega \mathcal{I}_{\{f>t\}}\, d\mu. \tag{I.3.1}$$

Thus, if $p = 1$ and $\mu = \delta_x$ then

$$f(x) = \int_0^\infty \mathcal{I}_{\{f>t\}}(x)\, dt.$$

Notation Let (X, d) be a metric space. For any subset A in X, let

int A denote the *interior of* A,
ext $A = \operatorname{int}(X \setminus A)$ denote the *exterior of* A,
cl A or the traditional $\overline{A}$ denote the *closure of* A,

∂A denote the *boundary of* A,
diam A denote the *diameter of* A,
$[A]_\epsilon = \{x \in X : d(x, A) \le \epsilon\}$ denote the *ϵ-thickening of* A.

For any $x \in X$ and $r > 0$, let $B(x;r)$ denote the *open metric disk* and $D(x;r)$ the *closed metric disk in* X*, with center* x *and radius* r. We denote the *metric sphere bounding* $B(x;r)$ by $S(x;r)$.

We consider n-dimensional Euclidean space $\mathbb{R}^n$, $n \ge 1$, where we denote arbitrary points by $x = (x^1, \ldots, x^n)$, $y = (y^1, \ldots, y^n), \ldots \in \mathbb{R}^n$. The inner product is denoted by $x, y \mapsto x \cdot y$. The length of an element x is given by $|x| = \sqrt{x \cdot x}$, and the distance between x and y is given by $d(x, y) = |x - y|$.

In $\mathbb{R}^n$, we write $\mathbb{B}^n(x;r)$ for $B(x;r)$ and $\mathbb{D}^n(x;r)$ for $D(x;r)$. We omit the superscript n when the context is sufficiently clear. We also write $\mathbb{S}^{n-1}(x;r)$ for $S(x;r)$, and drop the superscript when no problem will caused by doing so. When x is the origin $\mathbf{o}$ we write

$$\mathbb{B}^n(r) = \mathbb{B}^n(\mathbf{o};r), \qquad \mathbb{D}^n(r) = \mathbb{D}^n(\mathbf{o};r), \qquad \mathbb{S}^{n-1}(r) = \mathbb{S}^{n-1}(\mathbf{o};r);$$

and

$$\mathbb{B}^n = \mathbb{B}^n(\mathbf{o};1), \qquad \mathbb{D}^n = \mathbb{D}^n(\mathbf{o};1), \qquad \mathbb{S}^{n-1} = \mathbb{S}^{n-1}(\mathbf{o};1).$$

Definition Let $(X, \mathfrak{M})$ be a measure space. Then an $\mathbb{R}^k$-valued measure $\boldsymbol{\lambda}$ on $(X, \mathfrak{M})$ is a map $\boldsymbol{\lambda} : \mathfrak{M} \to \mathbb{R}^k$ such that

$$\boldsymbol{\lambda}(E) = \sum_{j=1}^{\infty} \boldsymbol{\lambda}(E_j) \qquad \text{(which converges)}$$

for all partitions $\{E_j\}$ of E, for all $E \in \mathfrak{M}$. (A *partition* $\{E_j\}$ of E is a countable family of subsets in $\mathfrak{M}$ that are pairwise disjoint, and whose union is E.)

If $\boldsymbol{\lambda}$ is an $\mathbb{R}^k$-valued measure on X, then its *total variation measure*, $|\boldsymbol{\lambda}|$, is defined by

$$|\boldsymbol{\lambda}|(E) = \sup \sum_{j=1}^{\infty} |\boldsymbol{\lambda}(E_j)|,$$

where the supremum is taken over all partitions of E.

One knows that $|\boldsymbol{\lambda}|(X) < +\infty$.

Definition An $\mathbb{R}^k$-valued measure $\boldsymbol{\lambda}$ is said to be *absolutely continuous with respect to a positive measure* μ *on* $(X, \mathfrak{M})$, $\boldsymbol{\lambda} \ll \mu$, if $\boldsymbol{\lambda}(E) = 0$ whenever $\mu(E) = 0$.

Proposition I.3.4 (Radon–Nikodym Theorem) *If $\boldsymbol{\lambda} \ll \mu$, then there exists an $\mathbb{R}^k$-valued function $\boldsymbol{\Phi}$ in $L^1(\mu)$, that is, $\int_X |\boldsymbol{\Phi}|\, d\mu < +\infty$, such that*

$$\boldsymbol{\lambda}(E) = \int_E \boldsymbol{\Phi}\, d\mu$$

for all measurable E.

Also, there exists a measurable map $\boldsymbol{\nu} : X \to \mathbb{S}^{k-1}$ such that

$$d\boldsymbol{\lambda} = \boldsymbol{\nu}\, d|\boldsymbol{\lambda}|.$$

Finally, let μ be a positive measure, $\boldsymbol{\xi} : X \to \mathbb{R}^k$ in $L^1(\mu)$, and set

$$\boldsymbol{\lambda}(E) = \int_E \boldsymbol{\xi}\, d\mu.$$

Then

$$|\boldsymbol{\lambda}|(E) = \int_E |\boldsymbol{\xi}|\, d\mu.$$

Definition Let X be a locally compact Hausdorff topological space, and μ a positive Borel measure on X, that is, its σ-algebra contains $\mathfrak{B}$, all Borel subsets of X. We say that μ is *locally finite* if $\mu(K)$ is finite for all compact K; a measurable E is *outer regular* if

$$\mu(E) = \inf\{\mu(V) : E \subseteq V,\ V \text{ open}\};$$

a measurable E is *inner regular* if

$$\mu(E) = \sup\{\mu(K) : K \subseteq E,\ K \text{ compact}\};$$

the measure μ is *regular* if every measurable E is both inner and outer regular. The measure μ is a *Radon measure* if it is locally finite and regular. An $\mathbb{R}^k$-valued measure $\boldsymbol{\lambda}$ is an $\mathbb{R}^k$-valued *Radon measure* if its total variation measure $|\boldsymbol{\lambda}|$ is a Radon measure.

Proposition I.3.5 (Riesz Representation Theorem) *Let X be a locally compact Hausdorff space, and $\boldsymbol{\Phi}$ a bounded linear functional on $C_c(X;\mathbb{R}^k)$ (relative to the sup norm). Then there exists a unique $\mathbb{R}^k$-valued Radon measure $\boldsymbol{\lambda}$ on X such that*

$$\boldsymbol{\Phi}(\boldsymbol{\xi}) = \int_X \boldsymbol{\xi} \cdot d\boldsymbol{\lambda}$$

for all $\boldsymbol{\xi} : X \to \mathbb{R}^k$ in $C_c(X;\mathbb{R}^k)$. Moreover,

$$\|\boldsymbol{\Phi}\| = |\boldsymbol{\lambda}|(X).$$

In $\mathbb{R}^n$, we denote the *standard Lebesgue measure* by $d\mathbf{v}_n(x)$. When considering a domain Ω in $\mathbb{R}^n$ with C^1 boundary $\partial\Omega$, we denote the Lebesgue measure on Ω by dV, and the *standard surface area measure on* $\partial\Omega$ by dA. For the special case of $\mathbb{B}^n$ we denote the surface area measure on $\mathbb{S}^{n-1}$ by $d\mu_{n-1}$. Recall that we use the notation

$$\boldsymbol{\omega}_{\mathbf{n}} = \int_{\mathbb{B}^n} d\mathbf{v}_n, \qquad \mathbf{c}_{\mathbf{n-1}} = \int_{\mathbb{S}^{n-1}} d\mu_{n-1}.$$

When $n = 2$, we replace dA by dL, and dV by dA.

When discussing an arbitrary k-dimensional submanifold of an n-dimensional Riemannian manifold, we use dV_k to denote the k-dimensional Riemannian measure on the submanifold.

When discussing changes of coordinates on $\mathbb{R}^n$, we refer to the new coordinates as *Euclidean coordinates* if they are obtained from the standard coordinate system by a Euclidean transformation.

Proposition I.3.6 (Lebesgue Density Theorem) *For $f \in L^1(\mathbb{R}^n)$ we have*

$$\lim_{r\downarrow 0} \frac{1}{\boldsymbol{\omega}_{\mathbf{n}} r^n} \int_{\mathbb{B}(x_o;r)} |f(x) - f(x_0)|\, d\mathbf{v}_n(x) = 0$$

for almost all $x_0 \in \mathbb{R}^n$.

In particular, for every measurable E in $\mathbb{R}^n$ with positive measure, we have

$$\lim_{r\downarrow 0} \frac{\mathbf{v}_n(E \cap \mathbb{B}(x;r))}{\boldsymbol{\omega}_{\mathbf{n}} r^n} = 1$$

for almost all $x \in E$.

Notation A *multi-index* α is given by $\alpha = (\alpha_1, \ldots, \alpha_n)$, where α_j, $j = 1, \ldots, n$, is a nonnegative integer, $|\alpha| = \alpha_1 + \cdots + \alpha_n$. For the multi-index α, D^α denotes the differentiation

$$D^\alpha = \left(\frac{\partial}{\partial x^1}\right)^{\alpha_1} \cdots \left(\frac{\partial}{\partial x^n}\right)^{\alpha_n}.$$

Definition We define a *mollifier on* $\mathbb{R}^n$ to be a function $j : \mathbb{R}^n \to [0, \infty) \in C_c^\infty(\mathbb{B}^n)$ satisfying

$$\int j(x)\, d\mathbf{v}_n(x) = 1.$$

We define, for all ϵ in $(0, 1)$,

$$j_\epsilon(x) = \epsilon^{-n} j(x/\epsilon);$$

and for any locally L^1 function f on $\mathbb{R}^n$, that is, $f \in L^1_{\text{loc}}$, we define

$$(j_\epsilon * f)(x) = \int j_\epsilon(y-x)f(y)\,d\mathbf{v}_n(y) = \int j_\epsilon(y)f(x+y)\,d\mathbf{v}_n(y).$$

Proposition I.3.7

(a) If $u \in L^1_{\text{loc}}$, then $\operatorname{supp} j_\epsilon * u = [\operatorname{supp} u]_\epsilon$.
(b) If $u \in C^0$, then $j_\epsilon * u \to u$ uniformly on compact subsets of $\mathbb{R}^n$ as $\epsilon \downarrow 0$.
(c) If $u \in L^1_{\text{loc}}$, then $j_\epsilon * u \in C^\infty$, and

$$(D^\alpha(j_\epsilon * u))(x) = (-1)^{|\alpha|}\int (D^\alpha j_\epsilon)(y-x)u(y)\,d\mathbf{v}_n(y)$$

for any multi-index α. Moreover, if $u \in C^k$, then for all $|\alpha| \le k$ one has

$$D^\alpha(j_\epsilon * u) = j_\epsilon * (D^\alpha u),$$

by integration by parts.
(d) If $u \in L^p$, then (i) $j_\epsilon * u \in L^p$, (ii) $\|j_\epsilon * u\|_p \le \|u\|_p$ *for all* $p \in [1,\infty]$, *and* (iii) $\|j_\epsilon * u - u\|_p \to 0$ *for all* $p \in [1,\infty)$, *as* $\epsilon \downarrow 0$.

Remark I.3.1 If $j(x) = j(-x)$ for all $x \in \mathbb{R}^n$, then

(I.3.2) $$\int (j_\epsilon * f)g\,d\mathbf{v}_n = \int f(j_\epsilon * g)\,d\mathbf{v}_n$$

for all $f, g \in L^1(\mathbb{R}^n)$. We shall henceforth *assume* the symmetry of all mollifiers, so (I.3.2) will always apply.

Notation Let $F : \Omega \to \mathbb{R}^k \in C^1$, where Ω is open in $\mathbb{R}^n$. For each $x \in \Omega$, we denote the associated Jacobian linear transformation of F by $J_F(x)$.

Definition Let $F : \Omega \to \mathbb{R}^k \in C^1$, where Ω is open in $\mathbb{R}^n$. A point $x \in \Omega$ is called a *critical point of F* if its associated Jacobian linear transformation $J_F(x)$ has rank $< k$.

A point $y \in \mathbb{R}^k$ is a *critical value* if there exists at least one point in its preimage which is a critical point. The complement in $\mathbb{R}^k$ of all critical values of F consists of the *regular values of F*. (This includes points not in the image of F.)

Proposition I.3.8 (Sard's Theorem) *Let $F : \Omega \to \mathbb{R}^k \in C^\ell$, where Ω is open in $\mathbb{R}^n$, $\ell > \max\{n-k, 0\}$. Let A denote the critical points of F. Then* $\mathbf{v}_k(F(A)) = 0$.

Remark I.3.2 The cases where we invoke Sard's theorem usually have $k = n$, in which case $\ell \geq 1$, or $k = 1$, in which case $\ell \geq n$.

Definition Consider the differential operator

$$\mathbf{L}u = \sum_{j,k=1}^{n} a^{jk}(x)\frac{\partial^2 u}{\partial x^j \partial x^k} + \sum_{j=1}^{n} b^j(x)\frac{\partial u}{\partial x^j}$$

defined on some domain D in $\mathbb{R}^n$. We say that $\mathbf{L}$ *is elliptic at* x if there exists a positive number $\mu(x)$ such that

$$\sum_{j,k=1}^{n} a^{jk}(x)p_j p_k \geq \mu(x)\sum_{j=1}^{n} {p_j}^2$$

for all $p = (p_1, \ldots, p_n) \in \mathbb{R}^n$. We say $\mathbf{L}$ is *elliptic on* D if $\mathbf{L}$ is elliptic at every point of D. We say $\mathbf{L}$ is *uniformly elliptic on* D if $\mathbf{L}$ is elliptic on D and there is a positive constant μ_0 such that $\mu(x) \geq \mu_0$ for all $x \in D$.

Proposition I.3.9 (Strong Maximum Principle) *Let* $\mathbf{L}$ *be uniformly elliptic on the domain* D *in* $\mathbb{R}^n$*, with uniformly bounded coefficients* $a^{jk}(x)$, $b^j(x)$, *and assume* $u : D \to \mathbb{R} \in C^2$ *satisfies the differential inequality*

$$\mathbf{L}u \geq 0$$

on all of D. If u attains a maximum at a point of D, then u is constant on D.

Furthermore, assume (i) $\partial D \in C^1$, (ii) $u : \overline{D} \to \mathbb{R} \in C^2(D) \cap C^0(\overline{D})$, (iii) $u \leq M$ *on all of* $\overline{D}$, (iv) $u = M$ *at some point* $w \in \partial D$, *and* (v) *the exterior normal derivative* $\partial u/\partial \boldsymbol{\nu}$ *is defined at* w. *Then*

$$\frac{\partial u}{\partial \boldsymbol{\nu}} > 0 \qquad \text{at } w,$$

unless $u = M$ *on all of* D.

We remind the reader that $\partial u/\partial \boldsymbol{\nu}$ at w is defined to be the limit of $(\operatorname{grad} u)(x)\cdot\boldsymbol{\nu}$, $x \in D$, as x approaches w along the line through w with direction $\boldsymbol{\nu}$.

I.3.1 Lipschitz Functions

Definition A *Lipschitz function on the metric space* (X, d) is a function u on X for which there exists a nonnegative number K such that

$$|u(z_1) - u(z_2)| \leq K d(z_1, z_2) \qquad \forall\, z_1, z_2 \in X. \tag{I.3.3}$$

The *Lipschitz constant* Lip u is the smallest positive K for which (I.3.3) is valid for all $z_1, z_2 \in X$.

Theorem I.3.1 *Let (X, d) be a metric space, $A \subset X$, and $f : A \to \mathbb{R}$ Lipschitz. Then there exists $g : X \to \mathbb{R}$ Lipschitz such that $g|A = f$.*

Proof Let $L = \text{Lip } f$, the Lipschitz constant of f, and consider

$$g(x) = \inf_{y \in A}\{f(y) + Ld(x, y)\}.$$

For all $x \in X$ and $y, z, \in A$ we have

$$f(z) - f(y) \le Ld(z, y) \le L\{d(x, y) + d(x, z)\},$$

which implies

$$f(z) - Ld(x, z) \le f(y) + Ld(x, y),$$

which implies

$$f(z) - Ld(x, z) \le g(x) \le f(y) + Ld(x, y),$$

which implies $g|A = f$.

To show that g is Lipschitz, note

$$\begin{aligned} g(x_1) - g(x_2) &= \sup_{y_2 \in A} \inf_{y_1 \in A} f(y_1) + Ld(x_1, y_1) - f(y_2) - Ld(x_2, y_2) \\ &\le \sup_{y_2 \in A} L\{d(x_1, y_2) - d(x_2, y_2)\} \\ &\le Ld(x_1, x_2). \end{aligned}$$

■

Theorem I.3.2 (Rademacher Theorem) *If $f : \mathbb{R}^m \to \mathbb{R}^n$ is Lipschitz, then f is differentiable a.e.-$[d\mathbf{v}_m]$.*

Proof We may assume $n = 1$. We may also assume that we know the theorem for $m = 1$ (since Lipschitz implies f is an absolutely continuous function of one real variable).

Fix $\xi \in \mathbb{S}^{n-1}$, and let $(D_\xi f)(x)$ denote the directional derivative of f at $x \in \mathbb{R}^n$ in the direction ξ (should it exist). Also, let

$$B_\xi = \{x \in \mathbb{R}^n : (D_\xi f)(x) \text{ does } \textit{not} \text{ exist}\}.$$

Then $\mathbf{v}_1(B_\xi \cap \{x + t\xi : t \in \mathbb{R}\}) = 0$ for all x (we know the theorem for $n = 1$). Therefore (by Fubini's theorem), $\mathbf{v}_n(B_\xi) = 0$ for every $\xi \in \mathbb{S}^{n-1}$. In particular, by successively picking $\xi = \mathbf{e}_1, \ldots, \mathbf{e}_n$ (the natural basis of $\mathbb{R}^n$), the gradient vector field grad f exists almost everywhere.

Next we wish to show that, for each ξ,

$$(D_\xi f)(x) = \xi \cdot (\text{grad } f)(x) \tag{I.3.4}$$

for almost every x. Indeed, given any $\phi \in C_c^\infty(\mathbb{R}^n)$ we have (by change of variables)

$$\int \frac{f(x+t\xi) - f(x)}{t} \phi(x)\, d\mathbf{v}_n(x) = -\int \frac{\phi(x) - \phi(x - t\xi)}{t} f(x)\, d\mathbf{v}_n(x),$$

which implies (using dominated convergence)

$$\begin{aligned}
\int (D_\xi f)\phi &= -\int (D_\xi \phi) f \\
&= -\int (\xi \cdot \text{grad } \phi) f \\
&= -\int \sum_j \left(\xi^j \frac{\partial \phi}{\partial x^j} \right) f \\
&= -\sum_j \int \xi^j \frac{\partial(\phi f)}{\partial x^j} + \sum_j \int \phi \xi^j \frac{\partial f}{\partial x^j} \\
&= \int \phi(\xi \cdot \text{grad } f),
\end{aligned}$$

which implies (I.3.4).

Finally, let $\{\xi_\alpha\}$ be a countable dense set in $\mathbb{S}^{n-1}$,

$$\begin{aligned}
A_\alpha &= \{x : \text{grad } f,\, D_{\xi_\alpha} f \text{ exist \& } D_{\xi_\alpha} f = \xi_\alpha \cdot \text{grad } f\}, \\
A &= \bigcap_{\alpha=1}^{\infty} A_\alpha.
\end{aligned}$$

Then $\mathbf{v}_n(\mathbb{R}^n \setminus A) = 0$. We now show that f is differentiable on A. For $x \in A$, $y = x + t\xi \in \mathbb{R}^n$, we have

$$\begin{aligned}
f(y) - f(x) - (y - x) \cdot (\text{grad } f)(x) &= f(x + t\xi) - f(x) - t\xi \cdot (\text{grad } f)(x) \\
&= f(x + t\xi) - f(x + t\xi_\alpha) + f(x + t\xi_\alpha) \\
&\quad - f(x) - t\xi_\alpha \cdot (\text{grad } f)(x) \\
&\quad + t(\xi_\alpha - \xi) \cdot (\text{grad } f)(x),
\end{aligned}$$

which implies

$$|f(y) - f(x) - (y - x) \cdot (\text{grad } f)(x)| \le 2L|\xi - \xi_\alpha||y - x| + o(|y - x|),$$

which implies the theorem. ∎

Theorem I.3.3 *Let u be a Lipschitz function on $\mathbb{R}^n$ with compact support. Subject u to mollification by j_ϵ as above. Then*

$$\|j_\epsilon * u - u\|_p \to 0, \qquad \|\operatorname{grad} j_\epsilon * u - \operatorname{grad} u\|_p \to 0$$

for all $p \in [1, \infty)$ as $\epsilon \downarrow 0$.

Proof Since u is Lipschitz, the restriction of u to any line is absolutely continuous; and thus all partial derivatives exist almost everywhere on $\mathbb{R}^n$, and are bounded with compact support. Thus, each partial derivative is an element of L^p for all $p \in [1, \infty]$. Also, the absolute continuity along lines and the compactness of the support of u imply the fundamental theorem of calculus for each partial derivative, which implies

$$\int_{\mathbb{R}^n} \frac{\partial u}{\partial x^k}(x)\, d\mathbf{v}_n(x) = 0$$

for $k = 1, \ldots, n$. Then we may use the argument of Proposition I.3.7(**c**) to show

$$\frac{\partial}{\partial x^k}(j_\epsilon * u) = j_\epsilon * \frac{\partial u}{\partial x^k}.$$

One easily obtains the conclusion of the theorem. ∎

Example I.3.1 Let K be a convex subset of $\mathbb{R}^n$ with nonempty interior. (See §I.3.3 below.) If $K \neq \mathbb{R}^n$, then K is contained in a half-space H. Let $\Pi = \partial H$, and $p : \mathbb{R}^n \to \Pi$ the orthogonal projection. Introduce Euclidean coordinates in $\mathbb{R}^n$ so that $\Pi = \{x^n = 0\}$. Define, for $w = (x^1, \ldots, x^{n-1}) \in p(K)$,

$$f(w) = \inf\{y : (w, y) \in K\}.$$

Then the graph of $x^n = f(w)$ is the "lower" part of the bounding hypersurface of K, and f is a *convex function*, that is,

$$f(\lambda u + (1-\lambda)v) \le \lambda f(u) + (1-\lambda) f(v) \qquad \forall\, u, v \in p(K),\ \lambda \in [0, 1].$$

Given $u, v \in \operatorname{int} p(K)$, let $\ell_{u,v}$ denote the line in Π through u, v, and set

$$F = f|\ell_{u,v}, \qquad (a, b) = \operatorname{int}(p(K) \cap \ell_{u,v}).$$

Then one checks that

$$\frac{F(t) - F(\tau)}{t - \tau} \qquad (t > \tau)$$

is increasing with respect to both t and τ, which implies the right and left hand derivatives of f, F'_+ and F'_- respectively, exist everywhere, and

$$\left|\frac{F(t) - F(\tau)}{t - \tau}\right| \leq \sup \left\{|F'_+(A)|, |F'_-(B)|\right\}$$

for all $a \leq A \leq \tau < t \leq B \leq b$, which implies F is locally Lipschitz on (a, b). By Rademacher's theorem, the boundary of K is a.e.-$[\Pi, d\mathbf{v}_{n-1}]$ differentiable.

Remark I.3.3 We recall for the reader the basic definition of *Sobolev spaces*. Let Ω be open in $\mathbb{R}^n$. (The definitions that follow are also the ones that are used when Ω denotes an open subset of a Riemannian manifold.) By the Meyers–Serrin theorem one has two equivalent characterizations of the Sobolev space $W^{1,1}(\Omega)$.

First one considers the collection H of C^∞ functions f on Ω for which both f, grad $f \in L^1$. Then one endows H with the pre-Banach norm

$$\|f\|_{1,1} = \|f\|_1 + \|\text{grad}\, f\|_1$$

and completes H, with respect to the norm $\|\ \|_{1,1}$, to the Banach space $W^{1,1}(\Omega)$.

The second characterization goes as follows: We say that $f \in L^1(\Omega)$ has a *weak L^1-derivative* if there exists an $L^1(\Omega)$-vector field $\boldsymbol{\xi}$ such that

$$\int_\Omega f \,\text{div}\, \eta \, d\mathbf{v}_n = -\int_\Omega \boldsymbol{\xi} \cdot \eta \, d\mathbf{v}_n$$

for all compactly supported C^∞ vector fields η on Ω. The vector field $\boldsymbol{\xi}$, should it exist, must be unique, and we denote it by $\boldsymbol{\xi} = \text{Grad}\, f$. Let $\mathfrak{H}$ be the Banach space (it is indeed a Banach space) consisting of functions $f \in L^1$ possessing weak L^1-derivatives, with the norm $\|\ \|_{\mathfrak{H}}$ given by

$$\|f\|_{\mathfrak{H}} = \|f\|_1 + \|\text{Grad}\, f\|_1, \qquad f \in \mathfrak{H}.$$

Then, by the Meyers–Serrin theorem, $\mathfrak{H} = W^{1,1}(\Omega)$.

Clearly, all $L^1(\Omega)$ Lipschitz functions with $L^1(\Omega)$-derivatives are elements of $W^{1,1}(\Omega)$.

Notation We let $\mathfrak{H}_c(\Omega)$ denote the completion of $C_c^\infty(\Omega)$ in $\mathfrak{H}(\Omega)$.

I.3.2 Co-area Formula for Smooth Mappings

In this section, we use the method of moving frames. One can find treatments of the method in the standard texts.

Let M^m, N^n be C^r, $r \geq 1$, Riemannian manifolds, $m \geq n$, and $\Phi : M \to N$ a C^1 map from M to N. We want to give an effective calculation of the volume disortion of the map, namely, of

$$\mathcal{J}_\Phi(x) = \sqrt{\det\ \Phi_* \circ (\Phi_*)^{\mathrm{adj}}},$$

where Φ_* denotes the associated Jacobian linear map of tangent spaces of M to those of N, and $(\ )^{\mathrm{adj}}$ denotes the adjoint map. (When M and N are Euclidean spaces, we use the notation J_Φ for Φ_*.)

Let $\{e_A\}$, $A = 1, \ldots, m$, be an orthonormal moving frame on M with dual co-frame $\{\omega^A\}$, and $\{E_j\}$, $j = 1, \ldots, n$, an orthonormal moving frame on N with dual co-frame $\{\theta^j\}$. It is well known that the local volume forms on M and N are given by $\omega^1 \wedge \cdots \wedge \omega^m$ and $\theta^1 \wedge \cdots \wedge \theta^n$, respectively.

There exist functions $\sigma^j{}_A$ on M such that

$$\Phi^*\theta^j = \sum_A \sigma^j{}_A \omega^A,$$

where Φ^* denotes the pullback of differential forms on N to M, associated with the map Φ. If at some $x \in M$ we have $\dim\ker \Phi_{*|x} > m - n$ (that is, x is a critical point of Φ), then

$$\Phi^*(\theta^1 \wedge \cdots \wedge \theta^n) = 0.$$

If at some $x \in M$ we have $\dim\ker \Phi_{*|x} = m - n$ (that is, x is a regular point of Φ), then Φ is a submersion on a neighborhood of x, and we may pick $e_{n+1}, \ldots, e_m$ to be tangent to the fibers of the local fibration about x associated with the submersion, which implies $e_1, \ldots, e_n$ is orthogonal to the fibration. We therefore have

$$\sigma^j{}_\alpha = 0, \qquad j = 1, \ldots, n, \qquad \alpha = n + 1, \ldots, m,$$

which implies

$$\Phi^*\theta^j = \sum_k \sigma^j{}_k \omega^k,$$

which implies

$$\Phi^*(\theta^1 \wedge \cdots \wedge \theta^n) = (\det \sigma^j{}_k)\omega^1 \wedge \cdots \wedge \omega^n.$$

One can easily check that

$$\det(\sigma^j{}_k) = \det\ \Phi_*|(\ker\Phi_*)^\perp.$$

Therefore, since

$$\mathcal{J}_\Phi(x) = \begin{cases} 0, & \operatorname{rank}\Phi_* < n, \\ \left|\det(\Phi_*|(\ker\Phi_*)^\perp)\right|, & \operatorname{rank}\Phi_* = n \end{cases}$$

(as one verifes rather easily), we have

$$\text{(I.3.5)}\quad \mathcal{J}_\Phi\, \omega^1 \wedge \cdots \wedge \omega^m = \pm\Phi^*(\theta^1 \wedge \cdots \wedge \theta^n) \wedge \omega^{n+1} \wedge \cdots \wedge \omega^m,$$

depending on whether $\det(\sigma^j{}_k)$ is positive or negative, which implies

Theorem I.3.4 (Co-area Formula for Smooth Mappings) *Let M, N be C^r Riemannian manifolds, with $m = \dim M \geq \dim N = n$, $r > m - n$, and let $\Phi : M \to N \in C^r$. Then for any measurable function $f : M \to \mathbb{R}$ that is everywhere nonnegative or is in $L^1(M)$, one has*

$$\text{(I.3.6)}\qquad \int_M f\,\mathcal{J}_\Phi\, dV_m = \int_N dV_n(y) \int_{\Phi^{-1}[y]} (f|\Phi^{-1}[y])\, dV_{m-n}.$$

Proof We have, from (I.3.5),

$$\int_M f\,\mathcal{J}_\Phi\, dV_m = \int_{\text{regval}\,\Phi\subset N} dV_n(y) \int_{\Phi^{-1}[y]} (f|\Phi^{-1}[y])\, dV_{m-n},$$

where regval Φ denotes the set of regular values of Φ in N. Since the map Φ is C^r, the critical values of Φ constitute, at most, a set of measure 0 in N (by Sard's theorem – it is also valid for manifolds), which implies (I.3.6). ∎

Corollary I.3.1 (Co-area Formula for Smooth Functions) *Let M^m be a C^m Riemannian manifold, and let $\Phi : M \to \mathbb{R} \in C^m$. Then for any measurable function $f : M \to \mathbb{R}$ that is everywhere nonnegative or is in $L^1(M)$, one has*

$$\text{(I.3.7)}\qquad \int_M f\,|\text{grad}\,\Phi|\, dV = \int_{\mathbb{R}} d\mathbf{v}_1(y) \int_{\Phi^{-1}[y]} (f|\Phi^{-1}[y])\, dA,$$

where grad Φ *denotes the Riemannian gradient vector field of Φ on M (see* §VI.1).

Corollary I.3.2 *Let M^{k-1} be a hypersurface in $\mathbb{R}^k$ given by the graph of a C^1 function $\phi : G \to \mathbb{R}$, where G is open in $\mathbb{R}^{k-1}$; so M is given by*

$$x^k = \phi(x^1, \ldots, x^{k-1}), \qquad (x^1, \ldots, x^{k-1}) \in G.$$

Then the surface area element on M, dA, is given by

$$\text{(I.3.8)}\qquad dA = \sqrt{1 + |\text{grad}_{k-1}\,\phi|^2}\, d\mathbf{v}_{k-1},$$

where grad_{k-1} *denotes the gradient of functions on* $\mathbb{R}^{k-1}$.

Proof Let H be a hyperplane in $\mathbb{R}^k$, with normal vector $\boldsymbol{\nu}$; $\mathbb{R}^{k-1}$ the hyperplane $\{x^k = 0\}$ in $\mathbb{R}^k$, with normal vector $\boldsymbol{\xi}$; and $p : H \to \mathbb{R}^{k-1}$ the projection. Then

$$\mathcal{J}_p = |\boldsymbol{\nu}\cdot\boldsymbol{\xi}|. \tag{I.3.9}$$

Then apply (I.3.6) with $f = 1/\mathcal{J}_p$, and calculate. ■

Corollary I.3.3 *If $\Omega \subset\subset \mathbb{R}^n$ is a domain with C^1 boundary, $\boldsymbol{\nu}$ the exterior unit normal vector field along $\partial\Omega$, and, for a given $\boldsymbol{\xi} \in \mathbb{S}^{n-1}$, $p_{\boldsymbol{\xi}} : \partial\Omega \to \boldsymbol{\xi}^{\perp}$ is the projection, then*

$$\int_{\partial\Omega} |\boldsymbol{\nu}_w\cdot\boldsymbol{\xi}|\, dA(w) = \int_{\boldsymbol{\xi}^{\perp}} \operatorname{card}(\partial\Omega \cap p_{\boldsymbol{\xi}}{}^{-1}[y])\, d\mathbf{v}_{\mathbf{n-1}}(y). \tag{I.3.10}$$

Proof Apply (I.3.6), with $f = 1$. ■

Corollary I.3.4 *If $\Omega \subset\subset \mathbb{R}^n$ is a domain with C^1 boundary, and for every $\boldsymbol{\xi} \in \mathbb{S}^{n-1}$, $p_{\boldsymbol{\xi}} : \partial\Omega \to \boldsymbol{\xi}^{\perp}$ denotes the projection, then*

$$A(\partial\Omega) = \frac{1}{2\boldsymbol{\omega}_{\mathbf{n-1}}} \int_{\mathbb{S}^{n-1}} d\boldsymbol{\mu}_{n-1}(\boldsymbol{\xi}) \int_{\boldsymbol{\xi}^{\perp}} \operatorname{card}(\partial\Omega \cap p_{\boldsymbol{\xi}}{}^{-1}[y])\, d\mathbf{v}_{n-1}(y). \tag{I.3.11}$$

Proof Integrate (I.3.10) over $\boldsymbol{\xi} \in \mathbb{S}^{n-1}$. Then

$$\begin{aligned}
&\int_{\mathbb{S}^{n-1}} d\boldsymbol{\mu}_{n-1}(\boldsymbol{\xi}) \int_{\boldsymbol{\xi}^{\perp}} \operatorname{card}(\partial\Omega \cap p_{\boldsymbol{\xi}}{}^{-1}[y])\, d\mathbf{v}_{n-1}(y)\\
&\quad= \int_{\mathbb{S}^{n-1}} d\boldsymbol{\mu}_{n-1}(\boldsymbol{\xi}) \int_{\partial\Omega} |\boldsymbol{\nu}_w\cdot\boldsymbol{\xi}|\, dA(w)\\
&\quad= \int_{\partial\Omega} dA(w) \int_{\mathbb{S}^{n-1}} |\boldsymbol{\nu}\cdot\boldsymbol{\xi}|\, d\boldsymbol{\mu}_{n-1}(\boldsymbol{\xi})\\
&\quad= 2A(\partial\Omega)\mathbf{c}_{\mathbf{n-2}} \int_0^{\pi/2} \cos\theta \sin^{n-2}\theta\, d\theta\\
&\quad= 2\frac{\mathbf{c}_{\mathbf{n-2}}}{n-1} A(\partial\Omega)\\
&\quad= 2\boldsymbol{\omega}_{\mathbf{n-1}} A(\partial\Omega),
\end{aligned}$$

which implies the corollary. ■

Corollary I.3.5 (Cauchy's Formula) *If Ω is convex, $\partial\Omega \in C^1$, then*

$$A(\partial\Omega) = \frac{1}{\boldsymbol{\omega}_{\mathbf{n-1}}} \int_{\mathbb{S}^{n-1}} \mathbf{v}_{n-1}(p_{\boldsymbol{\xi}}(\Omega))\, d\boldsymbol{\mu}_{n-1}(\boldsymbol{\xi}).$$

Proof If Ω is convex, then card $(\partial\Omega \cap p_{\xi}{}^{-1}[y]) = 2$ for almost-$[d\mathbf{v}_{n-1}(\xi^{\perp})]$ all $y \in p_{\xi}(\partial\Omega) = p_{\xi}(\Omega)$, $\xi \in \mathbb{S}^{n-1}$, which implies the claim. ∎

Corollary I.3.6 *If Ω is convex, $\partial\Omega \in C^1$, and Ω_0 is open containing Ω, $\partial\Omega_0 \in C^1$, then*

$$A(\partial\Omega) \leq A(\partial\Omega_0).$$

I.3.3 Some Geometric Preliminaries

Theorem I.3.5 *Suppose K is compact in $\mathbb{R}^n$, with the property that, for any $\xi \in \mathbb{S}^{n-1}$, K is symmetric with respect to some hyperplane perpendicular to ξ. Then the boundary of K consists of a union of concentric $(n-1)$-spheres.*

Proof Pick new Euclidean coordinates in $\mathbb{R}^n$ so that K is symmetric with respect to reflection in each of the coordinate planes. Then K is symmetric with respect to the origin. Since any direction has an orthogonal hyperplane of symmetry for K, and since K is already symmetric with respect to the origin, then K must be symmetric with respect to any plane through the origin. Since any half-line from the origin to ∂K can be taken to any other by a reflection in a hyperplane through the origin, each connected component of ∂K must be an $(n-1)$-sphere centered at the origin. ∎

We recall just a few notions about convex sets in $\mathbb{R}^n$.

Definition A set A in $\mathbb{R}^n$ is *convex* if $x, y \in A$ implies that $\lambda x + (1-\lambda)y \in A$ for all $\lambda \in (0, 1)$, that is, for any x and y in A the closed line segment $[x, y]$ in $\mathbb{R}^n$ joining them is contained in A.

Definition *A convex linear combination of elements $x_1, \ldots, x_k \in \mathbb{R}^n$* is the linear combination

$$\sum_{j=1}^{k} \lambda_j x_j,$$

where the coefficients satisfy

$$\sum_{j=1}^{k} \lambda_j = 1, \qquad \lambda_j \geq 0 \ \forall j.$$

One checks that if A is convex then any convex linear combination of elements of A is a point in A.

Definition Given $A \subset \mathbb{R}^n$, define the *convex hull of* A, conv A, to be the smallest convex set containing A.

Proposition I.3.10 *Given* A*, then* conv A *is given by the collection of all finite convex linear combinations of elements of* A.

If A *is compact then* conv A *is compact.*
If A *is bounded, then* conv A *is bounded and* $\operatorname{diam} \operatorname{conv} A = \operatorname{diam} A$.
If S *is convex then* $\overline{S}$ *is convex.*
If S *is convex compact then* $S = \operatorname{conv} \partial S$.
If C *is convex with nonempty interior, then* $\operatorname{int} C$ *is homeomorphic to* $\mathbb{R}^n$.

Definition Let A, B be subsets of $\mathbb{R}^n$, and H a hyperplane. We say *A and B are separated by H* if A and B lie in different closed half-spaces determined by H. If neither A nor B intersects H, we say *H strictly separates A and B*.

Definition Let A be a subset of $\mathbb{R}^n$, H a hyperplane, $x \in A$. We say *H is a supporting hyperplane of A at x* if $x \in H$ and H separates $\{x\}$ and A. Note that $x \in \partial A$.

Proposition I.3.11 *Let A be closed, with nonempty interior. Then A is convex if and only if every $x \in \partial A$ has a supporting hyperplane. Furthermore, if A is closed convex, then A is the intersection of all closed half-spaces containing A.*

I.4 Bibliographic Notes

§I.1 For surveys of the isoperimetric inequality in Euclidean space, replete with historical remarks and references, see Blaschke (1956), Hadwiger (1957), Osserman (1978), Talenti (1993), and the more recent book by Burago and Zalgaller (1988). For further remarks, including some discussion of experimental proofs of the isoperimetric inequality [one of which goes back to Courant and Robbins (1941)], see Dierkes, Hildebrandt, Kürster, and Wohlrab (1992, pp. 420–423).

On the issue of the proof of the isoperimetric inequality for general boundaries, the following paragraph of Osserman (1978, p. 1188) is worth repeating:

> First, a general remark. If we start with a relatively smooth boundary, adding "wiggles" to it will have very little effect on the volume enclosed, but will greatly increase the surface area. Thus, one has the somewhat ironic situation that the more irregular the boundary, the stronger will be the isoperimetric inequality, but the harder it is to prove. The fact is, the isoperimetric inequality holds in the greatest generality imaginable, but one needs suitable definitions even to state it.

The first unified solution of the isoperimetric problem in the model spaces of constant sectional curvature (that is, Euclidean, spherical, and hyperbolic spaces) was given by Schmidt (1948, 1949). In Chavel (1994) we gave separate proofs in each of the cases, to highlight a variety of methods. The Euclidean argument given there will be given in Theorem II.2.2 below. The isoperimetric argument for the sphere there is a new Riemannian theoretic argument by Gromov (1986), and the hyperbolic space argument follows the symmetrization argument in Figiel, Lindenstrauss, and Milman (1977). Different symmetrization arguments will be developed here later.

The study of isoperimetric inequalities on surfaces is alive and well. See the discussions in Osserman (1978, 1979). More recent results can be found (just to give a sample) in Adams and Morgan (1999), Benjamini and Cao (1996), Howards, Hutchings, and Morgan (1999), and Topping (1997, 1999). Extensive discussion of Bonnessen inequalities can be found in Osserman (1979).

§I.2 The complex variables proof follows Topping (1997) (the proof was inspired by F. Hélein). The Fourier series argument was first carried out by Hurwitz (1901). The name "Wirtinger's inequality" is apparently a misnomer. See the historical remarks in the previous surveys.

Theorem I.2.3 is from Howards, Hutchings, and Morgan (1999). Blaschke (1956, p. 1), in discussing the idea, refers back to Steiner (1881, p. 193ff.).

The example of Remark I.2.2 was supplied by an anonymous reader, to whom I extend my thanks.

Another line of reasoning was initiated by M. Gage (1984), wherein he proved that if a convex curve in the Euclidean plane is deformed along its inward normal at a rate proportional to its curvature at that point, then the isoperimetric ratio L^2/A decreases to the limit 4π, and the convex curves become asymptotically circular – provided that the shrinking curves do not develop singularities such as corners. The subject has developed quite extensively since then (including the discussion whether the singularities exist), both for surfaces and for higher dimensions, and one can find a partially updated bibliography in Gage (1991). Recent results, with applications to geometric inequalities, can be found in Topping (1998). Also, see Chou and Zho (scheduled to appear in 2001).

§I.3 For background in analysis, we generally follow Rudin (1966). More recent books that are helpful here are Lieb and Loss (1996) and Ziemer (1989). Proposition I.3.2 is from Federer (1969, p. 69 ff.)

For treatments of Sard's theorem, see Hirsch (1976) and Narasimhan (1968). For the strong maximum principle, see Gilbarg and Trudinger (1977) and Protter and Weinberger (1984). For Sobolev spaces, see Adams (1975) and Gilbarg and Trudinger (1977).

The general area and co-area formulae of geometric measure theory are discussed in §3.2 of Federer's treatise (1969). It seemed helpful to provide the general co-area formula in the C^1 category in order to see how much it contains, beyond the usual case of real-valued functions. We present the area formula for Lipschitz mappings of $\mathbb{R}^m \to \mathbb{R}^n$, $m \leq n$, in §IV.2 below. The method of moving frames is discussed in Chavel (1994).

One can find introductory material on convexity (including the Hahn–Banach theorem and the Cauchy formula for areas of boundaries) in Berger's two-volume work on geometry (1987, Chapters 11, 12).

II

Differential Geometric Methods

In this chapter we present the differential geometric arguments. By their very nature, they presuppose a certain smoothness of the boundary. Our method will be to start with assuming that the boundary of the domain in which we are interested is C^2, and then weaken the assumption to just C^1. We extend the isoperimetric inequality to more general domains in the next chapter.

We use classical arguments to show that if a domain provides a solution to the C^2 isoperimetric problem, then the domain is a disk. Then we strengthen the result a bit, to show that if the domain is just an extremal of the isoperimetric functional, then it must be a disk (see the definitions below). More specifically, we first introduce the standard local calculations of the differential geometry of hypersurfaces in Euclidean space, and the first variations of volume and area of domains and their boundaries under a 1-parameter family of diffeomorphisms. Then we give Almgren's characterization of the solution to the isoperimetric problem, followed by Alexandrov's characterization of an extremal for the isoperimetric functional.

Thus, the C^2 theory yields a characterization of the minimum (should it exist) in its category. To prove the isoperimetric inequality for domains with C^1 boundary, we present a proof based on Stokes's theorem, following M. Gromov. The only drawback of the method is that, to characterize equality, one requires that the domain be also assumed to be convex. We solve this latter problem with Steiner symmetrization in the next chapter.

II.1 The C^2 Uniqueness Theory

II.1.1 The Variation of Volume and Area

We first review the local calculations of classical differential geometry pertaining to an $(n-1)$-dimensional regular hypersurface Γ in $\mathbb{R}^n$. Although we present the definitions of geometric quantities (such as the curvatures,

Riemannian divergence, etc.) relative to specific choices of local coordinates, standard arguments establish the invariance of these geometric quantities relative to changes of coordinates.

Assume Γ is given locally by the C^1 mapping $\mathbf{x} : G \to \mathbb{R}^n$ of everywhere maximal rank, where G is an open subset of $\mathbb{R}^{n-1}$. So $\mathbf{x} = \mathbf{x}(u)$; and the vectors

$$\partial \mathbf{x}/\partial u^1, \ldots, \partial \mathbf{x}/\partial u^{n-1}$$

are linearly independent and span the tangent space to Γ at every $\mathbf{x}(u)$. We let $\mathbf{n}(u)$ denote a choice of continuous normal unit vector field along Γ. When considering a hypersurface Γ which is the boundary of a domain in $\mathbb{R}^n$, the choice for $\mathbf{n}$ will always be the *exterior* normal unit vector field – unless otherwise indicated.

The Riemannian metric of Γ (that is, the first fundamental form) is given locally by the positive definite matrix $G(u)$, where

$$\text{(II.1.1)} \qquad G = (g_{jk}), \qquad g_{jk} = \frac{\partial \mathbf{x}}{\partial u^j} \cdot \frac{\partial \mathbf{x}}{\partial u^k}, \quad j, k = 1, \ldots, n-1.$$

We also use the notation

$$G^{-1} = (g^{jk}), \qquad g = \det G.$$

The associated surface area on Γ is given locally by

$$dA = \sqrt{g}\, du^1 \cdots du^{n-1}.$$

Definition Let Ω be a bounded domain in $\mathbb{R}^n$, with C^k boundary, $k \geq 1$. We say that Ω is a *solution to the C^k isoperimetric problem* if, for any domain D with C^k boundary and volume equal to that of Ω, we have $A(\partial D) \geq A(\partial\Omega)$.

Let Ω be a bounded domain in $\mathbb{R}^n$, with C^k boundary. We say that Ω is a *C^k extremal of the isoperimetric functional* if, for any 1-parameter family of C^k diffeomorphisms $\Phi_t : \mathbb{R}^n \to \mathbb{R}^n$ satisfying $V(\Phi_t(\Omega)) = V(\Omega)$ for all t, we have

$$\left. \frac{d}{dt} A(\Phi_t(\partial\Omega)) \right|_{t=0} = 0.$$

Assume that Γ is a C^2 hypersurface in $\mathbb{R}^n$ (so the Riemannian metric is C^1). For any tangent vector field

$$\zeta = \sum_{j=1}^{n-1} \zeta^j \frac{\partial \mathbf{x}}{\partial u^j}$$

along Γ, we have its Riemannian divergence (see §VII.2.1) given by

$$\operatorname{div}_\Gamma \boldsymbol{\zeta} = \frac{1}{\sqrt{g}} \sum_{j=1}^{n-1} \frac{\partial(\zeta^j \sqrt{g})}{\partial u^j},$$

with an attendent intrinsic Riemannian divergence theorem for $(n-1)$-dimensional domains in Γ with C^1 $(n-2)$-dimensional boundaries; namely, given an $(n-1)$-domain $\Lambda \subset\subset \Gamma$ with C^1 boundary $\partial\Lambda$, and unit normal exterior (Γ-tangent) vector field $\boldsymbol{\nu}$ along $\partial\Lambda$, then

(II.1.2) $$\int_\Lambda \operatorname{div}_\Gamma \boldsymbol{\zeta}\, dV_{n-1} = \int_{\partial\Lambda} \boldsymbol{\zeta}\cdot\boldsymbol{\nu}\, dV_{n-2}.$$

The second fundamental form of Γ in $\mathbb{R}^n$ is given locally by

(II.1.3) $$\mathcal{B} = (b_{jk}), \qquad b_{jk} = \frac{\partial^2 \mathbf{x}}{\partial u^j \partial u^k}\cdot\mathbf{n}, \quad j,k = 1,\ldots,n-1.$$

Then

$$b_{jk} = \frac{\partial^2 \mathbf{x}}{\partial u^j \partial u^k}\cdot\mathbf{n} = \frac{\partial}{\partial u^j}\left(\mathbf{n}\cdot\frac{\partial \mathbf{x}}{\partial u^k}\right) - \frac{\partial \mathbf{n}}{\partial u^j}\cdot\frac{\partial \mathbf{x}}{\partial u^k} = -\frac{\partial \mathbf{n}}{\partial u^j}\cdot\frac{\partial \mathbf{x}}{\partial u^k},$$

that is,

(II.1.4) $$b_{jk} = -\frac{\partial \mathbf{n}}{\partial u^j}\cdot\frac{\partial \mathbf{x}}{\partial u^k}.$$

If $\mathcal{L} = (L_j{}^k)$ denotes the matrix of the G-self-adjoint linear transformation associated with $\mathcal{B}$, then

$$\mathcal{L} = G^{-1}\mathcal{B}.$$

The *mean curvature* H *of* Γ *in* $\mathbb{R}^n$ is the trace of $\mathcal{L}$, that is, the trace of $\mathcal{B}$ relative to G, given by

$$H = \operatorname{tr} G^{-1}\mathcal{B},$$

and the *Gauss–Kronecker curvature* K *of* Γ *in* $\mathbb{R}^n$ is the determinant of $\mathcal{L}$, that is, the determinant of $\mathcal{B}$ relative to G, given by

$$K = \det G^{-1}\mathcal{B}.$$

We may think of $\mathbf{n}$ as a map that associates with every point $\mathbf{x} \in \Gamma$ its normal vector $\mathbf{n} \in \mathbb{S}^{n-1}$ – commonly referred to as the *Gauss map* $\mathfrak{G} : \Gamma \to \mathbb{S}^{n-1}$. The Jacobian linear transformation $\mathfrak{G}_*$ of the Gauss map is then determined by

$$\frac{\partial \mathbf{x}}{\partial u^j} \mapsto \frac{\partial \mathbf{n}}{\partial u^j}.$$

Then (II.1.4) implies that the matrix of $\mathfrak{G}_*$ is given by $-\mathcal{L}$, which we write as

$$\mathfrak{G}_* = -G^{-1}\mathcal{B}. \tag{II.1.5}$$

Example II.1.1 If Γ is a relatively open subset of $\mathbb{S}^{n-1}(r)$, the $(n-1)$-sphere of radius r, then for the normal unit vector field along Γ exterior to $\mathbb{B}^n(r)$, we have

$$\mathbf{n} = \frac{1}{r}\mathbf{x},$$

which implies [by (II.1.4)]

$$\mathcal{B} = -\frac{1}{r}G;$$

therefore,

$$H = -\frac{(n-1)}{r}, \qquad K = (-1)^{n-1}\frac{1}{r^{n-1}}.$$

Theorem II.1.1 *Let Ω be a bounded domain in $\mathbb{R}^n$, with C^2 boundary Γ. Given any C^2 time-dependent vector field $X : \mathbb{R}^n \times \mathbb{R} \to \mathbb{R}^n$ on $\mathbb{R}^n$, let $\Phi_t : \mathbb{R}^n \to \mathbb{R}^n$ denote the 1-parameter flow determined by X, that is, Φ_t and X are related by*

$$\frac{d}{dt}\Phi_t(x) = X(x,t), \qquad \Phi_0 = \mathrm{id}.$$

Set

$$\boldsymbol{\xi}(x) = X(x,0), \qquad \eta = \boldsymbol{\xi}|\Gamma.$$

Then

$$\left.\frac{d}{dt}V(\Phi_t(\Omega))\right|_{t=0} = \iint_\Omega \operatorname{div}\boldsymbol{\xi}\, d\mathbf{v}_n = \int_\Gamma \eta\cdot\mathbf{n}\, dA, \tag{II.1.6}$$

where $\mathbf{n}$ is chosen to be the exterior unit normal vector field along Γ, and

$$\left.\frac{d}{dt}A(\Phi_t(\Gamma))\right|_{t=0} = \int_\Gamma \{\operatorname{div}_\Gamma \eta^T - H\eta\cdot\mathbf{n}\}\, dA = -\int_\Gamma H\eta\cdot\mathbf{n}\, dA. \tag{II.1.7}$$

where η^T denotes the tangential component of η.

Proof First,

$$V(\Phi_t(\Omega)) = \iint_\Omega \det J_{\Phi_t}(x)\, d\mathbf{v}_n(x),$$

where J_{Φ_t} denotes the Jacobian matrix of Φ_t. Then

$$\frac{d}{dt}V(\Phi_t(\Omega)) = \iint_\Omega \left(\frac{d}{dt}\det J_{\Phi_t}(x)\right) d\mathbf{v}_n(x).$$

The formula for differentiating determinants states that for any differentiable matrix function $t \mapsto \mathcal{A}(t)$, where $\mathcal{A}(t)$ is nonsingular, one has

$$\frac{d}{dt}\det \mathcal{A} = \det \mathcal{A} \cdot \operatorname{tr}\left(\mathcal{A}^{-1}\frac{d\mathcal{A}}{dt}\right).$$

Therefore,

$$\frac{d}{dt}V(\Phi_t(\Omega)) = \iint_\Omega (\det J_{\Phi_t}) \cdot \operatorname{tr}\left({J_{\Phi_t}}^{-1}\frac{d}{dt}J_{\Phi_t}\right) d\mathbf{v}_n(x).$$

Now

$$(J_{\Phi_t})_A{}^B = \frac{\partial {\Phi_t}^B}{\partial x^A}, \quad A, B = 1, \dots, n;$$

so for $t = 0$ we have, since $\Phi_0 = \mathrm{id}$,

$$\left.\frac{\partial {\Phi_t}^B}{\partial x^A}\right|_{t=0} = \delta_A{}^B,$$

the Kronecker delta. Furthermore,

$$\frac{d}{dt}(J_{\Phi_t})_A{}^B = \frac{\partial}{\partial t}\frac{\partial {\Phi_t}^B}{\partial x^A} = \frac{\partial}{\partial x^A}\frac{\partial {\Phi_t}^B}{\partial t}.$$

which implies, for $t = 0$,

$$\left.\frac{d}{dt}(J_{\Phi_t})_A{}^B\right|_{t=0} = \frac{\partial \xi^B}{\partial x^A},$$

which implies the first equality of (II.1.6). The second equality follows from the divergence theorem in $\mathbb{R}^n$.

Assume the surface Γ is given locally by $\mathbf{x} = \mathbf{x}(u)$, and set

$$\mathbf{y}(u, t) = \Phi_t(\mathbf{x}(u)).$$

For each fixed t we denote the Riemannian metric tensor on $\Phi_t(\Gamma)$ by

$$h_{jk} = \frac{\partial \mathbf{y}}{\partial u^j} \cdot \frac{\partial \mathbf{y}}{\partial u^k}, \quad j, k = 1, \dots, n-1,$$

which implies

$$\frac{d}{dt}A(\Phi_t(\Gamma)) = \int_\Gamma \frac{\partial}{\partial t}\sqrt{\det(h_{jk})}\, du^1 \cdots du^{n-1}.$$

For the derivative of $\sqrt{\det(h_{jk})}$, set $\mathcal{H} = (h_{jk})$, $\mathcal{H}^{-1} = (h^{jk})$; then

$$\begin{aligned}\frac{\partial}{\partial t}\sqrt{\det \mathcal{H}} &= \frac{1}{2}\left\{\sqrt{\det \mathcal{H}}\right\}^{-1} \det \mathcal{H} \cdot \left(\operatorname{tr} \mathcal{H}^{-1}\frac{\partial \mathcal{H}}{\partial t}\right) \\ &= \frac{1}{2}\sqrt{\det \mathcal{H}}\sum_{j,k} h^{jk}\frac{\partial h_{kj}}{\partial t} \\ &= \sqrt{\det \mathcal{H}}\sum_{j,k} h^{jk}\frac{\partial \mathbf{y}}{\partial u^k}\cdot\frac{\partial}{\partial u^j}\frac{\partial \mathbf{y}}{\partial t},\end{aligned}$$

that is,

$$\frac{\partial}{\partial t}\sqrt{\det \mathcal{H}} = \sqrt{\det \mathcal{H}}\sum_{j,k} h^{jk}\frac{\partial \mathbf{y}}{\partial u^k}\cdot\frac{\partial}{\partial u^j}\frac{\partial \mathbf{y}}{\partial t}. \tag{II.1.8}$$

Set $t = 0$; then $\boldsymbol{\eta}(u) = (\partial \mathbf{y}/\partial t)(u, 0)$. Write

$$\boldsymbol{\eta} = \sum_{\ell=1}^{n-1} \eta^\ell \frac{\partial \mathbf{x}}{\partial u^\ell} + \phi \mathbf{n}$$

along Γ. Then (II.1.8) implies, for $t = 0$,

$$\left.\frac{\partial}{\partial t}\sqrt{\det \mathcal{H}}\right|_{t=0} = \sqrt{\det G}\sum_{j,k} g^{jk}\frac{\partial \mathbf{x}}{\partial u^k}\cdot\frac{\partial \boldsymbol{\eta}}{\partial u^j}.$$

To calculate in more detail, one has

$$\frac{\partial \boldsymbol{\eta}}{\partial u^j} = \sum_{\ell=1}^{n-1}\left\{\frac{\partial \eta^\ell}{\partial u^j}\frac{\partial \mathbf{x}}{\partial u^\ell} + \eta^\ell\frac{\partial^2 \mathbf{x}}{\partial u^j \partial u^\ell}\right\} + \frac{\partial \phi}{\partial u^j}\mathbf{n} + \phi\frac{\partial \mathbf{n}}{\partial u^j},$$

which implies, since $\mathbf{n}$ is perpendicular to $\partial \mathbf{x}/\partial u^k$ for all $k = 1, \ldots, n-1$,

$$\begin{aligned}&\sum_{j,k} g^{jk}\frac{\partial \mathbf{x}}{\partial u^k}\cdot\frac{\partial \boldsymbol{\eta}}{\partial u^j} \\ &\quad= \sum_{j,k} g^{jk}\frac{\partial \mathbf{x}}{\partial u^k}\cdot\left(\sum_{\ell=1}^{n-1}\left\{\frac{\partial \eta^\ell}{\partial u^j}\frac{\partial \mathbf{x}}{\partial u^\ell} + \eta^\ell\frac{\partial^2 \mathbf{x}}{\partial u^\ell \partial u^j}\right\} + \phi\frac{\partial \mathbf{n}}{\partial u^j}\right) \\ &\quad= \sum_{j,k,\ell} g^{jk}g_{k\ell}\frac{\partial \eta^\ell}{\partial u^j} + \sum_{j,k,\ell} g^{jk}\eta^\ell\frac{\partial \mathbf{x}}{\partial u^k}\cdot\frac{\partial^2 \mathbf{x}}{\partial u^\ell \partial u^j} + \sum_{j,k}\phi g^{jk}\frac{\partial \mathbf{x}}{\partial u^k}\cdot\frac{\partial \mathbf{n}}{\partial u^j} \\ &\quad= \sum_j \frac{\partial \eta^j}{\partial u^j} + \sum_j \eta^j\frac{1}{2}\operatorname{tr}\left(G^{-1}\frac{\partial G}{\partial u^j}\right) + \sum_{j,k}\phi g^{jk}\frac{\partial \mathbf{x}}{\partial u^k}\cdot\frac{\partial \mathbf{n}}{\partial u^j} \\ &\quad= \sum_j \frac{1}{\sqrt{g}}\frac{\partial(\eta^j\sqrt{g})}{\partial u^j} + \sum_{j,k}\phi g^{jk}\frac{\partial \mathbf{x}}{\partial u^k}\cdot\frac{\partial \mathbf{n}}{\partial u^j} \\ &\quad= \operatorname{div}_\Gamma \boldsymbol{\eta}^T - \phi H.\end{aligned}$$

Therefore,

$$\left.\frac{d}{dt}A(\Phi_t(\Gamma))\right|_{t=0} = \int_\Gamma \{\operatorname{div}_\Gamma \eta^T - H\eta\cdot\mathbf{n}\}\,dA.$$

Then the intrinsic Riemannian divergence theorem (II.1.2) implies (since Γ is closed, with no boundary) the second equality (II.1.7). ∎

II.1.2 Uniqueness of a Solution to the C^2 Isoperimetric Problem

Let Ω be a bounded domain in $\mathbb{R}^n$, with C^2 boundary Γ. Recall that the normal unit vector along Γ is outward. Thus, if Ω were convex, then the second fundamental form $\mathcal{B}$ would be negative semidefinite.

Theorem II.1.2 *Assume that Ω is a solution to the C^2 isoperimetric problem, with volume of Ω equal that of the unit n-disk in $\mathbb{R}^n$. Then the mean curvature H of Γ satisfies*

$$-H \le n-1$$

on all of Γ.

Proof Consider the isoperimetric functional

$$J(D) = \frac{A(\partial D)}{V(D)^{1-1/n}},$$

where D varies over bounded domains in $\mathbb{R}^n$ possessing C^2 boundary. Let Φ_t denote a 1-parameter family of diffeomorphisms of $\mathbb{R}^n$, with associated time-dependent vector field $X = X(x,t)$ (as in the proof of Theorem II.1.1), $\xi(x) = X(x,0)$, $\eta = \xi|\Gamma$. Then direct calculation implies, by (II.1.6), (II.1.7), and the fact that Ω is an extremal of the isoperimetric functional,

$$\begin{aligned} 0 &= \left.\frac{d}{dt}J(\Phi_t(\Omega))\right|_{t=0} \\ &= -\frac{1}{V(\Omega)^{1-1/n}}\int_\Gamma H\eta\cdot\mathbf{n}\,dA + \left(\frac{1}{n}-1\right)\frac{A(\Gamma)}{V(\Omega)^{2-1/n}}\int_\Gamma \eta\cdot\mathbf{n}\,dA, \end{aligned}$$

which implies

$$-\frac{\int H\eta\cdot\mathbf{n}\,dA}{\int \eta\cdot\mathbf{n}\,dA} = \frac{n-1}{n}\frac{A(\Gamma)}{V(\Omega)} \qquad \text{(II.1.9)}$$

(both integrals are over Γ). But since $V(\Omega) = \omega_n$, and Ω is a solution of the isoperimetric problem, then $A(\Gamma) \le \mathbf{c_{n-1}}$, which implies by (I.1.3),

$$-\frac{\int H\eta\cdot\mathbf{n}\,dA}{\int \eta\cdot\mathbf{n}\,dA} \le n-1 \tag{II.1.10}$$

(both integrals are over Γ).

For any $w_0 \in \Gamma$, let ϕ be a nonnegative C^∞ function compactly supported on a neighborhood of w_0 in $\mathbb{R}^n$, and choose

$$X(x,t) = \phi(x)\mathbf{n}_{w_0}$$

(so this particular vector field is time-independent). Then by picking ϕ with sufficiently small support about w_o one obtains from (II.1.10) that $-H(w_0) \le n-1$, which is the claim of the theorem. ∎

Remark II.1.1 The argument of the last paragraph, applied to (II.1.9), implies that $H = \text{const.}$, namely $H = (n-1)A(\Gamma)/nV(\Omega)$.

Theorem II.1.3 *Assume Ω is a bounded domain in $\mathbb{R}^n$, with C^2 boundary Γ, and* **n** *the exterior normal unit vector field along Γ. Assume the mean curvature H of Γ satisfies*

$$-H \le n-1$$

along all of Γ. Then

$$A(\Gamma) \ge \mathbf{c_{n-1}},$$

with equality if and only if Ω is an n-disk.

Proof Start with $\overline{\Omega} = \Omega \cup \Gamma$, and let C denote the convex hull of $\overline{\Omega}$, $\Sigma = \partial C$ the boundary of C.

Let $w \in \Sigma \setminus (\Sigma \cap \Gamma)$, Π a supporting hyperplane of C at w. (See Figure II.1.1.) Then there exists an $x \in \Pi \cap (\Sigma \cap \Gamma)$ (if not, then C would not be the smallest convex set containing $\overline{\Omega}$), which implies (a) Π is the tangent hyperplane to Γ at x, and (b) the line segment wx is contained in Π. These two imply that Π is the unique supporting hyperplane to C at every point of wx, except possibly at w. If Π were not the unique supporting hyperplane of C at w, then Σ would have a conical singularity at w, which would imply that C is not the minimal convex set containing $\overline{\Omega}$. Therefore, every point of Σ has a unique supporting hyperplane, that is, Σ is everywhere differentiable.

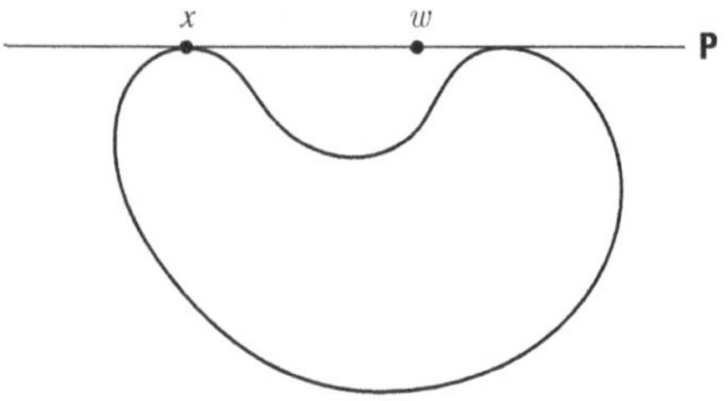

Figure II.1.1: The supporting hyperplane at x.

Since Σ is convex, everywhere differentiable, then every point of Σ has a unique exterior unit normal vector, and the Gauss map

$$\mathfrak{G} : \Sigma \to \mathbb{S}^{n-1}$$

that takes every point of $w \in \Sigma$ to its exterior unit normal vector $\mathsf{n}(w)$ is well defined. The Gauss map is onto; furthermore, the Gauss map restricted to $\Sigma \setminus (\Sigma \cap \Gamma)$ adds no new points to the image in $\mathbb{S}^{n-1}$ that are not already in the image of $\Sigma \cap \Gamma$, namely,

$$\mathfrak{G}(\Sigma \setminus (\Sigma \cap \Gamma)) \subseteq \mathfrak{G}(\Sigma \cap \Gamma).$$

Therefore, to calculate the image of the area of the Gauss map, that is, the area of $\mathbb{S}^{n-1}$, we restrict the integral to $\Sigma \cap \Gamma$.

In a coordinate system on Γ, as described in the calculations above, the Jacobian transformation of $\mathfrak{G}$, $\mathfrak{G}_*$, is given in local coordinates by the linear transformation described by $-G^{-1}\mathcal{B}$ [see (II.1.5)]. For points of $\Sigma \cap \Gamma$ we have $-\mathcal{B}$ positive semidefinite, which implies, by the arithmetic–geometric mean inequality,

$$\begin{aligned} \mathbf{c_{n-1}} &= \int_{\Sigma\cap\Gamma} \det(-G^{-1}\mathcal{B})\, dA \\ &\leq \int_{\Sigma\cap\Gamma} \left\{\frac{\operatorname{tr}(-G^{-1}\mathcal{B})}{n-1}\right\}^{n-1} dA \\ &= \int_{\Sigma\cap\Gamma} \left(\frac{-H}{n-1}\right)^{n-1} dA \\ &\leq A(\Sigma \cap \Gamma) \\ &\leq A(\Gamma), \end{aligned}$$

which implies the inequality.

If we have equality, then $\Sigma = \Gamma$. Also, we have equality in the arithmetic–geometric mean inequality, which implies all points are umbilics (that is, $G^{-1}\mathcal{B}$

is a scalar matrix at every point of Γ), which implies $-G^{-1}\mathcal{B}$ is the identity. Therefore (II.1.4) implies there exists a constant vector **b** such that

$$\mathbf{n} = \mathbf{x} - \mathbf{b}$$

on all of Γ. Thus

$$|\mathbf{x} - \mathbf{b}| = |\mathbf{n}| = 1,$$

which implies Γ is a unit sphere. ■

II.1.3 Uniqueness of an Extremal of the C^2 Isoperimetric Functional

Theorem II.1.4 *Let Ω be a bounded domain in $\mathbb{R}^n$, with C^2 boundary, that is an extremal of the C^2 isoperimetric functional. Then $\partial\Omega$ has constant mean curvature.*

Proof Use the notation of Theorem II.1.1, and set

$$\phi = \eta \cdot \mathbf{n}.$$

If Ω is an extremal for the isoperimetric functional then, for any choice of vector field X for which

$$V(\Phi_t(\Omega)) = \text{const.} \qquad \forall\, t,$$

(such a flow is called *incompressible*) we have, by (II.1.6),

$$\int_\Gamma \phi\, dA = 0,$$

and, by (II.1.7),

$$\int_\Gamma \phi H\, dA = 0.$$

Thus,

$$\int_\Gamma \phi H\, dA = 0 \qquad \forall\, \phi \;\text{ such that } \int_\Gamma \phi\, dA = 0.$$

A standard argument [see the proof of (I.2.3)] implies that $H = \text{const.}$ ■

We now write the equation for H as a second order elliptic quasilinear partial differential operator of divergence form. We assume the hypersurface Γ is given in nonparametric form, that is, the equation is described by

$$x^n = \phi(x^1, \ldots, x^{n-1}), \qquad \phi \in C^2,$$

over some domain in $\mathbb{R}^{n-1}$. So, in the parametric language,

$$x^j = u^j, \quad j = 1, \ldots, n-1, \qquad x^n = \phi(u^1, \ldots, u^{n-1}).$$

Then

$$\frac{\partial \mathbf{x}}{\partial u^j} = \mathbf{e}_j + \frac{\partial \phi}{\partial u^j}\mathbf{e}_n, \quad j = 1, \ldots, n-1, \qquad \mathbf{n} = \frac{-\text{grad}_{n-1}\phi + \mathbf{e}_n}{\sqrt{1 + |\text{grad}_{n-1}\phi|^2}},$$

(where grad_{n-1} denotes the gradient in $\mathbb{R}^{n-1}$), from which one calculates the metric tensor

$$g_{jk} = \delta_{jk} + \frac{\partial \phi}{\partial u^j}\frac{\partial \phi}{\partial u^k},$$

which we write as

$$G = I + A, \qquad A_{jk} = \frac{\partial \phi}{\partial u^j}\frac{\partial \phi}{\partial u^k}.$$

Note that

$$A^2 = |\text{grad}_{n-1}|^2 A,$$

which implies

$$G^{-1} = I - \frac{A}{1 + |\text{grad}_{n-1}|^2}.$$

Set

$$\Phi = \sqrt{1 + |\text{grad}_{n-1}|^2}, \qquad \mathcal{S}_{jk} = \frac{\partial^2 \phi}{\partial u^j \partial u^k}.$$

Then for the second fundamental form we have

$$b_{jk} = \frac{\partial^2 \mathbf{x}}{\partial u^j \partial u^k}\cdot\mathbf{n} = \frac{\partial^2 \phi}{\partial u^j \partial u^k}\mathbf{e}_n\cdot\mathbf{n} = \Phi^{-1}\mathcal{S}_{jk};$$

so $\mathcal{B} = \Phi^{-1}\mathcal{S}$. Then $H = \text{tr}G^{-1}\mathcal{B}$ implies

$$H = \Phi^{-1}\text{tr}(\{I - \Phi^{-2}A\}\mathcal{S}).$$

In what follows in this section (only), we substitute the notation ∇ for grad_{n-1}. Suppose we are given two fixed hypersurfaces

$$x^n = \phi(x^1, \ldots, x^{n-1}), \quad \phi \in C^2, \qquad x^n = \psi(x^1, \ldots, x^{n-1}), \quad \psi \in C^2,$$

over some domain in $\mathbb{R}^{n-1}$. Set

$$\Phi = \sqrt{1 + |\nabla\phi|^2}, \qquad \Psi = \sqrt{1 + |\nabla\psi|^2},$$

and let H_ϕ, H_ψ denote the mean curvatures associated with ϕ, ψ, respectively. Similarly, we use the notations G_ϕ and G_ψ, A_ϕ and A_ψ, $\mathcal{S}_\phi$ and $\mathcal{S}_\psi$, to indicate the quantities on each of the surfaces. Then direct calculation implies

$$H_\phi - H_\psi = \Phi^{-1}\mathrm{tr}(I - \Phi^{-2}A_\phi)\,\mathcal{S}_{\phi-\psi} + (\Phi^{-1} - \Psi^{-1})\,\mathrm{tr}(I - \Phi^{-2}A_\phi)\,\mathcal{S}_\psi$$
$$- \Psi^{-1}(\Psi^{-2} - \Phi^{-2})\,\mathrm{tr}A_\psi\mathcal{S}_\psi - \Psi^{-1}\Phi^{-2}\mathrm{tr}(A_\phi - A_\psi)\mathcal{S}_\psi.$$

Now

$$|\nabla\phi|^2 - |\nabla\psi|^2 = \nabla(\phi - \psi)\cdot\nabla(\phi + \psi),$$

which implies

$$\Phi^{-1} - \Psi^{-1} = -\frac{\nabla(\phi + \psi)}{\Phi^2\Psi^2(\Phi + \Psi)}\cdot\nabla(\phi - \psi)$$
$$\Phi^{-2} - \Psi^{-2} = -\frac{\nabla(\phi + \psi)}{\Phi^2\Psi^2}\cdot\nabla(\phi - \psi).$$

Also,

$$\begin{aligned}\mathrm{tr}(A_\phi - A_\psi)\mathcal{S}_\psi &= \sum_{j,k=1}^{n-1}(\partial_j\phi\partial_k\phi - \partial_j\psi\partial_k\psi)\mathcal{S}_{jk}\\ &= \sum_{j,k=1}^{n-1}\{(\partial_j\phi - \partial_j\psi)\partial_k\phi\mathcal{S}_{jk} + \partial_j\phi(\partial_k\phi - \partial_k\psi)\mathcal{S}_{jk}\}\\ &= \nabla(\phi + \psi)\mathcal{S}\cdot\nabla(\phi - \psi),\end{aligned}$$

where $\nabla(\phi + \psi)$ is written as a row vector.

Therefore there exist a vector field $\mathbf{B}$ on $\mathbb{R}^{n-1}$ such that

$$H_\phi - H_\psi = \mathcal{A}(\phi - \psi) + \mathbf{B}\cdot\nabla(\phi - \psi),$$

where the operator $\mathcal{A}$ is given by

$$\mathcal{A}w = \Phi^{-1}\mathrm{tr}(I - \Phi^{-2}A_\phi)\mathcal{S}_w.$$

We immediately have

Lemma II.1.1 *Consider the hypersurfaces in $\mathbb{R}^{n-1}$ determined by the functions ϕ and ψ, define $\mathcal{A}$ and $\mathbf{B}$ as above, and assume the two hypersurfaces have identical mean curvature. Consider the second order linear elliptic differential operator on $\mathbb{R}^{n-1}$ given by*

$$\mathbf{L}w = \mathcal{A}w + \mathbf{B}\cdot\nabla w.$$

Then the function $w = \phi - \psi$ satisfies the second order linear equation

$$\mathbf{L}w = 0.$$

Theorem II.1.5 *Let Ω be a bounded domain in $\mathbb{R}^n$, with C^2 boundary Γ, and assume Γ has constant mean curvature. Then Ω is an n-disk in $\mathbb{R}^n$.*

Proof Set $K = \overline{\Omega} = \Omega \cup \Gamma$.

Fix any unit vector $\boldsymbol{\xi} \in \mathbb{S}^{n-1}$, and pick Euclidean coordinates in $\mathbb{R}^n$ so that

$$\inf_{x \in K} x^n = 0$$

(where x^n is the nth coordinate of x). Then there exists $w_o \in \Gamma \cap \{\mathbb{R}^{n-1} \times \{0\}\}$ with tangent plane of Γ at w_o equal to $\mathbb{R}^{n-1} \times \{0\}$.

For each real α let Π_α denote the hyperplane

$$\Pi_\alpha = \mathbb{R}^{n-1} \times \{\alpha\},$$

and set

$$K_\alpha^- = K \cap \{x^n \leq \alpha\}, \qquad K_\alpha^+ = K \cap \{x^n \geq \alpha\},$$
$$\Gamma_\alpha^- = \Gamma \cap \{x^n < \alpha\}, \qquad \Gamma_\alpha^+ = \Gamma \cap \{x^n > \alpha\}.$$

Let

$$\text{rfl } K_\alpha^- = \text{reflection of } K_\alpha^- \text{ in } \Pi_\alpha,$$
$$\text{rfl } \Gamma_\alpha^- = \text{reflection of } \Gamma_\alpha^- \text{ in } \Pi_\alpha.$$

(See Figure II.1.2.) Then for sufficiently small $\alpha > 0$, Γ_α^- has a nonparametric representation over Π_α, and

$$\text{rfl } K_\alpha^- \subseteq K_\alpha^+.$$

Set

$$\beta = \sup \{\alpha_0 : \text{rfl } K_\alpha^- \subseteq K_\alpha^+ \ \forall \alpha \leq \alpha_0\}.$$

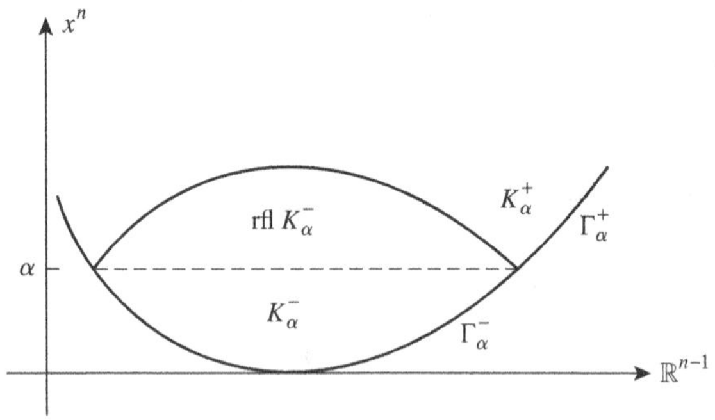

Figure II.1.2: The reflection in Π_α.

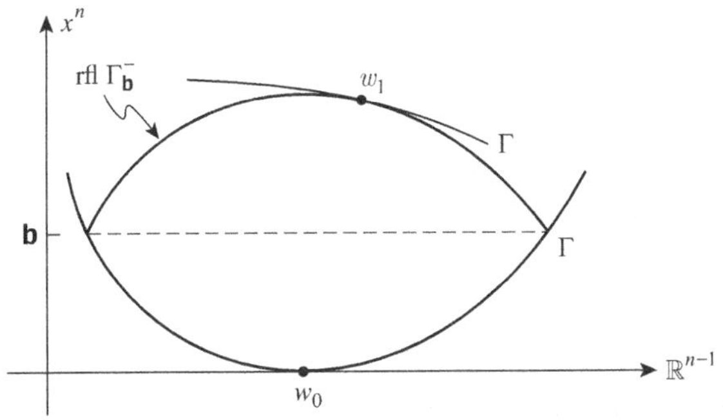

Figure II.1.3: Possibility (a).

Then one has two possibilities:

(a) There exists $w_1 \in \text{rfl}\,\Gamma_\beta^- \cap \Gamma_\beta^+$, in which case rfl Γ_β^- and Γ_β^+ are tangent at w_1. (See Figure II.1.3.)
(b) There exists $w_2 \in \Gamma \cap \Pi_\beta$ such that the tangent hyperplane to Γ at w_2 is perpendicular to Π_β. (See Figure II.1.4.)

In either case (a) or (b) one may introduce Euclidean coordinates $y = (y^1, \ldots, y^n)$ in $\mathbb{R}^n$ about w – which is either w_1 or w_2 as the case may be – so that Γ_β^+ and rfl Γ_β^- are described by

$$\left.\begin{array}{rl} \Gamma_\beta^+ : & y^n = \phi(\overline{y}) \\ \text{rfl}\,\Gamma_\beta^- : & y^n = \psi(\overline{y}) \end{array}\right\}, \qquad \overline{y} = (y^1, \ldots, y^{n-1}),$$

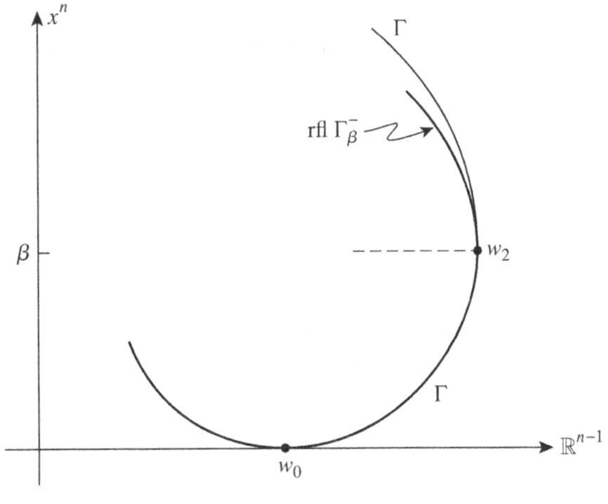

Figure II.1.4: Possibility (b).

satisfying

$$\phi(\overline{\mathsf{w}}) = \psi(\overline{\mathsf{w}}) = 0, \qquad \nabla\phi(\overline{\mathsf{w}}) = \nabla\psi(\overline{\mathsf{w}}) = 0.$$

[so $\mathsf{w} = (\overline{\mathsf{w}}, 0)$]. Thus, the hyperplane $y^n = 0$ is the tangent plane of the two hypersurfaces at w. In case (a) we have ϕ and ψ defined on some open set $U \subseteq \mathbb{R}^{n-1}$ about $\overline{\mathsf{w}}$, with

$$\phi \geq \psi \text{ on } U, \qquad \phi(\overline{\mathsf{w}}) = \psi(\overline{\mathsf{w}}).$$

The strong maximum principle (Proposition I.3.9), applied to Lemma II.1.1, then implies $\phi = \psi$ on a neighborhood of $\overline{\mathsf{w}}$, and therefore on all of U. Thus K is symmetric with respect to Π_β. Similarly, in case (b) we have ϕ and ψ defined on an open set U in $\mathbb{R}^{n-1}$ with $\overline{\mathsf{w}} \in \partial U$, and

$$\phi \geq \psi \text{ on } U, \qquad \phi(\overline{\mathsf{w}}) = \psi(\overline{\mathsf{w}}), \qquad \left.\frac{\partial\phi}{\partial\boldsymbol{\nu}_U}\right|_{\overline{\mathsf{w}}} = \left.\frac{\partial\psi}{\partial\boldsymbol{\nu}_U}\right|_{\overline{\mathsf{w}}},$$

where $\boldsymbol{\nu}_U$ denotes the exterior normal (with respect to U) unit vector at $\overline{\mathsf{w}}$. Again, the strong maximum principle implies $\phi = \psi$ on all of U. Thus K is, again, symmetric with respect to Π_β.

We conclude that given any $\boldsymbol{\xi} \in \mathbb{S}^{n-1}$, then K is symmetric with respect to some hyperplane perpendicular to $\boldsymbol{\xi}$. But then Theorem I.3.5 implies ∂K is a finite union of concentric spheres. Since the mean curvature is constant on *all* of Γ, one concludes that K is a closed disk. ∎

II.2 The C^1 Isoperimetric Inequality

We first establish the equivalence of the geometric isoperimetric inequality on $\mathbb{R}^n$ to an analytic L^1-Sobolev inequality.

Definition Define the *isoperimetric constant* $\mathfrak{I}$ *of* $\mathbb{R}^n$ by

$$\mathfrak{I} = \inf_\Omega \frac{A(\partial\Omega)}{V(\Omega)^{1-1/n}},$$

where Ω varies over bounded domains in $\mathbb{R}^n$ with C^1 boundary.

Define the *Sobolev constant* $\mathfrak{S}$ *of* $\mathbb{R}^n$ by

$$\mathfrak{S} = \inf_f \frac{\|\text{grad } f\|_1}{\|f\|_{n/(n-1)}},$$

where f varies over $C_c^\infty(\mathbb{R}^n)$.

Theorem II.2.1 (The Federer–Fleming Theorem) *The isoperimetric and Sobolev constants are equal, that is,*

$$\mathfrak{I} = \mathfrak{S}. \tag{II.2.1}$$

Proof Let Ω be any bounded domain in $\mathbb{R}^n$ with C^1 boundary. For sufficiently small $\epsilon > 0$ consider the function

$$f_\epsilon(x) = \begin{cases} 1, & x \in \Omega, \\ 1 - (1/\epsilon)d(x, \partial\Omega), & x \in \mathbb{R}^n \setminus \Omega, \quad d(x, \partial\Omega) < \epsilon, \\ 0, & x \in \mathbb{R}^n \setminus \Omega, \quad d(x, \partial\Omega) \geq \epsilon. \end{cases}$$

Then f_ϵ is Lipschitz for every ϵ, and (by Theorem I.3.3) we may approximate f_ϵ by functions $\phi_{\epsilon,j} \in C_c^\infty(\mathbb{R}^n)$ for which

$$\|\phi_{\epsilon,j} - f_\epsilon\|_{n/(n-1)} \to 0, \qquad \|\operatorname{grad} \phi_{\epsilon,j} - \operatorname{grad} f_\epsilon\|_1 \to 0$$

as $j \to \infty$. One has

$$\mathfrak{S} \leq \frac{\|\operatorname{grad} f_\epsilon\|_1}{\|f_\epsilon\|_{n/(n-1)}}.$$

One checks that

$$\lim_{\epsilon \downarrow 0} \int_{\mathbb{R}^n} |f_\epsilon|^{n/(n-1)} \, dV = V(\Omega).$$

Also,

$$|\operatorname{grad} f_\epsilon| = \begin{cases} 1/\epsilon, & x \in \mathbb{R}^n \setminus \Omega, \quad d(x, \partial\Omega) < \epsilon, \\ 0 & \text{otherwise}, \end{cases}$$

which implies

$$\lim_{\epsilon \downarrow 0} \int_{\mathbb{R}^n} |\operatorname{grad} f_\epsilon| \, dV = \lim_{\epsilon \downarrow 0} \frac{V(\{x \notin \Omega : d(x, \partial\Omega) < \epsilon\})}{\epsilon} = A(\partial\Omega).$$

Thus,

$$\mathfrak{S} \leq \lim_{\epsilon \downarrow 0} \frac{\|\operatorname{grad} f_\epsilon\|_1}{\|f_\epsilon\|_{n/(n-1)}} = \frac{A(\partial\Omega)}{V(\Omega)^{1-1/n}}$$

for all such Ω, from which we conclude

$$\mathfrak{S} \leq \mathfrak{I}.$$

It remains to prove the opposite inequality, that is,

$$\int_{\mathbb{R}^n} |\operatorname{grad} f| \, dV \geq \mathfrak{I} \left\{ \int_{\mathbb{R}^n} |f|^{n/(n-1)} \, dV \right\}^{1-1/n} \tag{II.2.2}$$

for all $f \in C_c^\infty(\mathbb{R}^n)$. ∎

Lemma II.2.1 *Given any* $\phi \in C_c^\infty(\mathbb{R}^n)$, *then*

$$\text{(II.2.3)} \qquad |\operatorname{grad} |\phi|\,| = |\operatorname{grad} \phi|$$

almost everywhere on $\mathbb{R}^n$.

Proof On the open set $\{\phi > 0\}$ we have $|\phi| = \phi$, and on the open set $\{\phi < 0\}$ we have $|\phi| = -\phi$, and (II.2.3) is certainly valid. Also, $\{\phi = 0\} \cap \{\operatorname{grad} \phi \neq 0\}$ is an $(n-1)$-submanifold of $\mathbb{R}^n$, which implies $\mathbf{v}_n(\{\phi = 0\} \cap \{\operatorname{grad} \phi \neq 0\}) = 0$.

It remains to consider what happens when $\phi = 0$ and $\operatorname{grad} \phi = 0$. Set, for any $\epsilon > 0$,

$$\phi_\epsilon = \sqrt{\phi^2 + \epsilon^2}.$$

Then $\phi_\epsilon \to |\phi|$ as $\epsilon \downarrow 0$, and $\operatorname{grad} \phi_\epsilon \to \pm\operatorname{grad} \phi$ as $\epsilon \downarrow 0$ when $\phi \neq 0$ (depending on whether ϕ is positive or negative at the point in question). When $\phi = 0$, then $\operatorname{grad} \phi_\epsilon = 0$; therefore $\operatorname{grad} \phi_\epsilon \to \operatorname{grad} \phi$ when $\phi = 0$ and $\operatorname{grad} \phi = 0$. Integration by parts implies $\operatorname{grad} \phi_\epsilon$ converges to the weak derivative of $|\phi|$ in $W^{1,1}$, which is $\operatorname{grad} |\phi|$. ∎

Proof of (II.2.2) Given $f \in C_c^\infty(\mathbb{R}^n)$, let

$$\Omega(t) = \{x : |f|(x) > t\}, \qquad V(t) = V(\Omega(t)).$$

Then the co-area formula (Corollary I.3.1) implies

$$\int_{\mathbb{R}^n} |\operatorname{grad} f|\, dV = \int_0^\infty A(|f|^{-1}[t])\, dt \geq \mathfrak{I} \int_0^\infty V(t)^{1-1/n}\, dt,$$

and, by Cavalieri's principle (Proposition I.3.3),

$$\int_{\mathbb{R}^n} |f|^{n/(n-1)}\, dV = \frac{n}{n-1} \int_0^\infty t^{1/(n-1)} V(t)\, dt.$$

So to prove (II.2.2) it suffices to show

$$\text{(II.2.4)} \qquad \int_0^\infty V(t)^{1-1/n}\, dt \geq \left\{ \frac{n}{n-1} \int_0^\infty t^{1/(n-1)} V(t)\, dt \right\}^{1-1/n}.$$

To establish (II.2.4) set

$$F(s) = \int_0^s V(t)^{1-1/n}\, dt, \qquad G(s) = \left\{ \frac{n}{n-1} \int_0^s t^{1/(n-1)} V(t)\, dt \right\}^{1-1/n}.$$

First,

$$F(0) = G(0);$$

second, since $V(s)$ is a decreasing function of s, we have

$$\begin{aligned} G'(s) &= \frac{n-1}{n}\left[\frac{n}{n-1}\right]^{1-1/n}\left\{\int_0^s t^{1/(n-1)}V(t)\,dt\right\}^{-1/n} s^{1/(n-1)}V(s) \\ &\le \left[\frac{n}{n-1}\right]^{-1/n}\left\{\int_0^s t^{1/(n-1)}\,dt\right\}^{-1/n} s^{1/(n-1)}V(s)^{1-1/n} \\ &= V(s)^{1-1/n} \\ &= F'(s). \end{aligned}$$

Then (II.2.4) follows immediately. ∎

Theorem II.2.2 (Isoperimetric Inequality for C^1 Surface Area) *Let Ω be a bounded domain in $\mathbb{R}^n$, with C^1 boundary $\partial\Omega$. Then*

(II.2.5)
$$\frac{A(\partial\Omega)}{V(\Omega)^{1-1/n}} \ge n\boldsymbol{\omega_n}^{1/n} = \frac{A(\mathbb{S}^{n-1})}{V(\mathbb{B}^n)^{1-1/n}}.$$

Proof By the Federer–Fleming theorem (Theorem II.2.1), it suffices to show that

(II.2.6)
$$\int |\text{grad } f|\,dV \ge n\boldsymbol{\omega_n}^{1/n}\|f\|_{n/(n-1)}$$

for all $f \in C_c^\infty(\mathbb{R}^n)$.

Set $\mathcal{C} = [0,1]^n$, the unit n-cube in $\mathbb{R}^n$. To prove (II.2.6) we first consider, for any nonnegative function $\mu \in C_c^1(\mathbb{R}^n)$, the maps $u^j : \mathbb{R}^n \to [0,\infty)$ defined by

$$u^j(x) = \int_{-\infty}^{+\infty} d\xi^j \int_{\mathbb{R}^{n-j}} \mu(x^1,\dots,x^{j-1},\xi^j,\xi^{j+1},\dots,\xi^n)\,d\xi^{j+1}\cdots d\xi^n,$$

$j = 1,\dots,n$, and the function $\eta : \mathbb{R}^n \to \mathcal{C}$ defined by

$$\eta^j(x) = \frac{1}{u^j(x)}\int_{-\infty}^{x^j} d\xi^j \int_{\mathbb{R}^{n-j}} \mu(x^1,\dots,x^{j-1},\xi^j,\xi^{j+1},\dots,\xi^n)\,d\xi^{j+1}\cdots d\xi^n.$$

Then $\boldsymbol{\eta} = (\eta^1,\dots,\eta^n)$ maps $\mathbb{R}^n$ *onto* $\mathcal{C}$, and

$$\eta^j - \eta^j(x^1,\dots,x^j), \qquad \frac{\partial\eta^j}{\partial x^j} = \frac{u^{j+1}}{u^j}$$

[where we set $u^{n+1}(x) = \mu(x)$], which implies

$$\det J_\eta(x) = \prod_{j=1}^n \frac{\partial\eta^j}{\partial x^j}(x) = \left\{\int_{\mathbb{R}^n} \mu(x)\,dx\right\}^{-1}\mu(x),$$

where $J_\eta(x)$ denotes the Jacobian linear transformation of $\boldsymbol{\eta}$ at x.

The best way to appreciate the mapping is to interpret it when Ω is a bounded domain in $\mathbb{R}^n$ with C^1 boundary, and $\mu = \mathcal{I}_\Omega$, the indicator function of the domain Ω [despite the fact that, in this case, $\mu \notin C_c^1(\mathbb{R}^n)$]. Then

$$u^1(x) = \mathbf{v}_n(\Omega)$$
$$\eta^1(x) = \frac{\mathbf{v}_n(\Omega \cap \{\xi \in \mathbb{R}^n : \xi^1 < x^1\})}{\mathbf{v}_n(\Omega)},$$

and

$$u^2(x) = \mathbf{v}_{n-1}(\Omega \cap \{\xi \in \mathbb{R}^n : \xi^1 = x^1\}),$$
$$\eta^2(x) = \frac{\mathbf{v}_{n-1}(\Omega \cap \{\xi \in \mathbb{R}^n : \xi^1 = x^1,\ \xi^2 < x^2\})}{\mathbf{v}_{n-1}(\Omega \cap \{\xi \in \mathbb{R}^n : \xi^1 = x^1\})},$$

and, more generally,

$$u^j(x) = \mathbf{v}_{n-j+1}(\Omega \cap \{\xi : \xi^1 = x^1, \dots, \xi^{j-1} = x^{j-1}\}),$$
$$\eta^j(x) = \frac{\mathbf{v}_{n-j+1}(\Omega \cap \{\xi : \xi^1 = x^1, \dots, \xi^{j-1} = x^{j-1},\ \xi^j < x^j\})}{\mathbf{v}_{n-j+1}(\Omega \cap \{\xi : \xi^1 = x^1, \dots, \xi^{j-1} = x^{j-1}\})}.$$

(See Figure II.2.1.)

If Ω is convex, then $\boldsymbol{\eta} \in C^1(\Omega) \cap C^0(\overline{\Omega})$, and $J\boldsymbol{\eta}(x)$ extends to a continuous matrix function on $\overline{\Omega}$. A special case is $\Omega = \mathbb{B}^n$, in which case we denote $\boldsymbol{\eta}$ corresponding to $\mathcal{I}_{\mathbb{B}^n}$ by Φ. Then $\Phi|\overline{\mathbb{B}^n} : \overline{\mathbb{B}^n} \to C$ is a diffeomorphism, with

$$\det J_\Phi = \boldsymbol{\omega_n}^{-1}.$$

We return to the general $\mu \in C_c^1(\mathbb{R}^n)$, set

$$\boldsymbol{\zeta} = \Phi^{-1} \circ \boldsymbol{\eta} : \mathbb{R}^n \to \overline{\mathbb{B}^n},$$

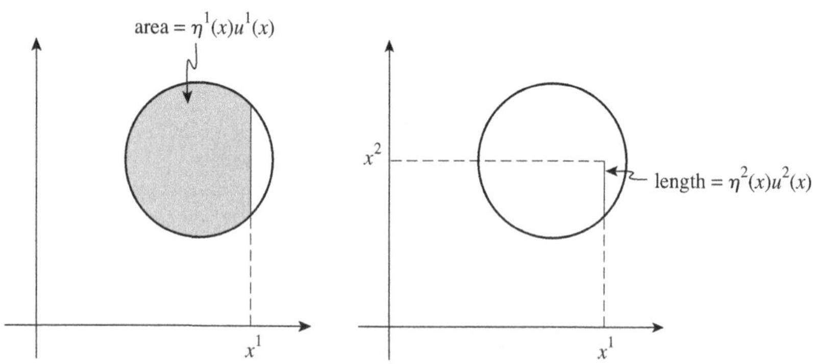

Figure II.2.1: When $\mu = \mathcal{I}_{\mathbb{B}^2}$.

and think of ζ as a vector field on $\mathbb{R}^n$ with $|\zeta| \le 1$, as well as a mapping of $\mathbb{R}^n$ onto $\overline{\mathbb{B}^n}$. Then

$$\left\{\int_{\mathbb{R}^n} \mu(x)\,dx\right\}^{-1} \omega_n \mu(x) = \det J_\zeta(x) \le \left\{\frac{\operatorname{div}\zeta}{n}\right\}^n (x) \tag{II.2.7}$$

by the arithmetic–geometric mean inequality.

To apply the above, given any *nonnegative* $f \in C_c^\infty$, consider

$$\mu = f^{n/(n-1)} = f^1 f^{1/(n-1)} = f\mu^{1/n}.$$

Then (II.2.7) implies

$$\begin{aligned}
\mu(x) &= f(x)\mu(x)^{1/n} \\
&\le \{n\omega_n{}^{1/n}\}^{-1}\left\{\int_{\mathbb{R}^n} \mu\right\}^{1/n} (f \operatorname{div}\zeta)(x) \\
&= \frac{\|f\|_{n/(n-1)}{}^{1/(n-1)}}{n\omega_n{}^{1/n}}\{\operatorname{div} f\zeta - \operatorname{grad} f \cdot \zeta\}(x),
\end{aligned}$$

which implies

$$\|f\|_{n/(n-1)} \le \frac{-1}{n\omega_n{}^{1/n}}\int_{\mathbb{R}^n} \operatorname{grad} f \cdot \zeta \, dV \le \frac{1}{n\omega_n{}^{1/n}}\int_{\mathbb{R}^n} |\operatorname{grad} f|\, dV, \tag{II.2.8}$$

which implies (II.2.6) for $f \ge 0$.

For arbitrary $f \in C_c^1$, let (j_ϵ) be a mollifier on $\mathbb{R}^n$ and set $F_\epsilon = j_\epsilon * |f|$. Then (II.2.8) implies

$$\|F_\epsilon\|_{n/(n-1)} \le \frac{1}{n\omega_n{}^{1/n}}\int_{\mathbb{R}^n} |\operatorname{grad} F_\epsilon|\, dV.$$

Let $\epsilon \downarrow 0$; then, by Lemma II.2.1,

$$\|f\|_{n/(n-1)} \le \frac{1}{n\omega_n{}^{1/n}}\int_{\mathbb{R}^n} |\operatorname{grad}|f||\, dV = \frac{1}{n\omega_n{}^{1/n}}\int_{\mathbb{R}^n} |\operatorname{grad} f|\, dV. \qquad \blacksquare$$

To consider the case of equality in (II.2.5), we consider, more closely, what happens when $\mu = I_\Omega$, for the bounded domain Ω with C^1 boundary. But we must *assume that* Ω *is convex* to guarantee that the mapping $\zeta \in C^1(\Omega) \cap C^0(\overline{\Omega})$, and to guarantee that $J_\zeta(x)$ extends to a continuous matrix function on $\overline{\Omega}$. First, we repeat the argument for the inequality. For all $x \in \Omega$, (II.2.7) implies

$$1 \le \frac{V(\Omega)^{1/n}}{n\omega_n{}^{1/n}} \operatorname{div}\zeta(x), \tag{II.2.9}$$

which implies, by the standard divergence theorem,

$$V(\Omega) \leq \frac{V(\Omega)^{1/n}}{n{\omega_n}^{1/n}} \int_{\partial\Omega} \zeta \cdot \nu \, dA \leq \frac{V(\Omega)^{1/n}}{n{\omega_n}^{1/n}} A(\partial\Omega), \tag{II.2.10}$$

which is the original version of the isoperimetric inequality.

If we have equality, then we have equality in (II.2.7) and (II.2.9) – which implies

$$\frac{\partial \zeta^j}{\partial x^j} = \left\{ \frac{\omega_n}{V(\Omega)} \right\}^{1/n}, \quad j = 1, \ldots, n, \tag{II.2.11}$$

on all of Ω; and we have equality in (II.2.10) – which implies

$$\zeta|_{\partial\Omega} = \left\{ \frac{\omega_n}{V(\Omega)} \right\}^{1/n} \nu \tag{II.2.12}$$

on all of $\partial\Omega$. For convenience we assume $V(\Omega) = \omega_n$; so $\omega_n / V(\Omega) = 1$.

From (II.2.11) we conclude, since $\zeta^j = \zeta^j(x^1, \ldots, x^j)$, that

$$\zeta^j = x^j + \alpha^j(x^1, \ldots, x^{j-1})$$

for all $j = 1, \ldots, n$ (for $j = 1$, $\alpha^1 = \text{const.}$). On the boundary of Ω we have (since $\zeta|\partial\Omega : \partial\Omega \to \mathbb{S}^{n-1}$)

$$\begin{aligned} 0 &= \sum_{j=1}^{n} \zeta^j \, d\zeta^j \\ &= \sum_{j=1}^{n} \zeta^j \left\{ dx^j + \sum_{k=1}^{j-1} \frac{\partial \alpha^j}{\partial x^k} \, dx^k \right\} \\ &= \sum_{j=1}^{n} \left\{ \zeta^j + \sum_{k=j+1}^{n} \zeta^k \frac{\partial \alpha^k}{\partial x^j} \right\} dx^j. \end{aligned}$$

From (II.2.12) we conclude that there exists a constant λ (depending at most on the point of the boundary of Ω) such that

$$\zeta^j + \sum_{k=j+1}^{n} \zeta^k \frac{\partial \alpha^k}{\partial x^j} = \lambda \zeta^j$$

for all $j = 1, \ldots, n$. From the case $j = n$ we have $\lambda = 1$ for all $\zeta^n \neq 0$, which implies

$$\sum_{k=j+1}^{n} \zeta^k \frac{\partial \alpha^k}{\partial x^j} = 0$$

for all points of $\partial\Omega$ for which $\zeta^n \neq 0$.

Fix $x^1, \ldots, x^{n-2}$. Then

$$\frac{\partial \alpha^n}{\partial x^{n-1}} = 0,$$

which implies that the locus $\{x^\tau = \text{const.}_\tau : \tau = 1, \ldots, n-2\} \cap \partial\Omega$ is a circle. Thus the intersection of any 2-plane with $\partial\Omega$ is a translate of a circle on the sphere. Thus the intersection of Ω with any 2-plane is a 2-disk of radius less than or equal to 1.

Now the map ζ takes Ω *onto* $\mathbb{B}^n$. Therefore there must exist at least one 2-plane for which its intersection with Ω is a 2-disk of radius equal to 1. Pick two antipodal points m_1 and m_2 on the boundary of this 2-disk of radius equal to 1. Then for *any* 2-plane through m_1 and m_2 the intersection of this new 2-plane with Ω is a 2-disk; since this new 2-disk contains m_1 and m_2, it must also have radius equal to 1. We conclude that for every 2-plane through m_1 and m_2 the intersection of the 2-plane with Ω is a 2–disk with radius equal to 1. But then Ω itself is an n-disk of radius equal to 1 and center at the midpoint of m_1 and m_2.

II.3 Bibliographic Notes

§II.1.1 The calculations of this section are classical, and one can find the necessary background in the standard books. Nowadays, it is not fashionable to do the calculations in local coordinates, but for a hypersurface in $\mathbb{R}^n$ they are still extremely useful.

§II.1.2 The paper of F. Almgren (1986), on which this section is based, is far more ambitious than the result presented here. In fact, our formulation constitituted only the heuristic for his results. Almgren starts with a $(k-1)$-dimensional submanifold Γ (not necessarily a hypersurface) of $\mathbb{R}^n$, $k \leq n$, and then seeks a k-submanifold Ω with boundary Γ that minimizes k-dimensional measure among all k-submanifolds bounded by Γ. For the pair (Ω, Γ), he then proves the k-dimensional isoperimetric inequality. The first version of this theorem (the isoperimetric inequality for minimal surfaces) goes back to Carleman (1921). But Carleman's theorem assumes the existence of the minimal surface, and that it is topologically a disk. Here, the existence must be proven as well. Since it is known that one cannot expect (in sufficiently high dimensions) an everywhere smooth area-minimizing submanifold Ω spanning Γ, one must broaden the competition of submanifolds Ω to more general geometric objects. But then one can include more general Γ. We also note that methods are sufficiently powerful to yield characterization of equality in the isoperimetric inequality, namely, it is realized if and only if Γ is a $(k-1)$-sphere bounding a k-disk Ω.

§II.1.3 Our proof of Alexandrov's theorem (Theorem II.1.5) follows his original (1962). A subsequent analytic proof was given by Reilly (1977). A more recent proof was given in Ros (1988); see the presentation in Oprea (1997), Chapter 4. Earlier results can be found in Liebmann (1900) and Hopf (1950).

§II.2 Federer–Fleming theorem was first proven independently in Federer and Fleming (1960) and Maz'ya (1960); and the isoperimetric inequality for C^1 surface area is from Gromov (1986), based on an idea of Knothe (1957).

III

Minkowski Area and Perimeter

In this chapter, we take some first steps in geometric measure theory, namely, we consider general compact sets, irrespective of the smoothness of their boundaries; and our prime method for proving the isoperimetric inequality for them (after we give suitable definitions of the area of the boundary) is by Steiner symmetrization. The symmetrization argument has great appeal to geometric intuition (although its first presentation by Steiner was incomplete), so we work with it in some detail. There are many other symmetrization schemes (see the bibliographic notes at the end of the chapter); but Steiner symmetrization lends itself, especially in the case of Minkowski area, to very simple arguments for the isoperimetric inequality. Moreover, we are able, in the case of C^1 boundary, to stay within the arguments of Minkowski area and Steiner symmetrization to characterize the disk as the only case of equality. The argument of this last characterization seems to be new, albeit unnecessary – because we shall characterize equality in the more general setting of finite perimeter. However, it is a useful preparation for the general argument. Also, the C^1 isoperimetric inequality suffices to prove the Faber–Krahn inequality (§III.3 below) with characterization of equality.

There are two types of possible definitions of areas of the boundary ∂K of an arbitrary compact set K:

1. The area of ∂K views ∂K as a subset of $\mathbb{R}^n$ without, necessarily, any reference to K. Two examples are the $(n - 1)$-dimensional Riemannian measure of a C^1 hypersurface of $\mathbb{R}^n$, and the $(n - 1)$-dimensional Hausdorff measure of *any arbitrary* subset F of $\mathbb{R}^n$ (to be considered below in Chapter IV). In each example, the two definitions are simply applied to ∂K.
2. The second type of definition of area is a functional defined on the collection of compact subsets of $\mathbb{R}^n$, which assigns to each compact subset a number meant to describe the area of its boundary. The examples we consider

below are Minkowski area and perimeter. Our most general formulation of the isoperimetric inequality for compact sets uses perimeter instead of Minkowski area; and we are able, in the category of finite perimeter, to characterize the case of equality. So, for us, this will be the optimal solution to the isoperimetric problem.

Practical differences between Minkowski area and perimeter include the following:

1. The perimeter functional is lower semicontinuous with respect to convergence of indicator functions in L^1, whereas no such semicontinuity exists, in general, for the Minkowski area functional with respect to Hausdorff convergence of compact sets. Nevertheless, one does have a restricted version of lower semicontinuity for Minkowski area (see the end of the proof of Theorem III.2.8 below).
2. It is possible for a compact subset of $\mathbb{R}^n$ to have n-dimensional Lebesgue measure equal to 0 but its boundary to have positive Minkowski area, which is not the case with perimeter.
3. Minkowski area has no natural localization for a relative neighborhood in the boundary (but see a partial attempt in the proof of Theorem III.2.4), whereas perimeter does.
4. Finally, the isoperimetric inequality for perimeter is sharper than the isoperimetric inequality for Minkowski area.

We note that a third example of the first type of area is the $(n-1)$-*dimensional integral geometric area* $IGA^{n-1}(F)$ of a subset F of $\mathbb{R}^n$, given by Corollary I.3.4,

$$IGA^{n-1}(F) = \frac{1}{2\omega_{n-1}} \int_{\mathbb{S}^{n-1}} d\mu_{n-1}(\xi) \int_{\xi^\perp} \operatorname{card}\left(F \cap p_\xi^{-1}[y]\right) d\mathbf{v}_{n-1}(y),$$

where p_ξ denotes projection onto the hyperplane $\xi^\perp$. But we do not go into any details here.

III.1 The Hausdorff Metric on Compacta

Definition Let (X, d) be a metric space. For any compact subset K of X, define the *circumradius* of K, $r(K)$, by

$$r(K) = \inf\{\rho > 0 : K \subset D(x;\rho) \text{ for some } x \in X\}.$$

Notation (Recall.) $D(x;\rho)$ denotes the *closed* n-disk in $\mathbb{R}^n$ of radius ρ, and centered at x.

Notation (Recall.) Let (X, d) be a metric space. For any subset A in X, and $\epsilon > 0$, $[A]_\epsilon$ denotes the *ϵ-thickening of A*, defined by

$$[A]_\epsilon = \{x \in X : d(x, A) \leq \epsilon\}.$$

Remark III.1.1 If $A \subset [B]_r$ and $B \subset [C]_\epsilon$, then $A \subset [C]_{r+\epsilon}$. Also, if K is compact, then

$$K = \bigcap_{\epsilon>0} [K]_\epsilon.$$

Definition Let $\mathbf{X}$ denote the collection of compact subsets of X. Given E and F in $\mathbf{X}$, define their *Hausdorff distance*, $\delta(E, F)$, by

$$\delta(E, F) = \inf\{\epsilon > 0 : E \subset [F]_\epsilon,\ F \subset [E]_\epsilon\}.$$

One checks that $(E, F) \mapsto \delta(E, F)$ satisfies the axioms of a metric space.

Lemma III.1.1 *Let X be an arbitrary metric space. Then the functions*

$$K \mapsto r(K), \qquad K \mapsto \operatorname{diam} K$$

are continuous on $\mathbf{X}$.

If μ is a positive measure on X that is finite on compact subsets, then

$$K \mapsto \mu(K)$$

is upper semicontinuous on $\mathbf{X}$, that is,

$$\limsup \mu(E_j) \leq \mu(\lim E_j)$$

for any convergent sequence (E_j) in $\mathbf{X}$.

Proof If $\epsilon > 0$, and $E, F \in \mathbf{X}$ satisfy $\delta(E, F) < \epsilon$, then, for any $x \in X$ and $R > 0$ for which we have $E \subset D(x; R)$, we must also have $F \subset D(x; R + \epsilon)$. This implies

$$r(F) \leq r(E) + \epsilon.$$

By switching the roles of E and F, we obtain

$$|r(E) - r(F)| \leq \epsilon,$$

which implies the claim for the circumradius.

Similarly, if $E, F \in \mathbf{X}$, and $E \subset [F]_\epsilon$, then

$$\operatorname{diam} E \leq \operatorname{diam} [F]_\epsilon \leq \operatorname{diam} F + 2\epsilon.$$

One easily has that $\delta(E, F) < \epsilon$ implies $|\operatorname{diam} E - \operatorname{diam} F| < 2\epsilon$.

If $E_j \to E$, then, given any $\epsilon > 0$, we have $E_j \subset [E]_\epsilon$ for sufficiently large j, which implies $\mu(E_j) \leq \mu([E]_\epsilon)$ for sufficiently large j, which implies

$$\limsup \mu(E_j) \leq \mu([E]_\epsilon)$$

for all $\epsilon > 0$, which implies the claim. ∎

Theorem III.1.1 (Blaschke Selection Theorem) *Assume that (X, d) has the property that closed and bounded subsets are compact. Then $\mathbf{X}$, the space of compact subsets of X, is complete. Furthermore, if X is compact then $\mathbf{X}$ is compact.*

Proof Let (E_j) be a Cauchy sequence in $\mathbf{X}$. Then consider

$$E = \bigcap_{j=1}^{\infty} \operatorname{cl} \bigcup_{i=j}^{\infty} E_i .$$

We claim that E_j converges to E in the Hausdorff metric.

The Cauchy hypothesis implies that

$$\bigcup_{i \geq j} E_i$$

is bounded for each j, which implies that

$$\operatorname{cl} \bigcup_{i \geq j} E_i$$

is compact for each j, which implies that E is nonempty compact.

Given $\epsilon > 0$, there exists $N_\epsilon > 0$ such that

$$\delta(E_i, E_j) < \epsilon \qquad \forall\, i, j \geq N_\epsilon,$$

which implies

$$\operatorname{cl} \bigcup_{i \geq j} E_i \subset [E_j]_\epsilon \qquad \forall\, j \geq N_\epsilon,$$

which implies

$$E \subset [E_j]_\epsilon \qquad \forall\, j \geq N_\epsilon .$$

To prove that

$$E_j \subset [E]_\epsilon \qquad \forall\, j \geq N_\epsilon,$$

we argue as follows. If $x \in E_j$, then $x \in [E_i]_\epsilon$ for all $i \geq j$, which implies

$$x \in \bigcup_{i \geq k} [E_i]_\epsilon = \left[\bigcup_{i \geq k} E_i \right]_\epsilon \qquad \forall\, k \geq j,$$

which implies that for each $k \geq j$ there exists $y_k \in \bigcup_{i \geq k} E_i$ such that $d(x, y_k) \leq \epsilon$. Since (y_k) is a bounded sequence, it has a subsequence converging to some point x_0. Since $y_\ell \in \bigcup_{i \geq k} E_i$ for all $\ell \geq k$, we have $x_0 \in \operatorname{cl} \bigcup_{i \geq k} E_i$ for all $k \geq j$, which implies $x_0 \in E$, and $d(x, x_0) \leq \epsilon$, which implies

$$x \in [E]_\epsilon.$$

Therefore **X** is complete.

Assume X is compact. Let (E_j) be a sequence in **X**, all E_j pairwise distinct. Given $\epsilon > 0$, there exists a cover of X by open metric disks $B(x_i; \epsilon)$, $i = 1, \ldots, M$. Then there exists at least one subset

$$\{x_{i_1}, \ldots, x_{i_\ell}\} \subseteq \{x_1, \ldots, x_M\}$$

for which the cardinality

$$\operatorname{card} \left\{ E_s : E_s \subset \bigcup_{k=1}^{\ell} B(x_{i_k}; \epsilon),\ E_s \cap B(x_{i_k}; \epsilon) \neq \emptyset\ \forall\, i_k \right\} = \infty.$$

Then, for each such E_s we have

$$E_s \subset \bigcup_{k=1}^{\ell} B(x_{i_k}; \epsilon) \subset [E_s]_{2\epsilon}.$$

Set $E_{i;0} = E_i$, the original sequence. Assume one has the subsequence $(E_{i;N})$ of (E_i). Pick the subsequence $(E_{i;N+1})$ of $(E_{i;N})$ by applying the above argument to $(E_{i;N})$ with $\epsilon = 1/(N+1)$. Set

$$G_{N+1} = \bigcup_{k=1}^{\ell_{N+1}} B(x_{i_{k;N+1}}; 1/(N+1)).$$

Then

$$E_{i;N+1} \subset G_{N+1} \subset [E_{i;N+1}]_{2/(N+1)},$$

which implies

$$\delta(E_{i;N+1}, G_{N+1}) < \frac{2}{N+1},$$

which implies

$$\delta(E_{i;N+1}, E_{j;N+1}) < \frac{4}{N+1}$$

for all i, j. Set $F_N = E_{N;N}$. Then

$$\delta(F_N, F_{N'}) < \frac{4}{\min\{N, N'\}},$$

which implies (F_N) is a Cauchy sequence, which, by completeness, has a limit. ■

Let (X, d) be a metric space such that closed and bounded subsets of X are compact. Then for each compact $K \subset X$ there exists $x_0 \in X$ such that $K \subset D(x_0; r(K))$, where $r(K)$ denotes the circumradius of K. We refer to this closed metric disk as a *circumdisk of* K, and denote it by D_K. In Euclidean space $\mathbb{R}^n$, D_K is unique.

Lemma III.1.2 *Assume also that μ is a positive measure on X, and that (X, d, μ) is homogeneous with respect to the metric and measure, that is, for every $x, y \in X$ there exists a bijection $\phi : X \to X$, preserving the distance d and the measure μ, and taking x to y.*

Given $K \in \mathbf{X}$, define

$$\mathbf{M}(K) = \{F \in \mathbf{X} : \mu(F) = \mu(K),\ \mu([F]_\epsilon) \le \mu([K]_\epsilon)\ \forall \epsilon > 0\}.$$

Then there exists $E \in \mathbf{M}(K)$ such that

$$r(E) = \min_{F \in \mathbf{M}(K)} r(F).$$

Similarly, given $K \in \mathbf{X}$, define

$$\mathbf{N}(K) = \{F \in \mathbf{X} : \mu(F) \ge \mu(K),\ \operatorname{diam} F \le \operatorname{diam} K\}.$$

Then there exists $E' \in \mathbf{N}(K)$ such that

$$r(E') = \min_{F \in \mathbf{N}(K)} r(F).$$

Proof Because we wish to minimize $r(F)$ in both statements of the lemma, it suffices to consider only those $F \in \mathbf{X}$ for which $r(F) \le r(K)$. Furthermore, because of the homogeneity of the metric and measure, it suffices to consider only those F contained in D_K, a circumdisk of K. Let

$$\mathbf{H} = \{F \in \mathbf{X} : F \subset D_K\}.$$

Then we already know that $\mathbf{H}$ is compact.

In the first statement of the lemma, we let

$$\mathbf{M}_0(K) = \mathbf{H} \cap \mathbf{M}(K).$$

and

$$\alpha = \inf_{F \in \mathbf{M}(K)} r(F) = \inf_{F \in \mathbf{M}_0(K)} r(F).$$

There exists a sequence (F_j) in $\mathbf{M}_0(K)$ with $r(F_j) \to \alpha$ as $j \to \infty$. Since $\mathbf{M}_0(K) \subset \mathbf{H}$, there exists a convergent subsequence $E_k \to E$, $E \in \mathbf{H}$. By the continuity of $F \mapsto r(F)$, we have $r(E) = \alpha$. Therefore, it suffices to show $E \in \mathbf{M}(K)$.

Since $\mu(E_k) = \mu(K)$ for all k, we have $\mu(K) \leq \mu(E)$. Given any $\epsilon > 0$, there exists $J > 0$ such that $j \geq J$ implies

$$E \subset [E_j]_\epsilon,$$

which implies

$$[E]_h \subset [E_j]_{\epsilon+h}$$

for all $h > 0$, which implies

$$\mu([E]_h) \leq \mu([E_j]_{\epsilon+h}) \leq \mu([K]_{\epsilon+h}).$$

Thus,

$$\mu([E]_h) \leq \mu([K]_{\epsilon+h})$$

for all $\epsilon, h > 0$. First let $\epsilon \to 0$ and then $h \to 0$. One then has $E \in \mathbf{M}(K)$, which implies the first claim.

In the second statement of the lemma, we let

$$\mathbf{N}_0(K) = \mathbf{H} \cap \mathbf{N}(K).$$

and

$$\beta = \inf_{F \in \mathbf{N}(K)} r(F) = \inf_{F \in \mathbf{N}_0(K)} r(F).$$

Since $\mathbf{H}$ is compact, and $\mathbf{N}(K)$ is closed, we conclude that $\mathbf{N}_0(K)$ is compact, and the second claim follows from the continuity of the functions $F \mapsto r(F)$ and $F \to \operatorname{diam} F$. ■

III.2 Minkowski Area and Steiner Symmetrization

Definition Given a compact subset K of $\mathbb{R}^n$, we define its *Minkowski area* Mink (K) by

$$\operatorname{Mink}(K) = \liminf_{h \downarrow 0} \frac{\mathbf{v}_n([K]_h) - \mathbf{v}_n(K)}{h}.$$

Remark III.2.1 Note that $K \mapsto \text{Mink}(K)$ is a functional defined on the collection of compacts sets K – not on the boundary of K.

Remark III.2.2 When $n = 1$, then

(III.2.1) $$\text{Mink}(K) = 2 + 2(\text{card}\{\text{components of } \mathbb{R} \setminus K\} - 2).$$

In particular, the Minkowski area of a point x_0 is 2, at the same time that $\mathbf{v}_1(\{x_0\}) = 0$.

Remark III.2.3 When K is the closure of a domain Ω in $\mathbb{R}^n$, with C^1 boundary $\partial\Omega$, then it is a standard exercise in differential geometry that

(III.2.2) $$\text{Mink}(K) = A(\partial\Omega),$$

where A denotes the standard Riemannian $(n-1)$-dimensional area.

Definition Let $X = \mathbb{R}^n$, $\mathbf{X}$ the collection of compact subsets of $\mathbb{R}^n$ with Hausdorff metric, and Π a hyperplane in $\mathbb{R}^n$.

For every $x \in \mathbb{R}^n$, let ℓ^x denote the line in $\mathbb{R}^n$ through x that is perpendicular to Π. For every compact subset K of $\mathbb{R}^n$ define the *Steiner symmetrization of K with respect to* Π, $\text{st}_\Pi K$, by

$$\text{st}_\Pi K = \bigcup_{w \in \Pi} \{w\} \times I^w,$$

where I^w denotes the closed interval in $\mathbb{R}$,

$$I^w = [-\sigma_w, \sigma_w], \qquad \sigma_w = \frac{1}{2}\mathbf{v}_1(\ell^w \cap K).$$

So $\mathbf{v}_1(I^w) = \mathbf{v}_1(\ell^w \cap K)$. (See Figure III.2.1.)

1. One verifies that $\text{st}_\Pi K$ is compact if K is compact.
2. Also, since

   (III.2.3) $$\mathbf{v}_n(K) = \int_\Pi \mathbf{v}_1(\ell^w \cap K)\, d\mathbf{v}_{n-1}(w),$$

   we have

   (III.2.4) $$\mathbf{v}_n(\text{st}_\Pi K) = \mathbf{v}_n(K)$$

   for all $K \in \mathbf{X}$.
3. If τ is a translation of $\mathbb{R}^n$, then $\text{st}_{\tau(\Pi)}(K) = \tau(\text{st}_\Pi(K))$.

Lemma III.2.1 *Let $K \in \mathbf{X}$, and assume $K \neq D_K$, that is, K is not equal to its circumdisk. Then there exist a finite number of Steiner symmetrizations in*

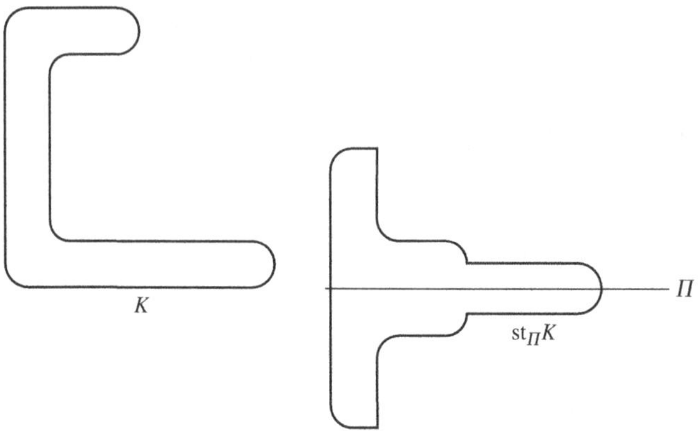

Figure III.2.1: Steiner symmetrization.

hyperplanes $\Pi_1, \ldots \Pi_k$ *such that*

$$r(\mathrm{st}_{\Pi_1} \circ \cdots \circ \mathrm{st}_{\Pi_k} K) < r(K).$$

Proof The proof of the lemma rests on the following fact: Given a closed n-disk D in $\mathbb{R}^n$, with boundary $(n-1)$-sphere S. Let o denote the center of D, and Π a hyperplane through o. For $x \in D$, let $\{x_1, x_2\} = \ell^x \cap S$. Then given $\epsilon > 0$, there exists $\delta = \delta(\epsilon; \Pi) > 0$ such that

$$D \setminus \mathrm{st}_\Pi (D \setminus \mathbb{B}(x; \epsilon)) \supset B_S(x_1; \delta) \cup B_S(x_2; \delta),$$

where B_S denotes the spherical cap in S with indicated center and radius. (See Figure III.2.2.) That is, if we remove a disk in D of radius ϵ centered at x, then after we Steiner symmetrize we are missing spherical caps in S of radius δ and centered at x_1 and x_2. If $x \in S$, then δ can be chosen independent of the plane Π.

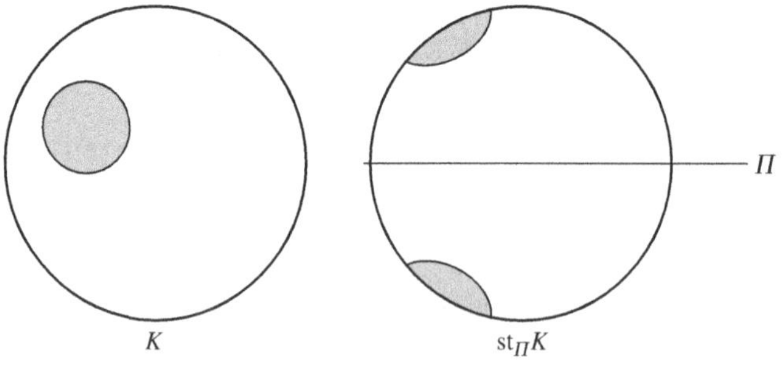

Figure III.2.2: Steiner symmetrization for disk with hole.

We now prove the lemma. We are given that $K \neq D_K$. Let S denote the boundary sphere of D_K. We may assume that $K \cap S \neq S$. For if not, there exists a disk in the interior of D_K, and in the complement of K. Then any symmetrization st with respect to a plane through the origin o of D_K will produce spherical neighborhoods in S contained in $D_K \setminus \mathrm{st}\, K$.

So we assume $K \cap S \neq S$. Since $D_K \setminus K$ is relatively open, it suffices to produce a finite number of planes $\Pi_1, \ldots, \Pi_k$ for which

(III.2.5) $$(\mathrm{st}_{\Pi_1} \circ \cdots \circ \mathrm{st}_{\Pi_k} K) \cap S = \emptyset.$$

Given $x \in S \setminus K$, there exists $\epsilon > 0$ such that the spherical cap $B_S(x;\epsilon) \subset S \setminus K$. Then the compactness of $K \cap S$ implies there exist a finite number of points $x_1, \ldots, x_k$ in $K \cap S$ such that the (relatively) open spherical caps $\{B_S(x_j;\epsilon) : j = 1, \ldots, k\}$ cover $K \cap S$, that is,

$$\bigcup_{j=1}^{k} B_S(x_j;\epsilon) \supset S \cap K.$$

For each j, consider the hyperplane Π_j bisecting the chord xx_j. Then this choice of Π_j, $j = 1, \ldots, k$ will satisfy (III.2.5), which implies the lemma. ∎

Theorem III.2.1 (Isodiametric Inequality) *For any $K \in \mathbf{X}$ we have*

$$\mathbf{v}_n(K) \leq \boldsymbol{\omega}_n \left\{ \frac{\mathrm{diam}\ K}{2} \right\}^n.$$

Thus, the disk minimizes the diameter of a compact set of given volume.

Proof We first show that given any $F \in \mathbf{X}$, we have

(III.2.6) $$\mathrm{diam}\ \mathrm{st}_\Pi F \leq \mathrm{diam}\ F,$$

for any hyperplane Π in $\mathbb{R}^n$.

Pick Euclidean coordinates in $\mathbb{R}^n$ where Π is identified by $x^n = 0$, and write $\mathbb{R}^n$ as $\mathbb{R}^{n-1} \times \mathbb{R}$ with $\mathbb{R}^{n-1} = \Pi$, and $\mathbb{R} = \ell^{\mathbf{o}}$, where $\mathbf{o}$ is the origin of $\mathbb{R}^n$. Let (w, α) and (v, β) maximize distance in $\mathrm{st}_\Pi F$, that is,

$$d((w,\alpha),(v,\beta)) - \mathrm{diam}\ \mathrm{st}_\Pi F,$$

where $w, v \in \mathbb{R}^{n-1}$, $\alpha, \beta \in \mathbb{R}$. Because

$$d((w,\alpha),(v,\beta)) = \sqrt{|w - v|^2 + (\alpha - \beta)^2},$$

we must have both cases

$$\alpha = -\sigma_w, \qquad \beta = \sigma_v \quad \text{and} \quad \alpha = \sigma_w, \qquad \beta = -\sigma_v.$$

Let

$$-\xi_w = \min \ell^w \cap F, \qquad \eta_w = \max \ell^w \cap F,$$
$$-\xi_v = \min \ell^v \cap F, \qquad \eta_v = \max \ell^v \cap F.$$

Then

$$\eta_w + \xi_w \geq 2|\alpha|, \qquad \eta_v + \xi_v \geq 2|\beta|,$$

which implies

$$\eta_w + \xi_w + \eta_v + \xi_v \geq 2\{|\alpha| + |\beta|\},$$

so at least one of $\eta_w + \xi_v$ or $\eta_v + \xi_w$ must be $\geq |\alpha| + |\beta|$, which implies (III.2.6).

To finish the proof, given K, we consider the compact set E' in $\mathbf{N}(K)$ that minimizes the circumradius (Lemma III.1.2). We claim that $D_{E'} = E'$, that is, E' is a disk. If not, then Steiner symmetrizations preserve $\mathbf{N}(K)$ [that is, $F \in \mathbf{N}(K)$ implies $\mathrm{st}_\Pi F \in \mathbf{N}(K)\ \forall\, \Pi$], and there exist a finite number of them whose composition lowers the circumradius (Lemma III.2.1), which contradicts the definition of E'. But $2r(E') \leq \operatorname{diam} K$, which implies

$$\omega_n \left\{\frac{\operatorname{diam} K}{2}\right\}^n \geq \omega_n \{r(E')\}^n = \mathbf{v}_n(E') \geq \mathbf{v}_n(K),$$

which implies the theorem. ∎

III.2.1 The Isoperimetric Inequality for Minkowski Area

Theorem III.2.2 (Brunn–Minkowski Inequality) *Given $K \in \mathbf{X}$, let D denote the closed n-disk of the same measure as K, that is, $\mathbf{v}_n(D) = \mathbf{v}_n(K)$. Then*

$$\mathbf{v}_n([D]_\epsilon) \leq \mathbf{v}_n([K]_\epsilon)$$

for all $\epsilon > 0$.

Proof First note that, by (III.2.1), the theorem is true if $n = 1$.

For general $n > 1$, we take the same approach as in the proof of the isodiametric inequality, namely, we first show that

$$\mathbf{v}_n([\mathrm{st}_\Pi F]_\epsilon) \leq \mathbf{v}_n([F]_\epsilon) \qquad \text{(III.2.7)}$$

for all $F \in \mathbf{X}$, $\epsilon > 0$, and all hyperplanes Π in $\mathbb{R}^n$.

Pick Euclidean coordinates in $\mathbb{R}^n$ so that $\Pi = \mathbb{R}^{n-1} \times \{0\}$ – so we identify Π with $\mathbb{R}^{n-1}$, let $\mathbf{o}$ be the origin of $\mathbb{R}^n$, and $\ell^{\mathbf{o}} = \mathbb{R}$, as in the proof of Theorem III.2.1. Let $q : \mathbb{R}^n \to \mathbb{R}$ denote projection onto the x^n-axis$= \ell^{\mathbf{o}}$. For any $w \in \Pi$, let $\tau_w : \mathbb{R}^n \to \mathbb{R}^n$ denote the translation by w. To help avoid confusion, we let $\mathbf{D}(w; \epsilon) = \mathbb{D}^{n-1}(w; \epsilon)$ denote the closed $(n-1)$-disk in Π centered at w and with radius ϵ. Given $w \in \Pi$, and $F \in \mathbf{X}$, then

$$\ell^w \cap [F]_\epsilon = \bigcup_{v \in \mathbf{D}(w;\epsilon)} \tau_w \left([q(\ell^v \cap F)]_{\sqrt{\epsilon^2 - |v-w|^2}} \right). \tag{III.2.8}$$

[We have written the formula in this manner to emphasize that we project $\ell^v \cap F$ onto the x^n-axis to obtain $q(\ell^v \cap F)$, carry out the 1-dimensional $\sqrt{\epsilon^2 - |v-w|^2}$-thickening *in the* x^n*-axis,* and then translate out to the line ℓ^w.] To prove (III.2.8) just note that $(w, \alpha) \in [F]_\epsilon$ if and only if there exists $(v, \beta) \in F$ such that $|w - v|^2 + (\beta - \alpha)^2 < \epsilon^2$, which is true if and only if $d_{\mathbb{R}^1}(\alpha, q(\ell^v \cap F)) < \sqrt{\epsilon^2 - |w - v|^2}$, which is the claim. Therefore

$$\begin{aligned}
\mathbf{v}_1(\ell^w \cap [F]_\epsilon) &\geq \sup_{v \in \mathbf{D}(w;\epsilon)} \mathbf{v}_1 \left(\tau_w \left([q(\ell^v \cap F)]_{\sqrt{\epsilon^2 - |v-w|^2}} \right) \right) \\
&= \sup_{v \in \mathbf{D}(w;\epsilon)} \mathbf{v}_1 \left([q(\ell^v \cap F)]_{\sqrt{\epsilon^2 - |v-w|^2}} \right) \\
&\geq \sup_{v \in \mathbf{D}(w;\epsilon)} \mathbf{v}_1 \left([q(I^v)]_{\sqrt{\epsilon^2 - |v-w|^2}} \right) \\
&= \mathbf{v}_1(\ell^w \cap [\mathrm{st}_\Pi\, F]_\epsilon),
\end{aligned}$$

where I^v is the 1-dimensional symmetrization of $\ell^v \cap F$ (to pass from the second to the third line we use the 1-dimensional isoperimetric inequality), that is,

$$\mathbf{v}_1(\ell^w \cap [\mathrm{st}_\Pi\, F]_\epsilon) \leq \mathbf{v}_1(\ell^w \cap [F]_\epsilon) \tag{III.2.9}$$

Then (III.2.3) and (III.2.9) imply (III.2.7).

To finish the proof, given K, we consider the compact set E in $\mathbf{M}(K)$ which minimizes the circumradius (Lemma III.1.2). We claim that $D_E = E$, that is, E is a disk. If not, then Steiner symmetrizations preserve $\mathbf{M}(K)$, and there exist a finite number of them whose composition lowers the circumradius (Lemma III.2.1), which contradicts the definition of E. ■

Theorem III.2.3 (Isoperimetric Inequality for Minkowski Area) *If K is a compact subset of $\mathbb{R}^n$, then Steiner symmetrization of K does not increase its*

Minkowski area, that is,

$$\text{Mink}\,(\text{st}_\Pi K) \le \text{Mink}\,(K) \tag{III.2.10}$$

for any hyperplane Π.

Furthermore, if the closed n-disk $\mathbb{D}^n(R)$ *has the same measure as* K, *then*

$$A(\mathbb{S}^{n-1}(R)) = \text{Mink}\,(\mathbb{D}^n(R)) \le \text{Mink}\,(K). \tag{III.2.11}$$

If K *is the closure of an open subset* Ω *in* $\mathbb{R}^n$ *with* C^1 *boundary* $\partial\Omega$, *then*

$$A(\mathbb{S}^{n-1}(R)) \le A(\partial\Omega), \tag{III.2.12}$$

which is expressed analytically as

$$\frac{A(\partial\Omega)}{V(\Omega)^{1-1/n}} \ge n\omega_n{}^{1/n}. \tag{III.2.13}$$

Proof Indeed, Theorem III.2.2 implies

$$\begin{aligned}\frac{\mathbf{v}_n(\mathbb{D}^n(R+h)) - \mathbf{v}_n(\mathbb{D}^n(R))}{h} &= \frac{\mathbf{v}_n([\mathbb{D}^n(R)]_h) - \mathbf{v}_n(\mathbb{D}^n(R))}{h} \\ &\le \frac{\mathbf{v}_n([K]_h) - \mathbf{v}_n(K)}{h}\end{aligned}$$

for all $h > 0$, which implies the claim. ∎

The problem with this approach is that while it quickly and elementarily gives the isoperimetric inequality, it cannot characterize the case of equality – unless one can sharpen the Brunn–Minkowski inequality to include an error term that persists to the limit, as $h \to 0$, and that vanishes if and only if K is a disk. In what follows, we shall give a variant of this argument.

Theorem III.2.4 *Assume* Ω *has* C^1 *boundary,* $K = \overline{\Omega}$ *compact. One has equality in* (III.2.12) [*equivalently,* (III.2.13)] *if and only if* Ω *is a disk.*

Proof First note that we have the theorem for $n = 1$.

One might wish to apply Steiner symmerization directly to prove

$$A(\partial\text{st}_\Pi\,\Omega) \le A(\partial\Omega),$$

and then characterize the case of equality. But one does not know in advance that if Ω has C^1 boundary, Steiner symmetrization will preserve this smoothness. (One can easily construct counterexamples. See Figure III.2.1.) Yet, there are obvious examples where some smoothness is preserved. We show that this is always true, and the smoothness that is preserved has enough information to provide the characterization of equality in (III.2.12) we are seeking.

Namely, we first give a local analogue of (III.2.7). Let Π be a hyperplane in $\mathbb{R}^n$, and coordinatize $\mathbb{R}^n$ as in the proof of Theorem III.2.1. For $z \in \mathbb{R}^{n-1}$ and $\epsilon > 0$, let (as above) $\mathbf{D}(z;\epsilon)$ denote the closed $(n-1)$-disk in $\mathbb{R}^{n-1}$ centered at z with radius ϵ, and let $\mathbf{C}(z;\epsilon)$ denote the solid n-cylinder in $\mathbb{R}^n$ over $\mathbf{D}(z;\epsilon)$, that is,

$$\mathbf{C}(z;\epsilon) = \mathbf{D}(z;\epsilon) \times \mathbb{R}.$$

For $z \in \mathbb{R}^{n-1}$, $\epsilon > 0$, and compact K in $\mathbb{R}^n$, we have

$$V(K \cap \mathbf{C}(z;\epsilon)) = \int_{\mathbf{D}(z;\epsilon)} \mathbf{v}_1(\ell^w \cap K)\, d\mathbf{v}_{n-1}(w) = V(\mathrm{st}_\Pi K \cap \mathbf{C}(z;\epsilon)),$$

We define a "localized" Minkowski area $\mathrm{Mink}_{\Pi;z;\epsilon}$ by

$$\mathrm{Mink}\,(K)_{\Pi;z;\epsilon} = \liminf_{h\downarrow 0} \frac{V(\{[K]_h \setminus K\} \cap \mathbf{C}(z;\epsilon))}{h}.$$

(See Figure III.2.3.) Then (III.2.9) implies

$$V\left([\mathrm{st}_\Pi K]_h \cap \mathbf{C}(z;\epsilon)\right) \le V\left([K]_h \cap \mathbf{C}(z;\epsilon)\right), \tag{III.2.14}$$

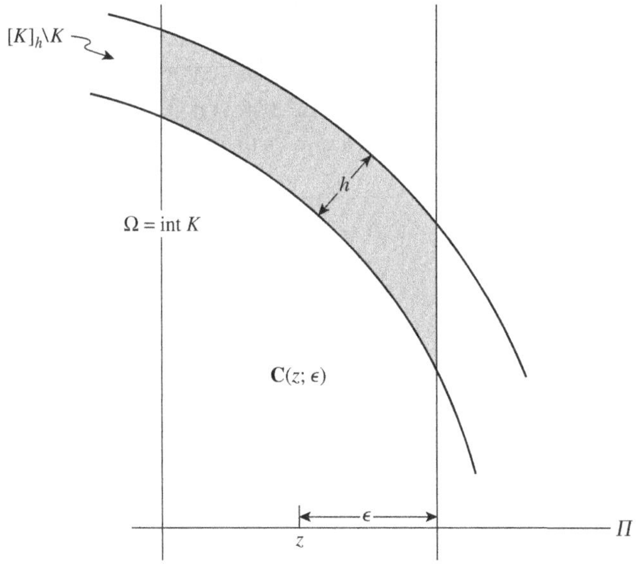

Figure III.2.3: $\{[K]_h \setminus k\} \cap \mathbf{C}(z;\epsilon)$.

which implies

$$\text{Mink}_{\Pi;z;\epsilon}(\text{st}_\Pi K) \leq \text{Mink}_{\Pi;z;\epsilon}(K), \tag{III.2.15}$$

Return to K equal to the closure of Ω with C^1 boundary $\partial\Omega$. Let

$$p : \mathbb{R}^n \to \mathbb{R}^{n-1},$$

denote the projection map. Fix $z \in \mathbb{R}^{n-1}$ and $\epsilon > 0$, and *assume that* $\mathbf{D}(z;\epsilon)$ *consists of regular values of the projection of* $\partial\Omega$, $p|\partial\Omega$. Let $\boldsymbol{\nu}$ denote the exterior unit normal vector field along $\partial\Omega$, and

$$E_h = \{v \in \mathbb{R}^n : v = u + \tau\boldsymbol{\nu}_u,\ u \in \partial\Omega \cap \mathbf{C}(z;\epsilon),\ \tau \in [0,h]\},$$

that is, E_h is the curved slab of thickness h determined by starting out at points $u \in \partial\Omega$ sitting over $\mathbf{D}(z;\epsilon)$ and going out h units in the direction $\boldsymbol{\nu}_u$.

Lemma III.2.2 *We have the differential geometric estimate*

$$|V(\{[K]_h \setminus K\} \cap \mathbf{C}(z;\epsilon)) - V([E_h])| = O(h^2) \tag{III.2.16}$$

as $h \downarrow 0$. *In particular,*

$$\text{Mink}_{\Pi;z;\epsilon}(K) = A(\partial\Omega \cap \mathbf{C}(z;\epsilon)).$$

Proof (See Figure III.2.4.) Let $\Gamma = \partial\Omega \cap \partial\mathbf{C}(z;\epsilon)$. Then Γ is a compact C^1 $(n-2)$-dimensional submanifold of $\mathbb{R}^n$, with fiber at $w \in \Gamma$ in the normal bundle of Γ, spanned by $\boldsymbol{\nu}_w$ (the unit normal exterior vector field along $\partial\Omega$) and $\mathbf{n}_w$ [the unit normal exterior vector field along $\partial\mathbf{C}(z;\epsilon)$]. The linear independence of $\boldsymbol{\nu}_w$ and $\mathbf{n}_w$ is guaranteed by the assumption that p is regular on all of $\mathbf{D}(z;\epsilon)$.

For each $w \in \Gamma$, let

$$\theta_w = \angle(\mathbf{n}_w, \boldsymbol{\nu}_w) \in (0,\pi) \quad \text{and} \quad \Psi = \inf_{w\in\Gamma} \sin\theta_w.$$

For any $r > 0$, $[\Gamma]_r$, the tubular neighborhood of Γ of radius r, satisfies

$$V([\Gamma]_r) = O(r^2)$$

as $r \downarrow 0$, since Γ has codimension 2. For any $\sigma > 1$, an application of Taylor's formula shows that the symmetric difference $\{\{[K]_h \setminus K\} \cap \mathbf{C}(z;\epsilon)\} \,\Delta\, E_h$ of $\{[K]_h \setminus K\} \cap \mathbf{C}(z;\epsilon)$ and E_h satisfies

$$\{\{[K]_h \setminus K\} \cap \mathbf{C}(z;\epsilon)\} \,\Delta\, E_h \subset [\Gamma]_{\sigma h/\Psi}$$

as $h \downarrow 0$, which implies (III.2.16). ∎

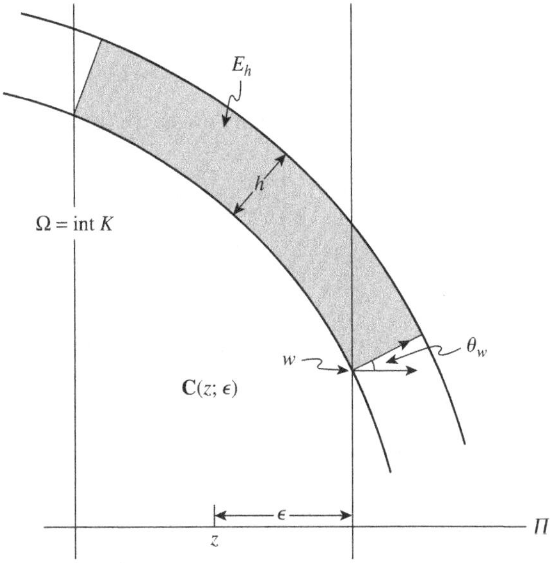

Figure III.2.4: E_h.

Lemma III.2.3 *Assume that* $\mathbf{D}(z;\epsilon)$ *consists of regular values of the projection of* $\partial\Omega$, $p|\partial\Omega$. *Let*

$$\alpha_1(w) < \beta_1(w) \leq \cdots \leq \alpha_k(w) < \beta_k(w)$$

be C^1 *functions on* $\mathbf{D}(z;\epsilon)$ *such that*

$$\ell^w \cap K = \bigcup_{j=1}^{k} [\alpha_j(w), \beta_j(w)]$$

for all $w \in \mathbf{D}(z;\epsilon)$. *Define*

$$\mathcal{A}(w) = \frac{1}{2} \sum_{j=1}^{k} \beta_j(w) - \alpha_j(w).$$

Then

$$\ell^w \cap \mathrm{st}_\Pi K - [-\mathcal{A}(w), \mathcal{A}(w)]$$

for all $w \in \mathbf{D}(z;\epsilon)$, *and*

(III.2.17) $$A(\partial(\mathrm{st}_\Pi \Omega) \cap \mathbf{C}(z;\epsilon)) \leq A(\partial\Omega \cap \mathbf{C}(z;\epsilon))$$

with equality if and only if $k = 1$, *and* $\Omega \cap \mathbf{C}(z;\epsilon)$ *is symmetric, up to translation, with respect to the hyperplane* Π.

Proof Note that $w \mapsto \mathcal{A}(w)$ is C^1, and

$$\operatorname{Mink}_{\Pi;z;\epsilon}(\mathrm{st}_\Pi K) = A(\partial \mathrm{st}_\Pi \Omega \cap \mathbf{C}(z;\epsilon)),$$

by Lemma III.2.2. To compare the boundary areas, we have

$$A(\partial\Omega \cap \mathbf{C}(z;\epsilon)) = \int_{\mathbf{D}(z;\epsilon)} \left\{ \sum_j \sqrt{1 + |\mathrm{grad}_{n-1} - \alpha_j|^2} + \sqrt{1 + |\mathrm{grad}_{n-1}\beta_j|^2} \right\} d\mathbf{v}_{n-1}$$

where $\mathrm{grad}_{n-1}\alpha_j$ is the $(n-1)$-vector $(\partial\alpha_j/\partial w_1, \ldots \partial\alpha_j/\partial w_{n-1})$, and the same for $\mathrm{grad}_{n-1}\beta_j$, $j = 1, \ldots, k$, and

$$A(\partial(\mathrm{st}_\Pi \Omega) \cap \mathbf{C}(z;\epsilon)) = 2\int_{\mathbf{D}(z;\epsilon)} \sqrt{1 + |\mathrm{grad}_{n-1}\mathcal{A}|^2}\, d\mathbf{v}_{n-1}.$$

Two applications of the triangle inequality imply

$$\sqrt{1^2 + |\mathrm{grad}_{n-1} - \alpha_j|^2} + \sqrt{1^2 + |\mathrm{grad}_{n-1}\beta_j|^2} \geq \sqrt{2^2 + |\mathrm{grad}_{n-1}\beta_j - \alpha_j|^2} \tag{III.2.18}$$

and

$$\sum_{j=1}^k \sqrt{2^2 + |\mathrm{grad}_{n-1}\beta_j - \alpha_j|^2} \geq \sqrt{(2k)^2 + \left| \sum_{j=1}^k \mathrm{grad}_{n-1}\beta_j - \alpha_j \right|^2} = \sqrt{(2k)^2 + 2^2|\mathrm{grad}_{n-1}\mathcal{A}|^2}.$$

Therefore

$$2\int_{\mathbf{D}(z;\epsilon)} \sqrt{k^2 + |\mathrm{grad}_{n-1}\mathcal{A}|^2}\, d\mathbf{v}_{n-1} \leq A(\partial\Omega \cap \mathbf{C}(z;\epsilon)), \tag{III.2.19}$$

which implies (III.2.17). We have equality in (III.2.17) if and only if $k = 1$, and the vectors $(\mathrm{grad}_{n-1} - \alpha_1(w), 1)$ and $(\mathrm{grad}_{n-1}\beta_1(w), 1)$ are linearly dependent. Hence (since their last coordinate is equal),

$$\alpha_1(w) = -\beta_1(w) + \text{const.}$$

on all of $\mathbf{D}(z;\epsilon)$, which implies $\Omega \cap \mathbf{C}(z;\epsilon)$ is symmetric, up to a translation, with respect to the hyperplane Π. ∎

Conclusion of the Proof of Theorem III.2.4. Sard's theorem (Proposition I.3.8) implies that z is a regular value of $p|\partial\Omega$ for almost all $z \in p(\partial\Omega)$. Given such a $z \in p(\partial\Omega)$, there exists $\epsilon > 0$ such that $\mathbf{D}(z;\epsilon)$ consists only of regular values, for which (III.2.17) follows, with its characterization of the case of equality.

If we have equality in (III.2.12), then Steiner symmetrization does not strictly decrease the Minkowski area of the boundary. In particular, we have equality in (III.2.15) for any hyperplane Π, $z \in \Pi$, and $\epsilon > 0$. Therefore, we have equality in (III.2.17) for all regular values of $p|\partial\Omega$. Therefore, by continuity, and by applying the argument to every possible hyperplane Π, we have that every line in $\mathbb{R}^n$ that intersects Ω does so in a segment, that is, Ω is convex. Furthermore, Ω is symmetric, up to a translation, with respect to any hyperplane Π. By the convexity of Ω and Theorem I.3.5, K is a disk. ■

III.2.2 Sequences of Steiner Symmetrizations

Our proof of the isoperimetric inequality used the fact that Steiner symmetrization did not increase Minkowski area, while it preserved volume. The minimizing property of the disk then required an existence argument, using the Blaschke theorem. A more intuitive approach to the minimizing property of the disk is to expect that the more symmetric a compact set, the "closer" it should be to realizing the minimum Minkowski area for fixed given volume. In this section we begin carrying out this approach, namely, we show that starting with any compact subset K of the plane, we may apply to it a sequence of rotations and Steiner symmetrizations so that it will converge, in the Hausdorff metric, to a closed disk. One can then use induction (on the dimension) to prove a similar result for higher dimensions.

So we see the Minkowski area decreasing, hopefully, toward the area of the disk. Note that Minkowski area is not necessarily lower semicontinuous, so we cannot (on general principles) pass to the limit to conclude that the area of the disk is less than or equal to all those in the sequence. But we shall be able to prove this result for our particular sequence. We postpone the characterization of equality to the discussion of perimeter.

Definition A Borel-measurable function $f : \mathbb{R}^n \to \mathbb{C}$ *vanishes at infinity* if

$$\mathbf{v}_n(|f| > t) < +\infty \qquad \forall\, t > 0.$$

Definition Let A be a Borel subset of $\mathbb{R}^n$, $\mathbf{v}_n(A) < +\infty$. We let A^* denote the *Schwarz symmetrization of* A, that is, A^* is the *open* disk in $\mathbb{R}^n$ with the same volume as A.

Also, define the *symmetric decreasing rearrangement of the function* $\mathcal{I}_A$, $\mathcal{I}_A^*$, by

$$\mathcal{I}_A^* = \mathcal{I}_{A^*} \ .$$

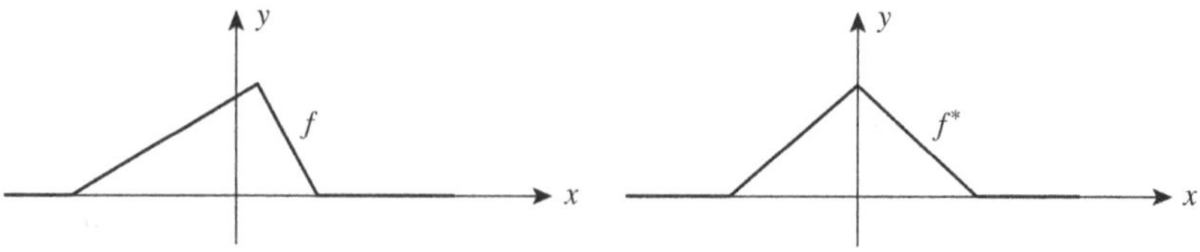

Figure III.2.5: Symmetric decreasing reaarangement.

For any Borel-measurable $f : \mathbb{R}^n \to \mathbb{C}$ that vanishes at infinity, define the *symmetric decreasing rearrangement of* f, f^*, by

$$f^*(x) = \int_0^\infty \mathcal{I}_{\{|f|>t\}}{}^*(x)\,dt.$$

(See Figure III.2.5.)

Then we have the following facts (recall Cavalieri's principle (Theorem I.3.3)):

1. $f^* \geq 0$.
2. f^* is spherically symmetric, that is, $|x| = |y|$ implies $f^*(x) = f^*(y)$.
3. f^* is lower semicontinuous (since A^* is open), which implies that f^* is measurable.
4. $\{f^* > t\} = \{|f| > t\}^*$ for every $t > 0$; of course, the two sets have the same volume.
5. If $\Phi : \mathbb{R}^+ \to \mathbb{R}^+$ is increasing, then

$$(\Phi \circ |f|)^* = \Phi \circ f^*.$$

6. One has

$$\int \phi \circ |f|\,d\mathbf{v}_n = \int \phi \circ f^*\,d\mathbf{v}_n$$

for any $\phi = \phi_1 - \phi_2$ such that each ϕ_j, $j = 1, 2$, is increasing and at least one of $\phi_1 \circ |f|$ or $\phi_2 \circ |f|$ is in $L^1(\mathbb{R}^n)$.

7. We always have

$$\|f\|_p = \|f^*\|_p \qquad \forall\ p \geq 1.$$

8. If $f \leq g$ then $f^* \leq g^*$.

Theorem III.2.5 *If f, g are nonnegative on $\mathbb{R}^n$, both vanishing at infinity, then*

(III.2.20)
$$\int fg\,d\mathbf{v}_n \leq \int f^* g^*\,d\mathbf{v}_n.$$

If $f = f^*$ and strictly *decreasing, then we have equality in* (III.2.20) *if and only if $g = g^*$ a.e.*

Proof By Cavalieri's principle (Theorem I.3.3), we want to show

$$\int_0^\infty ds \int_0^\infty dt \int_{\mathbb{R}^n} \mathcal{I}_{\{f>t\}}\mathcal{I}_{\{g>s\}}\, d\mathbf{v}_n$$
$$\leq \int_0^\infty ds \int_0^\infty dt \int_{\mathbb{R}^n} \mathcal{I}_{\{f>t\}}{}^*\mathcal{I}_{\{g>s\}}{}^*\, d\mathbf{v}_n;$$

it suffices to show that

$$\int \mathcal{I}_A\mathcal{I}_B\, d\mathbf{v}_n \leq \int \mathcal{I}_A^*\mathcal{I}_B^*\, d\mathbf{v}_n, \tag{III.2.21}$$

for all A, B of finite volume. Assume $\mathbf{v}_n(A) \leq \mathbf{v}_n(B)$. Then $A^* \subseteq B^*$, which implies that the right hand side of (III.2.21) is equal to $\mathbf{v}_n(A^*) = \mathbf{v}_n(A) \geq \mathbf{v}_n(A \cap B)$, which is the left hand side of (III.2.21). This proves (III.2.20).

Now assume equality in (III.2.20), $f = f^*$ strictly decreasing. Then

$$\int f\mathcal{I}_{\{g>s\}}\, d\mathbf{v}_n = \int f\mathcal{I}_{\{g^*>s\}}\, d\mathbf{v}_n$$

for almost all s. It suffices to show that

$$\mathcal{I}_{\{g>s\}} = \mathcal{I}_{\{g^*>s\}},$$

which would then imply $g = g^*$. Well, since f is spherically symmetric and strictly decreasing, there exists a *continuous* function $r = \rho(t)$ such that

$$\mathbb{B}^n(\rho(t)) = \{f > t\} \qquad \forall\, t > 0,$$

which implies

$$F_C(t) := \int \mathcal{I}_{\{f>t\}}\mathcal{I}_C\, d\mathbf{v}_n$$

is continuous for any measurable C. For any given s, pick $C = \{g > s\}$. Then $F_C(t) \leq F_{C^*}(t)$ for all t, by the original inequality (III.2.20), and $\int F_C(t)\,dt = \int F_{C^*}(t)\,dt$, by assumption. This implies (by continuity) that $F_C(t) = F_{C^*}(t)$ for all t, which implies

$$\mathbf{v}_n(\{f > t\} \cap C) = \mathbf{v}_n(\{f > t\} \cap C^*) = \min\{\mathbf{v}_n(f > t), \mathbf{v}_n(C^*)\}.$$

But this implies, for every t, that if C^* contains $\{f > t\}$ then C contains $\{f > t\}$, and if $\{f > t\}$ contains C^* then $\{f > t\}$ contains C. Therefore $C = C^*$, up to measure zero, which implies $\{g > s\}$ is a disk up to measure zero, for every s.

■

Corollary III.2.1 *For f and g in L^2 we have*

$$\|f^* - g^*\|_2 \leq \|f - g\|_2.$$

A generalization of Corollary III.2.1 goes as follows:

Theorem III.2.6 *Let $J : \mathbb{R} \to [0, \infty)$ be convex, $J(0) = 0$; and let f, g be nonnegative, vanishing at infinity. Then*

$$\int J \circ (f^* - g^*)\, d\mathbf{v}_n \leq \int J \circ (f - g)\, d\mathbf{v}_n. \tag{III.2.22}$$

If J is strictly convex, $f = f^*$ strictly decreasing, then equality in (III.2.22) *implies $g = g^*$ a.e.*

Proof Write $J = J_+ + J_-$, where

$$J_+(t) = \begin{cases} 0, & t \leq 0, \\ J(t), & t > 0, \end{cases}$$

and

$$J_-(t) = \begin{cases} J(t), & t \leq 0, \\ 0, & t > 0. \end{cases}$$

Then J_+ and J_- are convex.

Consider J_+, and its right hand derivative; then

$$J_+(t) = \int_0^t J_+{}'(s)\, ds,$$

which implies

$$\begin{aligned} J_+ \circ (f - g)(x) &= \int_{g(x)}^{f(x)} J_+{}'(f(x) - s)\, ds \\ &= \int_0^\infty J_+{}'(f(x) - s) \mathcal{I}_{\{g \leq s\}}(x)\, ds, \end{aligned}$$

which implies

$$\begin{aligned} \int_{\mathbb{R}^n} J_+ \circ (f - g)(x)\, d\mathbf{v}_n(x) &= \int_0^\infty ds \int_{\mathbb{R}^n} J_+{}'(f(x) - s) \mathcal{I}_{\{g \leq s\}}(x)\, d\mathbf{v}_n(x) \\ &\leq \int_0^\infty ds \int_{\mathbb{R}^n} J_+{}'(f^*(x) - s) \mathcal{I}_{\{g^* \leq s\}}(x)\, d\mathbf{v}_n(x) \\ &= \int_{\mathbb{R}^n} J_+{}' \circ (f^* - g^*)(x)\, d\mathbf{v}_n(x) \end{aligned}$$

(the second line follows from Theorem III.2.5). The same argument applies to J_-, which implies the inequality.

Now assume J is strictly convex, $f = f^*$ strictly decreasing, and equality in (III.2.22). Then we have equality, separately, for J_+ and J_-, which implies

$$\int J_+{}'(f - s)\mathcal{I}_{\{g \le s\}}\, d\mathbf{v}_n = \int J_+{}'(f - s)\mathcal{I}_{\{g^* \le s\}}\, d\mathbf{v}_n$$

a.e.-$[ds]$, which implies, by equality in (III.2.20), that $g = g^*$ a.e. ■

Corollary III.2.2 *For all* $p \ge 1$ *we have*

(III.2.23) $$\|f^* - g^*\|_p \le \|f - g\|_p.$$

Corollary III.2.3 *Let* $\mathcal{S}$ *denote Steiner symmetrization with respect to some hyperplane* Π *in* $\mathbb{R}^n$*, and* $\mathcal{S}f$ *the Steiner symmetrization of any function* f*, defined by*

$$\mathcal{S}f(x) = \int_0^\infty \mathcal{I}_{\mathcal{S}\{|f|>t\}}(x)\, dt.$$

Then

(III.2.24) $$\|\mathcal{S}f - \mathcal{S}g\|_p \le \|f - g\|_p$$

for all $f, g \in L^p$, $p \ge 1$.

Proof Use the 1-dimensional Corollary III.2.2 with Fubini's theorem. ■

Theorem III.2.7 (Helly Selection Theorem) *Given a uniformly bounded collection* Φ *of increasing functions on* $[a, b]$*, there exists a sequence in* Φ *that converges for every* $x \in [a, b]$.

Proof Let $r_1, r_2, \ldots$ be an enumeration of the rationals in $[a, b]$. Then there exists a sequence $\phi_{1,j}$ in Φ such that the sequence of numbers $\phi_{1,j}(r_1)$ converges, which, in turn, has a convergent subsequence $\phi_{2,j}(r_2)$, and so on. The sequence $\phi_{k,k}$ converges, as $k \to \infty$, for *all* rationals in $[a, b]$ to a function ϕ defined on the rationals. Extend ϕ to the irrationals by

$$\phi(x) = \lim_{r \uparrow x} \phi(r),$$

where r varies over rationals $\le x$. So ϕ is continuous from the left, and increasing.

We show that $\phi_{k,k}(x_0) \to \phi(x_0)$ for every x_0 at which ϕ is continuous. Pick $\rho_1 < x_0 < \rho_2$, ρ_j rational. Then

$$\phi_{k,k}(x_0) - \phi(x_0) = \{\phi_{k,k}(x_0) - \phi_{k,k}(\rho_1)\} + \{\phi_{k,k}(\rho_1) - \phi(\rho_1)\} \\ + \{\phi(\rho_1) - \phi(x_0)\}.$$

This implies

$$|\phi_{k,k}(x_0) - \phi(x_0)| \leq |\phi_{k,k}(\rho_2) - \phi_{k,k}(\rho_1)| + |\phi_{k,k}(\rho_1) - \phi(\rho_1)| \\ + |\phi(\rho_1) - \phi(x_0)|,$$

from which one has the claim. So $\phi_{k,k}(x) \to \phi(x)$ at points of continuity of ϕ.

Since ϕ is increasing, its points of discontinuity are countable. But one can use the above diagonal argument to obtain the convergence on these points of discontinuity of ϕ. ∎

Theorem III.2.8 (Isoperimetric Inequality for Minkowski Length in $\mathbb{R}^2$) *Let F be a compact subset of the plane $\mathbb{R}^2$. Let st_x denote Steiner symmetrization with respect to the x-axis, st_y Steiner symmetrization with respect to the y-axis, α an irrational multiple of 2π, and R_α the rotation of $\mathbb{R}^2$ by α radians. Finally, set*

$$T = \mathrm{st}_y\, \mathrm{st}_x\, R_\alpha.$$

Then a subsequence $T^{j_k} \cdot F$ of $T^j \cdot F$ converges to the Schwarz symmetrization of F, F^*, in the Hausdorff metric topology.

Also, $T^j \cdot F \to F^*$ in L^p, as $j \to \infty$, that is,

$$\|\mathcal{I}_{T^j \cdot F} - \mathcal{I}_{F^*}\|_p \to 0 \text{ as } j \to \infty, \qquad \forall\, p \geq 1.$$

Finally,

$$\text{(III.2.25)} \qquad \mathrm{Mink}\,(F^*) \leq \mathrm{Mink}\,(F).$$

Proof Note that (III.2.23) and (III.2.24) imply

$$\text{(III.2.26)} \qquad \|T \cdot f - T \cdot g\|_p \leq \|f - g\|_p.$$

Certainly, there is a closed disk D_0 such that F, and all $T^j \cdot F$, are contained in D_0. By Blaschke's selection theorem, one has a subsequence $T^{j_k} \cdot F \to D$, where D is a compact subset of D_0. We want to show that D is a disk.

Set $F_{j_k} = T^{j_k} \cdot F$. First note that, because $\mathrm{st}_y \mathrm{st}_{xy}\, \mathrm{st}_x F_{j_k} = F_{j_k}$ for all j_k, we have

$$\mathrm{st}_y\, \mathrm{st}_x D = D.$$

(Although it is not obvious that Steiner symmetrization is continuous in the Hausdorff metric – both Steiner and Schwarz symmetrization are continuous in L^p – it is nevertheless true that if $K_j \to K$ in the Hausdorff metric, and K_j is Steiner-symmetric with respect to some line for all j, then K is Steiner-symmetric with respect to that line.)

Next we show that

$$R_{2\alpha} D = D \quad \text{a.e.-}[d\mathbf{v}_2]. \tag{III.2.27}$$

Proof: Fix any quadrant Q of $\mathbb{R}^2$. Then $Q \cap F_{j_k}$ is described by a graph satisfying the hypotheses of the Helly selection theorem (except that the functions are decreasing), which implies there exists a subsequence – still denoted by j_k – for which the graph of F_{j_k} converges to the graph of D. Lebesgue's dominated convergence theorem then implies

$$\|\mathcal{I}_{F_{j_k}} - \mathcal{I}_D\|_p \to 0 \quad \text{as } k \to \infty, \tag{III.2.28}$$

for all $p \geq 1$. Let g be any spherically symmetric strictly decreasing function, such as

$$g(x, y) = e^{-\{x^2+y^2\}}, \qquad (x, y) \in \mathbb{R}^2.$$

Then $T^j \cdot g = g$ and (III.2.26) imply $\|g - \mathcal{I}_{T^j \cdot F}\| \downarrow$ limit (we drop the subscript p for now). Since $\|g - \mathcal{I}_{F_{j_k}}\| \to \|g - \mathcal{I}_D\|$, one has $\|g - \mathcal{I}_{T^j \cdot F}\| \to \|g - \mathcal{I}_D\|$. Also,

$$\|T(g - \mathcal{I}_D)\| = \|g - \mathcal{I}_{T \cdot D}\| = \lim \|g - \mathcal{I}_{T \cdot F_{j_k}}\| = \|g - \mathcal{I}_D\| = \|g - \mathcal{I}_{R_\alpha D}\|$$

(the last equality follows from the spherical symmetry of g), and

$$\|T(g - \mathcal{I}_D)\| = \|\mathrm{st}_x\, \mathrm{st}_y(R_\alpha g - \mathcal{I}_{R_\alpha D})\| = \|\mathrm{st}_x\, \mathrm{st}_y(g - \mathcal{I}_{R_\alpha D})\|,$$

which implies

$$\|\mathrm{st}_x\, \mathrm{st}_y(g - \mathcal{I}_{R_\alpha D})\| = \|g - \mathcal{I}_{R_\alpha D}\|,$$

which implies, by equality in (III.2.22) (since g is spherically symmetric strictly decreasing) that

$$\mathrm{st}_y\, \mathrm{st}_x\, R_\alpha D = R_\alpha D \quad \text{a.e.-}[d\mathbf{v}_2].$$

In particular, $R_\alpha D$, in addition to D, is symmetric with respect to the x- and y-axes. Let $\mathcal{R}_x$ denote reflection in the x-axis. Then

$$R_\alpha D = \mathcal{R}_x R_\alpha D = R_{-\alpha} \mathcal{R}_x D = R_{-\alpha} D,$$

which implies (III.2.27).

Now set

$$\mu(\theta) = \|\mathcal{I}_D - \mathcal{I}_{R_\theta D}\|;$$

then $\mu(\theta) = 0$ on a dense subset of $[0, 2\pi)$. We show that $\theta \mapsto \mu(\theta)$ is continuous when $p = 2$. It suffices to show that

$$r(\theta) := \int \mathcal{I}_D \mathcal{I}_{R_\theta D} \, d\mathbf{v}_n$$

is continuous. Pick a mollifier j_ϵ, $\epsilon > 0$; then

$$r_\epsilon(\theta) := \int (j_\epsilon * \mathcal{I}_D)\mathcal{I}_{R_\theta D} \, d\mathbf{v}_n = \int \mathcal{I}_D((j_\epsilon * \mathcal{I}_D) \circ R_{-\theta}) \, d\mathbf{v}_n$$

is continuous, and

$$\begin{aligned} |r_\epsilon(\theta) - r(\theta)| &= \left| \int (j_\epsilon * \mathcal{I}_D - \mathcal{I}_D)\mathcal{I}_{R_\theta D} \, d\mathbf{v}_n \right| \\ &\leq \|j_\epsilon * \mathcal{I}_D - \mathcal{I}_D\| \, \|\mathcal{I}_{R_\theta D}\| \\ &= \|j_\epsilon * \mathcal{I}_D - \mathcal{I}_D\| \, \|\mathcal{I}_D\| \end{aligned}$$

$\to 0$ uniformly in θ, as $\epsilon \downarrow 0$, which implies the claim.

Therefore $R_\theta D = D$ in L^2 for all θ, which implies that D is a disk. Also, since $\|\mathcal{I}_{T^j F} - \mathcal{I}_D\|_p \downarrow$, we have [by (III.2.28)] $\|\mathcal{I}_{T^j F} - \mathcal{I}_D\|_p \downarrow 0$.

It remains to show $D = F^*$ and (III.2.25). We certainly have (by Lemma III.1.1)

$$\mathbf{v}_n(F) = \limsup_{j \to \infty} \mathbf{v}_n(F_j) \leq \mathbf{v}_n(D).$$

Next, for every $\epsilon > 0$, we have $\mathbf{v}_n([F_j]_\epsilon) \downarrow$, with respect to j. Since the subsequence $F_{j_k} \to D$ in $\mathbf{X}$, then for every $\epsilon' < \epsilon$ there exists J such that

$$[D]_{\epsilon'} \subseteq [F_{j_k}]_\epsilon \qquad \forall \; k \geq J,$$

which implies

$$\mathbf{v}_n([D]_{\epsilon'}) \leq \mathbf{v}_n([F_{j_k}]_\epsilon) \leq \mathbf{v}_n([F]_\epsilon) \qquad \forall \; k \geq J,$$

which implies

$$\mathbf{v}_n(D) \leq \mathbf{v}_n([D]_{\epsilon'}) \leq \mathbf{v}_n([F]_\epsilon).$$

We first conclude that $\mathbf{v}_n(D) \leq \mathbf{v}_n(F)$, which implies the two volumes are equal. Therefore $D = F^*$. We also have

$$\mathbf{v}_n([D]_{\epsilon'}) \leq \mathbf{v}_n([F]_\epsilon) \qquad \forall \; \epsilon' < \epsilon,$$

which implies (III.2.25). ∎

Theorem III.2.9 (Isoperimetric Inequality for Minkowski Area in $\mathbb{R}^n$) *Let F be a compact subset of $\mathbb{R}^n$, $n \geq 3$. Then there exists a sequence T_j consisting of rotations, Steiner symmetrizations, and $(n-1)$-dimensional Schwarz symmetrizations, such that $T_j \cdot F \to F^*$ in the Hausdorff metric topology, and in L^p for all $p \geq 1$.*

Finally,

$$\text{Mink}\,(F^*) \leq \text{Mink}\,(F). \tag{III.2.29}$$

Proof Let $\mathcal{R}$ denote the rotation of $\mathbb{R}^n$ that takes $\mathbf{e}_n$ to $\mathbf{e}_{n-1}$ (and leaves the orthogonal $\mathbb{R}^{n-2}$ pointwise fixed), $\Pi = \mathbf{e}_n^\perp \cong \mathbb{R}^{n-1}$, $\mathcal{S}_1 = \text{st}_\Pi$ denote Steiner symmetrization *with respect to* Π, and $\mathcal{S}_2$ denote Schwarz symmetrization *in* Π; set

$$\mathcal{T} = \mathcal{S}_1\mathcal{S}_2\mathcal{R}.$$

Then the argument of the proof of (III.2.28) implies that there exists a subsequence (k_j),

$$T_j = \mathcal{T}^{k_j},$$

such that $\mathcal{I}_{T_j \cdot F} \to \mathcal{I}_D$ (in the various topologies) for some compact D, as $j \to \infty$. We also have $\mathcal{T}^j \cdot F \to D$ in L^p.

Since $\mathcal{S}_1$ and $\mathcal{S}_2$ commute, and $\mathcal{T}^j \cdot F \to D$ in L^p, we have

$$\mathcal{S}_2\mathcal{T}^j \cdot F = \mathcal{S}_2\mathcal{S}_1\mathcal{S}_2\mathcal{R}\mathcal{T}^{j-1} \cdot F = \mathcal{S}_1{\mathcal{S}_2}^2\mathcal{R}\mathcal{T}^{j-1} \cdot F = \mathcal{S}_1\mathcal{S}_2\mathcal{R}\mathcal{T}^{j-1} \cdot F = \mathcal{T}^j \cdot F$$

and

$$\mathcal{S}_2 D \leftarrow \mathcal{S}_2\mathcal{T}^j \cdot F = \mathcal{T}^j \cdot F \to D,$$

which implies

$$\mathcal{S}_2 D = D,$$

that is, D is rotationally symmetric with respect to the x^n-axis. Also, given $g = e^{-|x|^2}$, for example,

$$\|g - \mathcal{I}_{\mathcal{R}D}\| = \|\mathcal{R}(g - \mathcal{I}_D)\| = \|g - \mathcal{I}_D\| = \|g - \mathcal{I}_{\mathcal{T}D}\| = \|\mathcal{S}_1\mathcal{S}_2(g - \mathcal{I}_{\mathcal{R}D})\|,$$

that is,

$$\|g - \mathcal{I}_{\mathcal{R}D}\| = \|\mathcal{S}_1\mathcal{S}_2(g - \mathcal{I}_{\mathcal{R}D})\|,$$

which implies [by equality in (III.2.23)]

$$\mathcal{S}_2\mathcal{R}D = \mathcal{R}D.$$

So both D and $\mathcal{R}D$ are rotationally symmetric with respect to the x^n-axis, that is, D is rotationally symmetric with respect to both the x^n-axis and the x^{n-1}-axis. Let j_ϵ be a mollifier, depending only on distance from the origin, that is, spherically symmetric in $\mathbb{R}^n$,

$$\mathcal{I}_\epsilon = j_\epsilon * \mathcal{I}_D,$$

and let

$$\rho^2 = (x^1)^2 + \cdots + (x^{n-2})^2.$$

Then there exist functions f and g of two real variables, such that

$$\mathcal{I}_\epsilon(x^1, \ldots, x^n) = f(\sqrt{\rho^2 + (x^{n-1})^2}, x^n) = g(\sqrt{\rho^2 + (x^n)^2}, x^{n-1}).$$

Set $x^n = 0$. Then

$$f(\sqrt{\rho^2 + (x^{n-1})^2}, 0) = g(\sqrt{\rho^2}, x^{n-1}),$$

which implies

$$f(\sqrt{\rho^2 + (x^{n-1})^2 + (x^n)^2}, 0) = g(\sqrt{\rho^2 + (x^n)^2}, x^{n-1}) = \mathcal{I}_\epsilon(x^1, \ldots, x^n),$$

that is, the function $\mathcal{I}_\epsilon(x)$ is radial in x, for every ϵ, which implies $\mathcal{I}_D(x)$ is radial in x, which implies D is a disk.

The proof of $D = F^*$ and (III.2.29) is the same as in the previous theorem. ■

III.3 Application: The Faber–Krahn Inequality

We present an application of symmetrization to analysis, more particularly, an application of geometric isoperimetric inequalities to isoperimetric inequalities for eigenvalues. The result goes as follows:

Definition Given any open set Ω in $\mathbb{R}^n$, one considers the functional

$$F[\phi] = \frac{\|\text{grad}\,\phi\|^2}{\|\phi\|^2},$$

where ϕ ranges over $C_c^\infty(\Omega)$, and the associated infimum

$$\lambda^*(\Omega) = \inf_{\phi \in C_c^\infty(\Omega)} F[\phi].$$

We refer to $\lambda^*(\Omega)$ as the *fundamental tone of* Ω.

It is well known that $\lambda^*(\Omega)$ is the infimum of the spectrum of the Laplacian $-\Delta$ on Ω, subject to vanishing Dirichlet boundary data (see the discussions in §§VII.2 and VII.4 below). In addition, if Ω is a domain with compact closure and C^∞ boundary, then $\lambda^*(\Omega)$ is an eigenvalue of the Laplacian on Ω, that is, there exists a solution $\phi \in C^\infty(\overline{\Omega})$ to

$$\Delta \phi + \lambda^*(\Omega)\phi = 0, \qquad \phi|\partial\Omega = 0; \tag{III.3.1}$$

moreover, $\phi|\Omega \neq 0$ and $(\partial\phi/\partial\nu)|\partial\Omega \neq 0$, and the solution ϕ is unique up to a multiplicative constant.

Theorem III.3.1 *Let Ω be a bounded domain in $\mathbb{R}^n$, and let B be the open disk in $\mathbb{R}^n$ satisfying $V(\Omega) = V(B)$. Then*

$$\lambda^*(\Omega) \geq \lambda^*(B).$$

If Ω also has C^∞ boundary, then one has equality if and only if Ω is isometric to B.

Proof Fix a function $f \in C_c^\infty(\Omega)$, and set, for every $t > 0$,

$$\Omega_t = \{|f| > t\}, \qquad \Gamma_t = \partial\Omega_t \subseteq \{|f| = t\}.$$

We use the symmetric decreasing rearrangement (or Schwarz symmetrization) of functions on Ω, defined by

$$B_t = \Omega_t^* = \mathbb{B}^n(\rho(t)), \qquad f^*(x) = \int_0^\infty \mathcal{I}_{B_t}(x)\,dt.$$

So B_t is the disk of radius $\rho(t)$ whose volume equals that of Ω_t, and f^* is the function that on the sphere $\mathbb{S}^{n-1}(\rho(t))$ has the value t. Then

$$\|f\|_2 = \|f^*\|_2.$$

So to prove the first claim of the theorem, it suffices to show that

$$\|\text{grad } f\|_2 \geq \|\text{grad } f^*\|_2. \tag{III.3.2}$$

Set

$$V(t) := V(\Omega_t) = \boldsymbol{\omega}_{\boldsymbol{n}}\rho(t)^n.$$

Let (α, β) be an interval consisting of regular values of f, and for $t \in (\alpha, \beta)$ set

$$A(t) := A(\Gamma_t).$$

Then $t \mapsto V(t), \rho(t), A(t)$ are all C^∞ on (α, β) with both $V(t)$ and $\rho(t)$ strictly decreasing. On (α, β) we have

$$\frac{dV}{dt} = \mathbf{c}_{\mathbf{n-1}}\rho^{n-1}\frac{d\rho}{dt}, \qquad \frac{\partial f^*}{\partial r}\circ\rho = \frac{1}{d\rho/dt},$$

where r denotes distance from the origin. On the other hand,

$$\frac{dV}{dt} = -\int_{\Gamma_t} |\text{grad } f|^{-1}\, dA,$$

by the co-area formula (Corollary I.3.1).

Finally, set

$$T = \sup |f| \quad \text{and} \quad r_0 = \rho(0).$$

To prove (III.3.2) we have, on the one hand,

$$\begin{aligned}\iint_\Omega |\text{grad } f^*|^2\, dV &= \mathbf{c_{n-1}} \int_0^{r_0} \left(\frac{\partial f^*}{\partial r}\right)^2 r^{n-1}\, dr \\ &= -\mathbf{c_{n-1}} \int_0^T \frac{1}{(\rho')^2} \rho^{n-1} \rho'\, dt \\ &= -\mathbf{c_{n-1}} \int_0^T \frac{\rho^{n-1}}{\rho'}\, dt.\end{aligned}$$

On the other hand we have, by the Cauchy–Schwarz inequality,

$$A^2(t) = \left\{\int_{\Gamma_t} dA\right\}^2 \leq \left\{\int_{\Gamma_t} |\text{grad } f|\, dA\right\}\left\{\int_{\Gamma_t} |\text{grad } f|^{-1}\, dA\right\},$$

which implies

$$\begin{aligned}\iint_\Omega |\text{grad } f|^2\, dV &= \int_0^T dt \int_{\Gamma_t} |\text{grad } f|\, dA \\ &\geq \int_0^T A^2(t) \left\{\int_{\Gamma_t} |\text{grad } f|^{-1}\, dA\right\}^{-1} dt \\ &= -\int_0^T A^2(t)(V'(t))^{-1}\, dt \\ &\geq -\int_0^T A^2(\mathbb{S}^{n-1}(\rho(t)))(V'(t))^{-1}\, dt \\ &= -\mathbf{c_{n-1}} \int_0^T \frac{\rho^{n-1}}{\rho'}\, dt\end{aligned}$$

(the last inequality is the isoperimetric), which implies (III.3.2).

To consider the case of equality, we assume that Ω has a C^∞ boundary, in which case $\lambda^*(\Omega)$ has a positive eigenfunction f of $\lambda_1(\Omega)$. Of course, $f|\partial\Omega = 0$. The eigenfunction equation (III.3.1) implies

$$\frac{\partial^2 f}{\partial x^{j\,2}} < 0$$

for at least one of the $j \in \{1, \ldots, n\}$. Thus, by the implicit function theorem, the set of points for which $\operatorname{grad} f = 0$ is contained in an $(n-1)$-manifold. In particular, for every t, $V(\Gamma(t)) = 0$, which implies $V = V(t)$ is continuous with respect to t.

Now one can argue as above. If $\lambda^*(\Omega) = \lambda^*(B)$, then the argument shows that $A(\partial\Omega_t) = A(\partial B_t)$ for all regular values of t. But then for every such t one has Ω_t isometric to B_t, and the strong maximum principle implies that $\Omega = \Omega(0)$ is isometric to $B(0) = B$. ■

III.4 Perimeter

III.4.1 Geometric Perimeter

Definition We say that a collection of measurable subsets (E_j) of $\mathbb{R}^n$ *converges to E in* L^1 if $\mathbf{v}_n(E_j \,\Delta\, E) \to 0$ as $j \to \infty$, that is, if $\mathcal{I}_{E_j} \to \mathcal{I}_E$ in $L^1(\mathbb{R}^n)$:

$$\lim_{j\to\infty} \int |\mathcal{I}_{E_j} - \mathcal{I}_E| \, d\mathbf{v}_n = 0.$$

Definition The *geometric perimeter* geoper(E) *of a measurable subset E of* $\mathbb{R}^n$ is defined by

$$\operatorname{geoper}(E) = \inf\{\liminf A(\partial M_j)\},$$

where the infimum is taken over all sequences of open subsets M_j of $\mathbb{R}^n$, with C^∞ boundary, that converge to E in L^1. One easily checks that if $E_j \to E$ in L^1 then

$$\operatorname{geoper}(E) \le \liminf \operatorname{geoper}(E_j),$$

that is, $E \mapsto \operatorname{geoper}(E)$ is lower semicontinuous with respect to convergence in L^1.

Theorem III.4.1 *If K is compact, then K may be approximated in* L^1 *by open subsets with* C^∞ *boundary. Furthermore,*

$$\operatorname{geoper}(K) \le \operatorname{Mink}(K).$$

Proof Let $f(x) = d(x, K)$, (j_ϵ) a mollifier on $\mathbb{R}^n$, and set

$$f_\epsilon = j_\epsilon * f.$$

Then $\sup |\operatorname{grad} f_\epsilon| = \operatorname{Lip} f_\epsilon \le \operatorname{Lip} f = 1$.

For any small $h > 0$, let $\sigma = h^2$. Then there exists $\epsilon \in (0, \sigma)$ such that

$$f_\epsilon{}^{-1}[(\sigma, h - \sigma)] \subset\subset [K]_h \setminus K,$$

and there exists a regular value $\tau \in (\sigma, h - \sigma)$ of f_ϵ. By dominated convergence, K may be approximated in L^1 by $f_\epsilon^{-1}[(0, \tau)]$.

Furthermore, by the co-area formula (Corollary I.3.1), we have

$$\int_\sigma^{h-\sigma} A\big(f_\epsilon{}^{-1}[t]\big)\, dt = \int_{f_\epsilon{}^{-1}[(\sigma,h-\sigma)]} |\mathrm{grad}\, f_\epsilon|\, dV < \mathbf{v}_n([K]_h \setminus K),$$

which implies there exists a regular value $t_0 \in (\sigma, h - \sigma)$ such that

$$(h - 2\sigma) A(f_\epsilon{}^{-1}[t_0]) \le \mathbf{v}_n([K]_h \setminus K).$$

Also,

$$f_\epsilon{}^{-1}[t_0] = \partial(f_\epsilon{}^{-1}[0, t_0)), \qquad K \subset f_\epsilon{}^{-1}[0, t_0] \subset [K]_h.$$

Given any $\rho > 0$, there exists $h_j \to 0$ such that

$$\frac{\mathbf{v}_n([K]_{h_j} \setminus K)}{h_j} \le \mathrm{Mink}\,(K) + \rho,$$

which implies, for $\epsilon_j \in (0, \sigma_j)$, $\sigma_j = h_j^2$, and associated t_j,

$$A\big(f_{\epsilon_j}{}^{-1}[t_j]\big) \le \frac{h_j}{h_j - 2\sigma_j}\{\mathrm{Mink}\,(K) + \rho\},$$

which implies the theorem. ∎

III.4.2 Functions of Bounded Variation

Definition Given an open set Ω in $\mathbb{R}^n$. The collection of *functions on Ω of bounded variation*, $\mathrm{BV}(\Omega)$, consists of those functions $f \in L^1(\Omega)$ for which

$$\sup_{\boldsymbol{\xi}} \int f \,\mathrm{div}\, \boldsymbol{\xi}\, d\mathbf{v}_n < +\infty,$$

where $\boldsymbol{\xi}$ varies over $C_c^1(\Omega)$ vector fields on Ω satisfying $|\boldsymbol{\xi}| \le 1$ on all of Ω. We refer to these vector fields as *admissible vector fields on Ω*.

If the function $f \in \mathrm{BV}(\Omega)$, then the Riesz representation theorem (Proposition I.3.5) implies that there exists a vector-valued Radon measure δf on Ω defined by

$$\int f \,\mathrm{div}\, \boldsymbol{\xi}\, d\mathbf{v}_n = \int \boldsymbol{\xi} \cdot d\delta f$$

for all admissible vector fields $\boldsymbol{\xi}$ on Ω, and the total variation of δf satisfies

$$|\delta f|(\Omega) = \sup_{\boldsymbol{\xi}} \int_{\Omega} f \operatorname{div} \boldsymbol{\xi} \, d\mathbf{v}_n$$

where the supremum is taken over all admissible $\boldsymbol{\xi}$.

Example III.4.1 If $f \in C^1(\Omega) \cap L^1(\Omega)$, then $f \in \mathrm{BV}(\Omega)$ and

$$d\delta f(x) = -(\operatorname{grad} f)(x) \, d\mathbf{v}_n(x).$$

Indeed, for any admissible $\boldsymbol{\xi}$ we have

$$\int_{\Omega} f \operatorname{div} \boldsymbol{\xi} \, d\mathbf{v}_n = -\int_{\Omega} \operatorname{grad} f \cdot \boldsymbol{\xi} \, d\mathbf{v}_n$$

Remark III.4.1 The same result (Example III.4.1) is valid in $W^{1,1}(\Omega)$ (see Remark I.3.3 below).

Proposition III.4.1 *If* $\Omega' \subset \Omega$ *then* $(\delta f)|\Omega' = \delta(f|\Omega')$.

Theorem III.4.2 *Let* Ω *be open in* $\mathbb{R}^n$, (f_j) *a sequence of functions in* $\mathrm{BV}(\Omega)$ *such that* $f_j \to f$ *in* $L^1_{\mathrm{loc}}(\Omega)$. *Then* $f \in \mathrm{BV}(\Omega)$ *and*

$$|\delta f|(\Omega) \leq \liminf_{j \to \infty} |\delta f_j|(\Omega).$$

Proof For any admissible $\boldsymbol{\xi}$ on Ω we have

$$\begin{aligned}
\int f \operatorname{div} \boldsymbol{\xi} &= \int (\lim f_j) \operatorname{div} \boldsymbol{\xi} \\
&= \lim \int f_j \operatorname{div} \boldsymbol{\xi} \\
&= \lim \int \boldsymbol{\xi} \cdot d\delta f_j \\
&\leq \liminf |\delta f_j|(\Omega)
\end{aligned}$$

– the second line follows from the compactness of the support of $\boldsymbol{\xi}$, $f \in L^1_{\mathrm{loc}}$, and dominated convergence; and the last line uses $|\boldsymbol{\xi}| \leq 1$ on all of Ω. ■

Corollary III.4.1 *Given the open set* Ω *in* R^n, *endow the space* $\mathrm{BV}(\Omega)$ *with the norm*

$$\|f\|_{\mathrm{BV}} = \|f\|_1 + |\delta f|(\Omega).$$

Then $\mathrm{BV}(\Omega)$ *is complete relative to the norm* $\| \; \|_{\mathrm{BV}}$.

Proof If (f_j) is a Cauchy sequence in $\mathrm{BV}(\Omega)$, then (f_j) is a Cauchy sequence in $L^1(\Omega)$, which implies there exists $f \in L^1(\Omega)$ such that $f_j \to f$ as $j \to \infty$. Then, by Theorem III.4.2, we have

$$|\delta(f - f_j)|(\Omega) \leq \liminf_{k\to\infty} |\delta(f_k - f_j)|(\Omega) \leq \liminf_{k\to\infty} \|f_k - f_j\|_{\mathrm{BV}(\Omega)},$$

which implies the corollary. ■

Remark III.4.2 Note that for arbitrary $f \in \mathrm{BV}(\Omega)$ we do not expect to approximate f by C^∞ functions, namely that there exist $(f_j) \in C^\infty(\Omega)$ such that $\|f_j - f\|_1 \to 0$ and $|\delta(f_j - f)|_1(\Omega) \to 0$, for this would imply that $f \in W^{1,1}(\Omega)$ – which is not necessarily the case (see the case of indicator functions in Example III.4.2 below).

Theorem III.4.3 *Given the open sets Ω in $\mathbb{R}^n$, $A \subset\subset \Omega$, and $f \in \mathrm{BV}(\Omega)$, satisfying*

$$|\delta f|(\partial A) = 0. \tag{III.4.1}$$

Let (j_ϵ) be a mollifier on $\mathbb{R}^n$, $f_\epsilon = j_\epsilon * f$. Then

$$|\delta f|(A) = \lim_{\epsilon\downarrow 0} |\delta f_\epsilon|(A).$$

Proof One has $f_\epsilon \to f$ in $L^1(\Omega)$, which implies

$$|\delta f|(A) \leq \liminf_{\epsilon\downarrow 0} |\delta f_\epsilon|(A).$$

For $\boldsymbol{\xi}$ admissible on A we have

$$\int f_\epsilon \operatorname{div} \boldsymbol{\xi} = \int (j_\epsilon * f) \operatorname{div} \boldsymbol{\xi} = \int f \operatorname{div} (j_\epsilon * \boldsymbol{\xi}).$$

Now $\operatorname{supp} (j_\epsilon * \boldsymbol{\xi}) \subset [\operatorname{supp} \boldsymbol{\xi}]_\epsilon \subset [A]_\epsilon$ which implies

$$\int_A f_\epsilon \operatorname{div} \boldsymbol{\xi} \leq \int_{[A]_\epsilon} f \operatorname{div} (j_\epsilon * \boldsymbol{\xi}) \leq |\delta f|([A]_\epsilon).$$

Therefore

$$|\delta f_\epsilon|(A) \leq |\delta f|([A]_\epsilon),$$

which implies

$$\limsup_{\epsilon\downarrow 0} |\delta f_\epsilon|(A) \leq |\delta f|(\operatorname{cl} A) = |\delta f|(A)$$

by (III.4.1), which implies the theorem. ■

Theorem III.4.4 *If $f \in \mathrm{BV}(\mathbb{R}^n)$, (j_ϵ) a mollifier on $\mathbb{R}^n$, then*

$$|\delta f|(\mathbb{R}^n) = \lim_{\epsilon \downarrow 0} |\delta f_\epsilon|(\mathbb{R}^n). \tag{III.4.2}$$

Proof The proof is similar to the proof of Theorem III.4.3. ∎

Theorem III.4.5 *Let Ω be open in $\mathbb{R}^n$. Given any $f \in \mathrm{BV}(\Omega)$, there exists a sequence of C^∞ functions (f_j) in $\mathrm{BV}(\Omega)$ such that*

$$L^1\text{-}\lim_{j\to\infty} f_j = f, \qquad \lim_{j\to\infty} |\delta f_j|(\Omega) = |\delta f|(\Omega).$$

Proof Given $\epsilon > 0$, there exists an open $\Omega_0 \subset\subset \Omega$ such that

$$|\delta f|(\Omega \setminus \mathrm{cl}\,\Omega_0) < \epsilon.$$

Let $\Omega_0 \subset\subset \Omega_1 \subset\subset \Omega_2 \subset\subset \cdots$ be an exhaustion of Ω, with Ω_0 the one we chose above, and set

$A_1 = \Omega_2$,
$A_\ell = \Omega_{\ell+1} \setminus \mathrm{cl}\,\Omega_{\ell-1}, \qquad \ell = 2, 3, \ldots,$
(ϕ_k) a partition of unity of Ω subordinate to (A_k),
(j_ϵ) a mollifier on $\mathbb{R}^n$.

For each $k = 1, 2, \ldots,$ pick ϵ_k in $(0, \epsilon)$ such that

$\mathrm{supp}\, j_{\epsilon_k} * (f\phi_k) \subset \Omega_{k+2} \setminus \mathrm{cl}\,\Omega_{k-2}, \qquad \Omega_{-1} = \emptyset,$
$\int |j_{\epsilon_k} * (f\phi_k) - f\phi_k| < \epsilon/2^k,$
$\int |j_{\epsilon_k} * (f\,\mathrm{grad}\,\phi_k) - f\,\mathrm{grad}\,\phi_k| < \epsilon/2^k,$

and set

$$f_\epsilon = \sum_{k=1}^{\infty} j_{\epsilon_k} * (f\phi_k).$$

One easily sees that $f_\epsilon \to f$ in $L^1(\Omega)$ as $\epsilon \downarrow 0$, which implies, by Theorem III.4.2,

$$|\delta f| \le \liminf_{\epsilon \downarrow 0} |\delta f_\epsilon|.$$

Given an admissible ξ on Ω, we have

$$\begin{aligned}
\int f_\epsilon \operatorname{div}\xi &= \sum_k \int (j_{\epsilon_k} * (f\phi_k)) \operatorname{div}\xi \\
&= \sum_k \int f\phi_k \operatorname{div}(j_{\epsilon_k} * \xi) \\
&= \sum_k \int f \operatorname{div}(\phi_k (j_{\epsilon_k} * \xi)) - \sum_k \int f \operatorname{grad}\phi_k \cdot (j_{\epsilon_k} * \xi - \xi) \\
&\quad - \sum_k \int f \operatorname{grad}\phi_k \cdot \xi \\
&= \sum_k \int f \operatorname{div}(\phi_k (j_{\epsilon_k} * \xi)) \\
&\quad - \sum_k \int \{ j_{\epsilon_k} * (f \operatorname{grad}\phi_k) - f \operatorname{grad}\phi_k \} \cdot \xi \\
&\le \sum_k \int_{A_k} f \operatorname{div}(\phi_k (j_{\epsilon_k} * \xi)) + \epsilon
\end{aligned}$$

– the last equality uses the fact that (ϕ_k) is a partition of unity. Now

$$\int_{A_1} f \operatorname{div}(\phi_1 (j_{\epsilon_1} * \xi)) \le |\delta f|(\Omega);$$

$A_k \subset \Omega \setminus \operatorname{cl}\Omega_0$ for all $k \ge 2$, and every point in $\Omega \setminus \operatorname{cl}\Omega_0$ is covered by at least one of the collection (A_k) and at most two of the collection (A_k), which implies

$$\begin{aligned}
\sum_{k=2}^{\infty} \int_{A_k} f \operatorname{div}(\phi_k (j_{\epsilon_k} * \xi)) &\le 2 \int_{\Omega\setminus\Omega_0} \sum_{k=2}^{\infty} f \operatorname{div}(\phi_k (j_{\epsilon_k} * \xi)) \\
&\le 2|\delta f|(\Omega \setminus \operatorname{cl}\Omega_0) < 2\epsilon,
\end{aligned}$$

which implies

$$|\delta f_\epsilon|(\Omega) \le |\delta f|(\Omega) + 3\epsilon,$$

which implies the theorem. ∎

III.4.3 Caccioppoli Sets

We consider our most important example of functions of bounded variation – indicator functions of Borel sets in $\mathbb{R}^n$.

Notation For the indicator function of a set E for which $\mathcal{I}_E$ has bounded variation, we always write

$$\delta E := \delta \mathcal{I}_E.$$

Example III.4.2 Given Ω, let E be a domain in $\mathbb{R}^n$ with C^1 boundary, and $\mathcal{I}_E$ its indicator function, with $\mathbf{v}_n(E \cap \Omega) < +\infty$, that is, $\mathcal{I}_E|\Omega \in L^1(\Omega)$. Then it is standard that $\mathcal{I}_E|\Omega \notin W^{1,1}(\Omega)$. For any admissible $\boldsymbol{\xi}$ on Ω, we have

$$\int_\Omega \mathcal{I}_E \operatorname{div} \boldsymbol{\xi} = \int_{\partial E \cap \Omega} \boldsymbol{\xi} \cdot \boldsymbol{\nu}_E \, dA,$$

where $\boldsymbol{\nu}_E$ is the exterior unit vector field along ∂E. If $A(\partial E \cap \Omega) < +\infty$, then $\mathcal{I}_E|\Omega \in \mathrm{BV}(\Omega)$ and

$$d\boldsymbol{\delta} E|\Omega = \boldsymbol{\nu}_E \, dA_E|\Omega, \qquad |\boldsymbol{\delta} E|(\Omega) = A(\partial E \cap \Omega).$$

Remark III.4.3 For arbitrary measurable E with $\mathcal{I}_E|\Omega \in \mathrm{BV}(\Omega)$, we always have $\operatorname{supp} \boldsymbol{\delta} E \subset \partial E$. Indeed, if F is open in $\operatorname{int}(\Omega \setminus E)$, then $\mathcal{I}_E|F = 0$ implies $\int \mathcal{I}_E \operatorname{div} \boldsymbol{\xi} = 0$ for all admissible vector fields on F. If G is open in $\operatorname{int}(\Omega \cap E)$, then $\mathcal{I}_E|G = 1$ implies $\int \mathcal{I}_E \operatorname{div} \boldsymbol{\xi} = \int_G \operatorname{div} \boldsymbol{\xi} = 0$ for all admissible vector fields on G, which implies the claim.

We now extend Example III.4.2 to Lipschitz domains.

Definition A domain Ω in $\mathbb{R}^n$ is a *Lipschitz domain* if for each $w \in \partial\Omega$ there exist a neighborhood $U = U(w)$ in $\mathbb{R}^n$, Euclidean coordinates u, and a Lipschitz function

$$\phi : U \cap (\mathbb{R}^{n-1} \times \{0\}) \to \mathbb{R}$$

such that

$$\Omega \cap U = \{x : u^n(x) > \phi(u^1(x), \ldots, u^{n-1}(x))\},$$
$$\partial\Omega \cap U = \{x : u^n(x) = \phi(u^1(x), \ldots, u^{n-1}(x))\}.$$

For convenience, we shall write $\mathbb{R}^{n-1}$ for $\mathbb{R}^{n-1} \times \{0\}$ and $u = (u^1, \ldots, u^{n-1}, u^n) = (\overline{u}, u^n)$, $\overline{u} \in \mathbb{R}^{n-1}$.

Example III.4.3 Let Ω be a Lipschitz domain, $w \in \partial\Omega$, U, u, and ϕ as above, and assume ϕ is differentiable at $\overline{u}(w)$. [By Rademacher's theorem (Theorem I.1.3.2), ϕ is differentiable a.e.-$[d\mathbf{v}_{n-1}]$ on $U \cap \mathbb{R}^{n-1}$.] Then $\partial\Omega$ has a tangent hyperplane at w, with exterior unit vector $\boldsymbol{\nu}$ given by

$$\boldsymbol{\nu} = \frac{\operatorname{grad} \phi - \mathbf{e}_n}{\sqrt{|\operatorname{grad} \phi|^2 + 1}},$$

where $\mathbf{e}_n$ denotes the unit vector in the positive direction of the u^n-axis [the gradient here is $(n-1)$-dimensional]. Let (ϕ_j) approximate ϕ in the Sobolev space $W^{1,1}(U \cap \mathbb{R}^{n-1})$, $\phi_j \in C^\infty$ for all j (Theorem I.3.3). Then $\phi_j \to \phi$ uniformly on compacta. Let

$$\Lambda_j = \{x \in U : u^n(x) > \phi_j(\overline{u}(x))\},$$
$$\boldsymbol{\nu}_j = \text{exterior unit normal vector field along } \partial\Lambda_j.$$

Then

$$\int_U \mathcal{I}_{\Lambda_j} \operatorname{div} \boldsymbol{\xi} = \int_U \boldsymbol{\xi} \cdot \boldsymbol{\delta}\Lambda_j,$$

where

$$\boldsymbol{\delta}\Lambda_j = \boldsymbol{\nu}_j(\overline{u}, \phi(\overline{u}))\sqrt{|\operatorname{grad} \phi_j(\overline{u})|^2 + 1}\, d\mathbf{v}_{n-1}(\overline{u}) \otimes \delta^1_{\phi_j(\overline{u})}(u^n)\, du^n,$$

$\delta^1_\alpha(t)$ is the 1-dimensional delta function (of t) concentrated at $\alpha \in \mathbb{R}$, and

$$\int_U \boldsymbol{\xi} \cdot \boldsymbol{\delta}\Lambda_j = \int_{U \cap \mathbb{R}^{n-1}} (\boldsymbol{\xi} \cdot \boldsymbol{\nu}_j)(\overline{u}, \phi_j(\overline{u}))\sqrt{|\operatorname{grad} \phi_j(\overline{u})|^2 + 1}\, d\mathbf{v}_{n-1}(\overline{u}).$$

Let $j \to \infty$. Then

$$\int_U \mathcal{I}_{\Lambda_j} \operatorname{div} \boldsymbol{\xi} \to \int_U \mathcal{I}_\Omega \operatorname{div} \boldsymbol{\xi},$$
$$\int_U \boldsymbol{\xi} \cdot \boldsymbol{\delta}\Lambda_j \to \int_{U \cap \mathbb{R}^{n-1}} (\boldsymbol{\xi} \cdot \boldsymbol{\nu})\sqrt{|\operatorname{grad} \phi|^2 + 1}\, d\mathbf{v}_{n-1}.$$

Therefore, $\mathcal{I}_\Omega | U \in \mathrm{BV}(U)$ and

$$\boldsymbol{\delta}\Omega | U = \boldsymbol{\nu}(\overline{u}, \phi(\overline{u}))\sqrt{|\operatorname{grad} \phi(\overline{u})|^2 + 1}\, d\mathbf{v}_{n-1}(\overline{u}) \otimes \delta^1_{\phi(\overline{u})}(u^n)\, du^n,$$

which generalizes the classical formula (I.3.8).

Definition We start with a Borel set E. For any open Ω, we define the *perimeter of E in Ω, $P(E; \Omega)$*, to be $|\boldsymbol{\delta}E|(\Omega)$ when finite, and otherwise $+\infty$.

Remark III.4.4 E Borel guarantees that E is $|\boldsymbol{\delta}E|$-measurable.

Definition We also define

$$P(E) = P(E; \mathbb{R}^n).$$

E is called a *Caccioppoli subset of* $\mathbb{R}^n$ if $P(E; \Omega)$ is finite for all bounded open Ω in $\mathbb{R}^n$. A Caccioppoli set in $\mathbb{R}^n$ is also referred to as a *set with locally finite perimeter*.

Proposition III.4.2 *We always have:*

(a) $E \subset\subset \Omega \subset\subset \Omega_1 \Rightarrow P(E; \Omega) \leq P(E; \Omega_1)$.
(b) $P(E_1 \cup E_2; \Omega) \leq P(E_1; \Omega) + P(E_2; \Omega)$, *with equality if* $d(E_1, E_2) > 0$.

(c) $\mathbf{v}_n(E) = 0 \Rightarrow P(E) = 0.$
(d) $\mathbf{v}_n(E_1 \,\Delta\, E_2) = 0 \Rightarrow P(E_1) = P(E_2).$

Remark III.4.5 Note that (c) in the above proposition differs markedly from Minkowski area, where a set might have measure 0 but nonzero Minkowski area (Remark III.2.2).

Remark III.4.6 We restate the definition of Caccioppoli set slightly differently. A Borel set E is a Caccioppoli set in $\mathbb{R}^n$ if there exists a vector-valued Radon measure μ on $\mathbb{R}^n$ such that

$$\int_E \operatorname{div} \boldsymbol{\xi} \, d\mathbf{v}_n = \int \boldsymbol{\xi} \cdot d\mu$$

for all admissible $\boldsymbol{\xi}$ on $\mathbb{R}^n$. Indeed, for any bounded open Ω in $\mathbb{R}^n$ we have

$$P(E; \Omega) \leq |\mu|(\Omega) < +\infty,$$

which implies E is Caccioppoli. The measure μ must be unique; and since

$$\int \boldsymbol{\xi} \cdot d\mu = \int \boldsymbol{\xi} \cdot d\delta E$$

for all admissible $\boldsymbol{\xi}$ on $\mathbb{R}^n$, we have $\mu = \delta E$.

Theorem III.4.6 *Let E be a Borel set in $\mathbb{R}^n$ such that $\mathcal{I}_E \in \mathrm{BV}(\mathbb{R}^n)$ (that is, E has finite volume and perimeter), (j_ϵ) a mollifier on $\mathbb{R}^n$. Set $\varphi = \mathcal{I}_E$ and $\varphi_\epsilon = j_\epsilon * \varphi$. Then*

$$|\delta E|(A) = \lim_{\epsilon \downarrow 0} |\delta \varphi_\epsilon|(A) \tag{III.4.3}$$

for all $A \in \mathbb{R}^n$ with $|\delta E|(A) = 0$. Also,

$$|\delta E|(\mathbb{R}^n) = \lim_{\epsilon \downarrow 0} |\delta \varphi_\epsilon|(\mathbb{R}^n). \tag{III.4.4}$$

Proof This is a direct application of Theorems III.4.3 and III.4.4. ■

Theorem III.4.7 **(Co-area Formula for Perimeter)** *Given Ω open in $\mathbb{R}^n$, $f \in \mathrm{BV}(\Omega)$, let F_t denote the level domain*

$$F_t = \{x \in \Omega : f(x) > t\}.$$

Then

$$\delta f = \int_{-\infty}^{\infty} \delta F_t \, dt, \qquad |\delta f| = \int_{-\infty}^{\infty} |\delta F_t| \, dt. \tag{III.4.5}$$

Remark III.4.7 Thus, when f is C^∞, the second equality in (III.4.5) reduces (by Example III.4.1) to the co-area formula for smooth functions (Corollary I.3.1).

Proof For any $\alpha \in \mathbb{R}$, and

$$\Phi_\alpha(t) = \begin{cases} \mathcal{I}_{(-\infty,\alpha)}(t), & t > 0, \\ \mathcal{I}_{(-\infty,\alpha)}(t) - 1, & t \le 0, \end{cases}$$

we have

$$\alpha = \int_{-\infty}^{\infty} \Phi_\alpha(t)\, dt.$$

Note that

$$|\alpha - \beta| = \int_{-\infty}^{\infty} |\Phi_\alpha(t) - \Phi_\beta(t)|\, dt.$$

Then, for $f(x)$, we have $\mathcal{I}_{F_t}(x) = \mathcal{I}_{(-\infty, f(x))}(t)$, which implies that for any admissible $\boldsymbol{\xi}$ on Ω we have

$$\begin{aligned} \int \boldsymbol{\xi}\cdot\delta f &= \int f \operatorname{div} \boldsymbol{\xi} \\ &= \int_{-\infty}^{0} dt \int \{\mathcal{I}_{F_t} - 1\} \operatorname{div} \boldsymbol{\xi} + \int_0^\infty dt \int \mathcal{I}_{F_t} \operatorname{div} \boldsymbol{\xi} \\ &= \int_{-\infty}^{\infty} dt \int \mathcal{I}_{F_t} \operatorname{div} \boldsymbol{\xi} \\ &= \int_{-\infty}^{\infty} \boldsymbol{\xi}\cdot\delta F_t\, dt. \end{aligned}$$

Therefore,

$$\delta f = \int_{-\infty}^{\infty} \delta F_t\, dt, \qquad |\delta f| \le \int_{-\infty}^{\infty} |\delta F_t|\, dt.$$

By Theorem III.4.5, there exists a sequence $(f_j) \in C^\infty$ such that $f_j \to f$ in $L^1(\Omega)$ and $|\delta f_j| \to |\delta f|$. Then

$$\int |f_j - f|\, d\mathbf{v}_n = \int d\mathbf{v}_n \int |\mathcal{I}_{(F_j)_t} - \mathcal{I}_{F_t}|\, dt = \int dt \int |\mathcal{I}_{(F_j)_t} - \mathcal{I}_{F_t}|\, d\mathbf{v}_n$$

[where $(F_j)_t$ denotes the level domain of f_j associated with the value t], which implies (Proposition I.3.1) there exists a sequence (j_k) such that $\mathcal{I}_{(F_{j_k})_t} \to \mathcal{I}_{F_t}$

in $L^1(\Omega)$ for almost all t, as $k \to \infty$. Therefore,

$$\begin{aligned}\int |\delta F_t|\,dt &\leq \int \liminf |\delta(F_{j_k})_t|\,dt \\ &\leq \liminf \int |\delta(F_{j_k})_t|\,dt \\ &= \liminf |\delta f_{j_k}| \\ &= |\delta f|\end{aligned}$$

– the next to last equality follows from the co-area formula for C^∞ functions (Corollary I.3.1) – which implies the theorem. ■

Lemma III.4.1 *Let E be a Caccioppoli set in $\mathbb{R}^n$, (j_ϵ) a mollifier on $\mathbb{R}^n$, and $t \in (0, 1)$. Set $\varphi = \mathcal{I}_E$ and $\varphi_\epsilon = j_\epsilon * \varphi$, and*

$$(E_\epsilon)_t = \{x : \varphi_\epsilon(x) > t\}.$$

Then

$$\int |\mathcal{I}_{(E_\epsilon)_t} - \mathcal{I}_E|\,d\mathbf{v}_n \leq \frac{1}{\min(t, 1-t)} \int |\varphi_\epsilon - \varphi|\,d\mathbf{v}_n.$$

Proof If $x \in E\setminus(E_\epsilon)_t$ then $\varphi(x) = 1$ and $\varphi_\epsilon(x) \leq t$, which implies

$$(\varphi - \varphi_\epsilon)(x) \geq (1-t)\mathcal{I}_{E\setminus(E_\epsilon)_t}.$$

Similarly, if $x \in (E_\epsilon)_t \setminus E$ then $\varphi(x) = 0$ and $\varphi_\epsilon(x) > t$, which implies

$$(\varphi - \varphi_\epsilon)(x) < -t\mathcal{I}_{(E_\epsilon)_t\setminus E}.$$

Therefore

$$\begin{aligned}\int |\varphi_\epsilon - \varphi|\,d\mathbf{v}_n &\geq \int t\mathcal{I}_{(E_\epsilon)_t\setminus E} + (1-t)\mathcal{I}_{E\setminus(E_\epsilon)_t}\,d\mathbf{v}_n \\ &\geq \min(t, 1-t) \int |\mathcal{I}_{(E_\epsilon)_t} - \mathcal{I}_E|\,d\mathbf{v}_n,\end{aligned}$$

which implies the lemma. ■

Theorem III.4.8 *Let E be a Caccioppoli set in $\mathbb{R}^n$. Then there exists a sequence of domains (E_ℓ) in $\mathbb{R}^n$, with C^∞ boundary, such that*

$$L^1\text{-}\lim_{\ell\to\infty} \mathcal{I}_{E_\ell} = \mathcal{I}_E, \qquad \lim_{j\to\infty} |\delta E_\ell| = |\delta E|.$$

In particular, the perimeter of E is equal to its geometric perimeter.

Proof Start with $\varphi = \mathcal{I}_E$, the mollifier (j_ϵ) on $\mathbb{R}^n$, $\varphi_\epsilon = j_\epsilon * \varphi$, and $(E_\epsilon)_t$ as above. Then the co-area formula (Theorem III.4.7) implies

$$|\delta\varphi_\epsilon| = \int_0^1 |\delta(E_\epsilon)_t|\,dt.$$

For any bounded open Ω, Lemma III.4.1 implies that $\mathcal{I}_{(E_\epsilon)_t} \to \mathcal{I}_E$ in $L^1(\Omega)$ as $\epsilon \downarrow 0$, for every $t \in (0, 1)$. Therefore, by Theorem III.4.2,

$$|\delta E| \leq \liminf_{\epsilon\downarrow 0} |\delta(E_\epsilon)_t|$$

for all t, which implies,

$$\begin{aligned}
|\delta E| &= \int_0^1 |\delta E|\,dt \\
&\leq \int_0^1 \liminf_{\epsilon\downarrow 0} |\delta(E_\epsilon)_t|\,dt \\
&\leq \liminf_{\epsilon\downarrow 0} \int_0^1 |\delta(E_\epsilon)_t|\,dt \\
&= \lim_{\epsilon\downarrow 0} |\delta\varphi_\epsilon| \\
&= |\delta E|,
\end{aligned}$$

– the last equality follows from Theorem III.4.6 – which implies

$$|\delta E| = \liminf_{\epsilon\downarrow 0} |\delta(E_\epsilon)_t|$$

for almost all $t \in (0, 1)$.

Now pick $\epsilon_k \to 0$ as $k \to \infty$. Then the same argument for $E_k = E_{\epsilon_k}$ shows that

$$|\delta E| = \liminf_{k\to\infty} |\delta(E_k)_t|$$

for almost all $t \in (0, 1)$. Sard's theorem (Proposition I.3.8) implies

$$\mathbf{v}_1\left(\bigcup_{k=1}^{\infty} \operatorname{critval} \varphi_{\epsilon_k}\right) = 0,$$

where, for any function H, critval H denotes the set of critical values of H. Let t be in the complement of $\bigcup_k$ critval φ_{ϵ_k} in $[0, 1]$, that is, t is a regular value of

φ_{ϵ_k}, for all k. Then there is a subsequence $\sigma_\ell = \epsilon_{k_\ell}$ such that

$$|\delta E| = \lim_{\ell \to \infty} |\delta(E_{\sigma_\ell})_t|.$$

Then the sequence $(E_{\sigma_\ell})_t$ will do the job.

Thus geoper$(E) \leq P(E)$. For any sequence E_ℓ, $\partial E_\ell \in C^\infty$, with $E_\ell \to E$ in L^1, we have

$$\liminf_\ell A(\partial E_\ell) = \liminf_\ell P(\partial E_\ell) \geq P(E),$$

which implies $P(E) \leq$ geoper(E), which implies the two are equal. ∎

Remark III.4.8 Note that the sets $(E_{\sigma_\ell})_t$ converge to E in the Hausdorff metric on compact sets.

III.4.4 The Isoperimetric Inequality for Perimeter

Remark III.4.9 Note that Theorem III.4.1 now implies that the isoperimetric inequality (to be proved below) for perimeter is sharper than the isoperimetric inequality for Minkowski area.

Theorem III.4.9 *Steiner symmetrization does not increase perimeter.*

Proof Let $\mathcal{S}$ denote Steiner symmetrization with respect to some fixed hyperplane, K a finite disjoint union of compact domains with C^∞ boundary. Then

$$P(K) = A(\partial K) = \text{Mink }(K) \geq \text{Mink }(\mathcal{S}K) \geq P(\mathcal{S}K),$$

by Theorems III.2.3 and III.4.1. For arbitrary Borel E with finite perimeter, let $K_\ell \to E$ in L^1, $P(K_\ell) \to P(E)$, as $\ell \to \infty$, and $\partial K_\ell \in C^\infty$ for all ℓ. Then, by Corollary III.2.3,

$$\|\mathcal{I}_{\mathcal{S}K_\ell} - \mathcal{I}_{\mathcal{S}E}\|_1 \leq \|\mathcal{I}_{K_\ell} - \mathcal{I}_E\|_1 \to 0,$$

which implies $\mathcal{I}_{\mathcal{S}K_\ell} \to \mathcal{I}_{\mathcal{S}E}$ in L^1, which implies

$$P(\mathcal{S}E) \leq \liminf_{\ell \to \infty} P(\mathcal{S}K_\ell) \leq \liminf_{\ell \to \infty} P(K_\ell) = P(E),$$

which implies the claim. ∎

Theorem III.4.10 (Isoperimetric Inequality for Perimeter) *Let E be compact, and E^* denote the Schwarz symmetrization of E. Then*

$$P(E^*) \leq P(E), \tag{III.4.6}$$

with equality if and only if E is an n-disk, except for (at most) a set of $d\mathbf{v}_n$-measure equal to 0.

Proof (Geometric proof.) By Theorems III.2.8 and III.2.9, there exists a sequence T_j consisting of Steiner symmetrizations, $(n-1)$-dimensional Schwarz symmetrizations, and rotations such that $T_j \cdot E \to E^*$ in L^1 as $j \to \infty$. Then

$$P(E^*) \leq \liminf_{j\to\infty} P(T_j \cdot E) \leq P(E),$$

which implies the inequality.

(Analytic proof.) Given any $f \in \mathrm{BV}_c$ (where BV_c denotes the compactly supported functions in BV), we can approximate f by $f_k \in C_c^\infty$ such that

$$\|f_k - f\|_1 \to 0, \qquad \|\operatorname{grad} f_k\|_1 = \|\delta f_k\|_{\mathrm{BV}} \to \|\delta f\|_{\mathrm{BV}},$$

as $k \to \infty$. We already have

$$n\boldsymbol{\omega_n}^{1/n}\|f_k\|_{n/(n-1)} \leq \|\operatorname{grad} f_k\|_1 = \|\delta f_k\|_{\mathrm{BV}}.$$

Since $f_k \to f$ in L^1, then $f_k \to f$ pointwise a.e.-$[d\mathbf{v}_n]$, which implies, by Fatou's lemma,

$$n\boldsymbol{\omega_n}^{1/n}\|f\|_{n/(n-1)} \leq n\boldsymbol{\omega_n}^{1/n} \liminf_{k\to\infty} \|f_k\|_{n/(n-1)} \leq \liminf_{k\to\infty} \|\delta f_k\|_{\mathrm{BV}} = \|\delta f\|_{\mathrm{BV}}.$$

But this implies the isoperimetric inequality, by picking $f = \mathcal{I}_E$.

We now characterize the case of equality in (III.4.6). The reader is referred to the proof of Theorem III.2.4, where it is assumed that the boundary of the minimizer is C^1, since we will draw on that argument.

Fix E, compact, that satisfies equality in the isoperimetric inequality (III.4.6). Then $P(E) = P(\mathcal{S}E)$ for every Steiner symmetrization.

Let K_j be a sequence of compact domains with C^∞ boundary such that $K_j \to E$, in the Hausdorff metric and in L^1, and $P(K_j) \to P(E)$. Fix a hyperplane Π, and let $\mathcal{S}$ denote Steiner symmetrization with respect to Π.

Since $K_j \to E$ in $L^1(\mathbb{R}^n)$, we have $\mathcal{S}K_j \to \mathcal{S}E$ in $L^1(\mathbb{R}^n)$, and, by passing to subsequences if necessary, we may assume

$$|\delta K_j| \to |\delta E|, \qquad |\delta \mathcal{S}K_j| \to |\delta \mathcal{S}E|.$$

For each j, define the graph function $\mathcal{A}_j$ on Π associated with $\mathcal{S}K_j$, as in the proof of Lemma III.2.3. Then (III.2.19) implies

$$\begin{aligned} P(K_j) &\geq 2\int \sqrt{k_j(w)^2 + |\mathrm{grad}_{n-1}\mathcal{A}_j|(w)^2}\, d\mathbf{v}_{n-1}(w) \\ &\geq 2\int \sqrt{1 + |\mathrm{grad}_{n-1}\mathcal{A}_j|^2(w)}\, d\mathbf{v}_{n-1}(w) \\ &= P(\mathcal{S}K_j), \end{aligned}$$

which implies

$$\begin{aligned} P(E) &\geq \limsup 2\int \sqrt{k_j^2 + |\mathrm{grad}_{n-1}\mathcal{A}_j|^2}\, d\mathbf{v}_{n-1} \\ &\geq \limsup 2\int \sqrt{1 + |\mathrm{grad}_{n-1}\mathcal{A}_j|^2}\, d\mathbf{v}_{n-1} \\ &= \lim P(\mathcal{S}K_j) \\ &= P(\mathcal{S}E). \end{aligned}$$

We claim that $P(E) = P(\mathcal{S}E)$ implies

(III.4.7) $$\lim_{j\to\infty} \mathbf{v}_{n-1}(k_j > 1) = 0.$$

Proof Assume there exists $\epsilon > 0$ such that

$$\mathbf{v}_{n-1}(k_j > 1) > \epsilon \quad \text{for infinitely many } j.$$

Now

$$\begin{aligned} P(K_j) &\geq 2\int_{|\mathrm{grad}_{n-1}\mathcal{A}_j|\geq N} \sqrt{k_j(w)^2 + |\mathrm{grad}_{n-1}\mathcal{A}_j|(w)^2}\, d\mathbf{v}_{n-1}(w) \\ &\geq 2N\mathbf{v}_{n-1}(|\mathrm{grad}_{n-1}\mathcal{A}_j| \geq N), \end{aligned}$$

which implies that

$$\mathbf{v}_{n-1}(|\mathrm{grad}_{n-1}\mathcal{A}_j| \geq N) \leq \frac{P(E)}{N}$$

for sufficiently large j; so we may pick N sufficiently large so that

$$\mathbf{v}_{n-1}(|\mathrm{grad}_{n-1}\mathcal{A}_j| \geq N) \leq \frac{\epsilon}{2}$$

for sufficiently large j. Therefore

$$\mathbf{v}_{n-1}(\{|\mathrm{grad}_{n-1}\mathcal{A}_j| < N\} \cap \{k_j > 1\}) \geq \frac{\epsilon}{2},$$

for infinitely many j. Because

$$\begin{aligned}
\int \sqrt{k_j^2 + |\mathrm{grad}_{n-1}\mathcal{A}_j|^2}\, d\mathbf{v}_{n-1} &\geq \int \sqrt{1 + |\mathrm{grad}_{n-1}\mathcal{A}_j|^2}\, d\mathbf{v}_{n-1} \\
&\quad + \int_{|\mathrm{grad}_{n-1}\mathcal{A}_j|<N} \left[\left\{1 + \frac{k_j^2 - 1}{1+N^2}\right\}^{1/2} - 1\right] \sqrt{1 + |\mathrm{grad}_{n-1}\mathcal{A}_j|^2}\, d\mathbf{v}_{n-1} \\
&= \frac{P(\mathcal{S}K_j)}{2} \\
&\quad + \int_{\{|\mathrm{grad}_{n-1}\mathcal{A}_j|<N\}\cap\{k_j>1\}} \left[\left\{1 + \frac{k_j^2 - 1}{1+N^2}\right\}^{1/2} - 1\right] \cdot \\
&\quad \cdot \sqrt{1 + |\mathrm{grad}_{n-1}\mathcal{A}_j|^2}\, d\mathbf{v}_{n-1} \\
&\geq \frac{P(\mathcal{S}K_j)}{2} + \mathrm{const.}\frac{\epsilon}{2}
\end{aligned}$$

for infinitely many j, we have a contradiction to $P(E) = P(\mathcal{S}E)$, which implies (III.4.7).

Let E_L denote the *Lebesgue set of* E, that is, $q \in E_L$ if $q \in E$ and

$$\lim_{r\downarrow 0} \frac{1}{\boldsymbol{\omega}_n r^n} \int_{\mathbb{B}(q;r)} \mathcal{I}_E\, d\mathbf{v}_n = 1.$$

Then Lebesgue's density theorem (Proposition I.3.6) implies that $\mathbf{v}_n(E \setminus E_L) = 0$. Let E_0 denote the closure of E_L. Then $E_0 \subset E$, E_0 is compact, and $\mathbf{v}_n(E \setminus E_0) = 0$. So we replace E by E_0. Then $\mathbf{v}_n(E) = \mathbf{v}_n(E_0)$ and $P(E) = P(E_0)$.

Given the hyperplane $\Pi \cong \mathbb{R}^{n-1}$, let $p : \mathbb{R}^n \to \mathbb{R}^{n-1}$ denote the projection; and given any $w \in \mathbb{R}^{n-1}$, let ℓ^w denote the line through w perpendicular to $\mathbb{R}^{n-1}$.

Let $w \in p(E_0)$, and set

$$\alpha(w) = \inf \ell^w \cap E_0, \qquad \beta(w) = \sup \ell^w \cap E_0.$$

Assume there exists $t_0 \in (\alpha(w), \beta(w))$ such that $x_0 := (w, t_0) \notin E_0$. Then there exists $r > 0$ such that $\mathbb{B}(x_0; r) \subset \mathbb{R}^n \setminus E_0$; and there exist $t_1 \in [\alpha(w), t_0 - r]$ such that $x_1 := (w, t_1) \in E_0$, and $t_2 \in [t_0 + r, \beta(w)]$ such that $x_2 := (w, t_2) \in E_0$. Pick $\epsilon > 0$ so that, for any points $y_1 \in \mathbb{B}(x_1; \epsilon)$ and $y_2 \in \mathbb{B}(x_2; \epsilon)$, the line joining them must intersect $\mathbb{B}(x_0; r)$. Given such an $\epsilon > 0$, there exist $y_1 \in \mathbb{B}(x_1; \epsilon) \cap E_L$ and $y_2 \in \mathbb{B}(x_2; \epsilon) \cap E_L$, which implies that, for any given $\delta > 0$,

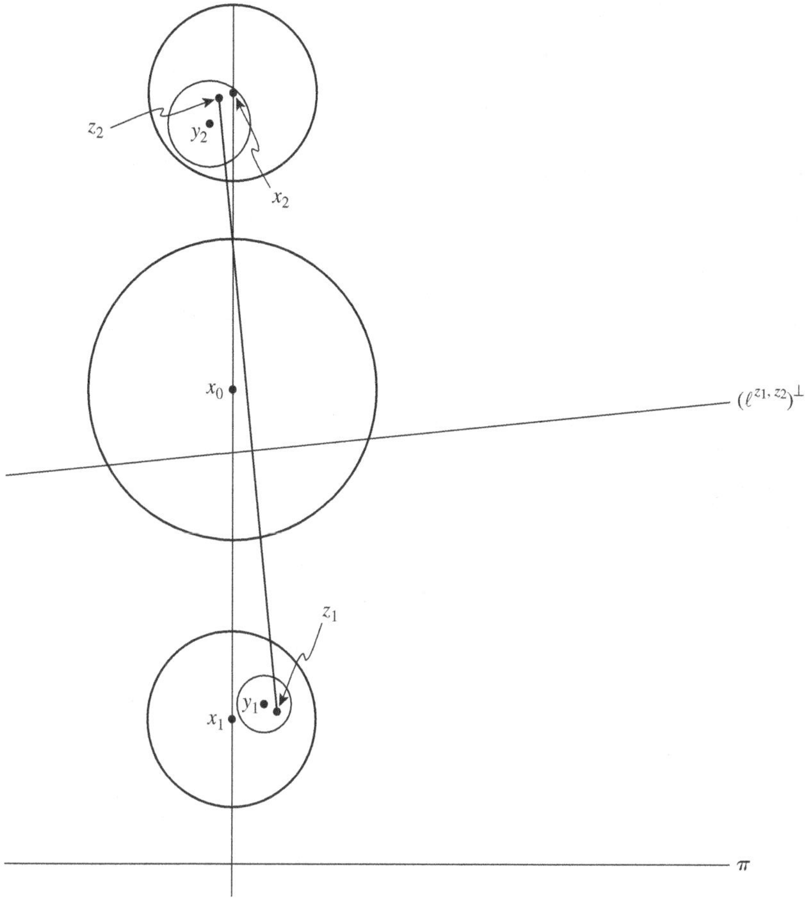

Figure III.4.1: The hole.

there exists $\epsilon' > 0$ such that

$$\mathbf{v}_n(\mathbb{B}(y_1;\epsilon') \cap E) > (1-\delta)\mathbf{v}_n(\mathbb{B}(y_1;\epsilon')), \qquad \mathbb{B}(y_1;\epsilon') \subset \mathbb{B}(x_1;\epsilon),$$
$$\mathbf{v}_n(\mathbb{B}(y_2;\epsilon') \cap E) > (1-\delta)\mathbf{v}_n(\mathbb{B}(y_2;\epsilon')), \qquad \mathbb{B}(y_2;\epsilon') \subset \mathbb{B}(x_2;\epsilon).$$

(See Figure III.4.1.) Since $K_j \to E$ in L^1 we have $J > 0$ such that

$$\mathbf{v}_n(\mathbb{B}(y_1;\epsilon') \cap K_j) > (1-2\delta)\mathbf{v}_n(\mathbb{B}(y_1;\epsilon')),$$
$$\mathbf{v}_n(\mathbb{B}(y_2;\epsilon') \cap K_j) > (1-2\delta)\mathbf{v}_n(\mathbb{B}(y_2;\epsilon'))$$

for all $j > J$. Since the boundaries of K_j are C^∞, we have $\mathbf{v}_n(K_j) = \mathbf{v}_n(\operatorname{int} K_j)$

for all $j > J$, which implies

$$\mathbf{v}_n(\mathbb{B}(y_1;\epsilon') \cap \operatorname{int} K_j) > (1 - 2\delta)\mathbf{v}_n(\mathbb{B}(y_1;\epsilon')),$$
$$\mathbf{v}_n(\mathbb{B}(y_2;\epsilon') \cap \operatorname{int} K_j) > (1 - 2\delta)\mathbf{v}_n(\mathbb{B}(y_2;\epsilon'))$$

for all $j > J$, which implies that for every $j > J$ there exist points z_1, z_2 and $\rho > 0$ such that

$$\mathbb{B}(z_1;\rho) \subset \mathbb{B}(y_1;\epsilon') \cap K_j, \qquad \mathbb{B}(z_2;\rho) \subset \mathbb{B}(y_2;\epsilon') \cap K_j.$$

Let ℓ^{z_1,z_2} denote the line determined by z_1 and z_2. Then for any line intersecting $\mathbb{B}(z_1;\epsilon'/2)$ and $\mathbb{B}(z_2;\epsilon'/2)$ must intersect $\mathbb{B}(x_0;r)$, which contradicts (III.4.7). Therefore, $\ell^w \cap E_0 = [\alpha(w), \beta(w)]$.

Thus, equality in the isoperimetric inequality implies $P(\mathcal{S}E_0) = P(E_0)$ for all symmetrizations $\mathcal{S}$, which implies E_0 must be convex. Then for any hyperplane Π, the corresponding functions $\alpha(w)$, $-\beta(w)$, $w \in p_\Pi(E_0)$, are convex, and hence locally Lipschitz. Set

$$\mathcal{A}(w) = \frac{\beta(w) - \alpha(w)}{2}.$$

Since E_0 is Lipschitz we have, by Example III.4.3,

$$\begin{aligned}|\delta E_0|(w, y) = d\mathbf{v}_{n-1}(w)& \\ &\otimes \left\{\sqrt{1 + |\operatorname{grad}_{n-1} - \alpha|^2}(w)\delta^1_{\alpha(w)}(y)\right. \\ &\quad \left. + \sqrt{1 + |\operatorname{grad}_{n-1}\beta|^2}(w)\delta^1_{\beta(w)}(y)\right\} dy\end{aligned}$$

(where δ^1 denotes the Dirac delta function on the line $\mathbb{R}^1$), which implies, by (III.2.18),

$$\begin{aligned}P(E_0) &= \int \left\{\sqrt{1 + |\operatorname{grad}_{n-1} - \alpha|^2} + \sqrt{1 + |\operatorname{grad}_{n-1}\beta|^2}\right\} d\mathbf{v}_{n-1} \\ &\geq 2\int \sqrt{1 + |\operatorname{grad}_{n-1}\mathcal{A}|^2}\, d\mathbf{v}_{n-1} \\ &= P(\mathcal{S}E_0),\end{aligned}$$

with equality if and only if $\beta(w) + \alpha(w) = \text{const.}$ for almost all $w \in p_\Pi(E_0)$. Continuity of the functions $\alpha(w)$ and $\beta(w)$ implies $\beta(w) + \alpha(w) = \text{const.}$ for all $w \in p_\Pi(E_0)$. Then E_0 is convex and is symmetric, up to a translation, with respect to any hyperplane Π, which implies E_0 is a disk, by the convexity of E_0 and Theorem I.3.5. ∎

III.5 Bibliographic Notes

The fundamental treatise on geometric measure theory is Federer's (1969). An elementary introduction is Morgan (1988), and a more advanced treatment is Simon (1984).

§III.1 The treatment of the Hausdorff metric on compact sets, including Blaschke's theorem, follows Berger (1987, Chapter 9).

§III.2 Steiner symmetrization can be found in Steiner (1838). It initiated a whole school of techniques with applications in geometry and analysis. See the geometric discussion in Burago and Zalgaller (1988, Chapter 9), and analytic discussions in the classic Pólya and Szegö (1951), and the more recent Kawohl (1985), and Baernstein (1995).

Our use of the phrase *Schwarz symmetrization* is not universal. Except for Steiner symmetrization, none of the nomenclature seems to be universal; so when reading any presentation, *caveat emptor* applies.

The technique of using sequences of Steiner symmetrizations goes back, at least, to the important paper of Carathéodory and Study (1909). The paper is one of the early – if not the first – rigorous proofs of the isoperimetric inequality by Steiner symmetrization. See the discussion in Blaschke (1956, p. 43 ff.) Our treatment follows, rather closely, Lieb and Loss (1996, Chapter 3). Deeper results on approximations by sequences of symmetrizations can be found in Almgren and Lieb (1989) and Burchard (1997).

§III.3 Theorem III.3.1 answers in the affirmative a conjecture of Lord Rayleigh (1877, §210). It was first proven in Faber (1923), Krahn (1925).

The argument can be easily carried over to spheres and hyperbolic spaces; namely, as soon as one has metric disks as the solution to the geometric isoperimetric problem in these spaces, one automatically has the disks as solutions to the corresponding eigenvalue isoperimetric problem. See Chavel (1984, Chapter IV).

§III.4 The discussion of perimeter follows the exceptionally clear discussion of Giusti (1984, Chapter 1). Ziemer (1989) was also very helpful.

The discussion of the isoperimetric inequality for the perimeter of compact sets was influenced by Talenti (1993). We did not follow his argument closely, since we were able to use the previous arguments from Minkowski area.

IV
Hausdorff Measure and Perimeter

In this chapter we introduce the most general construction of measures in a metric space, Hausdorff measure. As emphasized earlier, it provides a measure on the boundary of any domain, in terms of that set itself (not as a boundary), irrespective of regularity of the set.

We show that, in Euclidean space, Hausdorff measure in the top dimension leads to the usual Lebesgue measure, and on smooth Riemannian manifolds it leads to the usual Riemannian theoretic volume measure. Finally, we show that if Ω is a Lipschitz domain in $\mathbb{R}^n$, then the $(n-1)$-dimensional Hausdorff measure on the boundary of Ω, $\partial\Omega$, coincides with the perimeter measure of Ω. Our proof requires the area formula for Lipschitz maps, which we prove here as well. It was tempting to include in this chapter the structure of the boundary of domains with locally finite perimeter, and the relation of perimeter to Hausdorff measure. But we would have had hardly anything to add to the excellent treatment in Chapters 2–4 of Giusti (1984); so we left the matter to the reader to explore there.

IV.1 Hausdorff Measure

Definition (Recall.) Given a set X, a *σ-algebra $\mathcal{S}$ in X* is a collection of subsets of X satisfying:

(i) $\phi, X \in \mathcal{S}$;
(ii) if $A \in \mathcal{S}$ then $X \setminus A \in \mathcal{S}$;
(iii) if $(A_j) \in \mathcal{S}$ then $\bigcup_j A_j \in \mathcal{S}$ for any countable sequence of subsets of X.

A *measure on $\mathcal{S}$* is a function $\mu : \mathcal{S} \to [0, +\infty]$ satisfying:

(i) $\mu(\emptyset) = 0$,
(ii) μ is *countably additive*, that is, $\mu(\bigcup_j A_j) = \sum_j \mu(A_j)$ for countable *pairwise disjoint* $(A_j) \in \mathcal{S}$.

Sometimes it is awkward to specify at the outset the σ-algebra $\mathcal{S}$. Instead, using Carathéodory's criterion, one starts with an outer measure on the full collection of subsets of X, 2^X, and creates the σ-algebra and measure from the outer measure. Namely,

Definition A function $\phi : 2^X \to [0, +\infty]$ is an *outer measure* if it satisfies:

(i) $\phi(\emptyset) = 0$;
(ii) ϕ is *monotone*, that is, if $A \subseteq B$ then $\phi(A) \leq \phi(B)$;
(iii) ϕ is *countably subadditive*, that is, $\phi(\bigcup_j A_j) \leq \sum_j \phi(A_j)$ for *any* countable sequence of subsets of X.

One then calls a subset E of X *ϕ-measurable* if

$$\phi(A) = \phi(A \cap E) + \phi(A \setminus E)$$

for all $A \in 2^X$. The collection of ϕ-measurable subsets of X, $\mathcal{S}_\phi$, is known to be a σ-algebra, and the restriction $\mu = \phi | \mathcal{S}_\phi$ of ϕ to $\mathcal{S}_\phi$ is a measure.

Definition An outer measure μ is called *regular* if to each $A \subset X$ there exists a μ-measurable $B \supset A$ such that $\mu(B) = \mu(A)$. (Note that this definition of regularity is for *outer* measures.) Note that if μ is regular, then any sequence (A_j) in X satisfying $A_j \subset A_{j+1}$, for all j, also satisfies

$$\mu\left(\bigcup_{j=1}^{\infty} A_j\right) = \lim_{j \to \infty} \mu(A_j).$$

If X is a topological space with outer measure μ, we say that μ is *Borel regular* if for each $A \subset X$ there exists a Borel set $B \supset A$ such that $\mu(B) = \mu(A)$.

If (X, d) is a metric space with an outer measure μ, we say that μ is a *metric outer measure* if

$$d(A, B) > 0 \implies \mu(A \cup B) = \mu(A) + \mu(B).$$

Theorem IV.1.1 (Carathéodory's Theorem) *If μ is a metric outer measure on X, then all Borel sets are μ-measurable.*

Proof It suffices to show

$$\mu(E) \geq \mu(E \cap C) + \mu(E \setminus C)$$

for all $E \subset X$ satisfying $\mu(E) < +\infty$, and for all *closed* subsets C. Let

$$C_j = \{x \in X : d(x, C) \leq 1/j\} = [C]_{1/j}$$

for $j = 1, 2, \ldots$. Then $E \setminus C \supset E \setminus C_j$, which implies

$$\mu(E) \geq \mu((E \setminus C_j) \cup (E \cap C)) = \mu(E \setminus C_j) + \mu(E \cap C).$$

Let $j \to \infty$; we shall show that $\mu(E \setminus C_j) \to \mu(E \setminus C)$. To this end, set

$$R_k = \left\{ x \in E : \frac{1}{k+1} < d(x, C) \leq \frac{1}{k} \right\}.$$

Then

$$E \setminus C = (E \setminus C_1) \cup \bigcup_{j=1}^{\infty} R_j = (E \setminus C_k) \cup \bigcup_{j=k}^{\infty} R_j,$$

which implies

$$\mu(E \setminus C_k) \leq \mu(E \setminus C) \leq \mu(E \setminus C_k) + \sum_{j=k}^{\infty} \mu(R_j).$$

Now the series $\sum_j \mu(R_j)$ converges; indeed,

$$\begin{aligned} \sum_{j=1}^{\infty} \mu(R_j) &= \sum_{j=1}^{\infty} \mu(R_{2j-1}) + \mu(R_{2j}) \\ &= \mu\left(\bigcup_{j=1}^{\infty} R_{2j-1}\right) + \mu\left(\bigcup_{j=1}^{\infty} R_{2j}\right) \\ &\leq 2\mu(E). \end{aligned}$$

Therefore $\sum_{j \geq k} \mu(R_j) \to 0$ as $k \to \infty$, which implies the theorem. ■

Definition Let X be a metric space, $\mathcal{F}$ a family of subsets of X, and $\zeta : \mathcal{F} \to [0, +\infty]$. A covering (E_j) in $\mathcal{F}$ of a subset A in X is called a *δ-cover* of A if $\operatorname{diam} E_j < \delta$ for all j.

Assume that $(X, \mathcal{F}, \zeta)$ satisfy:

1. for every $\delta > 0$, X has a δ-cover;
2. for every $\delta > 0$, there exists $E \in \mathcal{F}$ such that $\zeta(E) < \delta$.

Define the *δ-Hausdorff (outer) measure* ψ_δ on X by

$$\psi_\delta(A) = \inf \sum_j \zeta(E_j),$$

where (E_j) varies over all δ-covers of A by sets in $\mathcal{F}$, and the *Hausdorff (outer) measure* ψ on X by

$$\psi(A) = \lim_{\delta \downarrow 0} \psi_\delta(A).$$

Theorem IV.1.2 *We have*

(a) $\psi(\emptyset) = 0$;
(b) $\psi(A)$ *is well-defined, because* ψ_δ *is monotone decreasing with respect to* δ;
(c) $A \mapsto \psi(A)$ *is an outer measure.*

Proof We only have to check subadditivity: If (E_j) is a δ-cover of A, and (F_j) is a δ-cover of B, then (E_j, F_j) is a δ-cover of $A \cup B$, and

$$\psi_\delta(A \cup B) \le \sum_j \zeta(E_j) + \sum_j \zeta(F_j),$$

which implies

$$\psi_\delta(A \cup B) \le \psi_\delta(A) + \psi_\delta(B) \le \psi(A) + \psi(B),$$

which implies

$$\psi(A \cup B) \le \psi(A) + \psi(B).$$ ■

Theorem IV.1.3 *The Hausdorff measure* ψ *is a metric outer measure. Therefore,* ψ *is a Borel measure.*

Proof Given A, B with $d(A, B) > 0$, consider $\delta < d(A, B)/3$ and a δ-cover (E_j) of $A \cup B$. Then any given E_j cannot intersect both A and B. Therefore

$$\sum_j \zeta(E_j) \ge \sum_{A \cap E_j \neq \emptyset} \zeta(E_j) + \sum_{B \cap E_j \neq \emptyset} \zeta(E_j) \ge \psi_\delta(A) + \psi_\delta(B),$$

which implies

$$\psi(A \cup B) \ge \psi_\delta(A) + \psi_\delta(B).$$

One now easily has

$$\psi(A \cup B) = \psi(A) + \psi(B).$$ ■

Theorem IV.1.4 *If the members of* $\mathcal{F}$ *are Borel sets, then* ψ *is Borel regular.*

Proof Given $A \subset X$, pick *open* $(1/k)$-covers $(E_{k,j})$ of A, for every $k = 1, 2 \ldots,$ satisfying

$$\sum_j \zeta(E_{k,j}) \le \psi_{1/k}(A) + 1/k,$$

and set

$$B = \bigcap_{k=1}^{\infty} \bigcup_{j=1}^{\infty} E_{k,j}.$$

One checks that B is a Borel set containing A, and $\psi(A) = \psi(B)$. ■

Definition Let X be a metric space, $0 \le s < +\infty$, $\mathcal{F} = 2^X$ all subsets of X, and

$$\zeta(E) = (\text{diam } E)^s,$$

with the conventions $0^0 = 1$ and $(\text{diam } \emptyset)^s = 0$. The resulting δ-Hausdorff and Hausdorff measures are denoted by $\mathcal{H}^s_\delta$, $\mathcal{H}^s$, respectively, and referred to as the *standard δ-Hausdorff* and the *standard Hausdorff measures on X*. When no explicit mention of $\mathcal{F}$ and ζ is made, we are always speaking of the standard Hausdorff measures.

Note that $\mathcal{H}^0(A) = \text{card } A$.

Proposition IV.1.1 *If X is a separable metric space, we obtain the same measure $\mathcal{H}^s$ if we pick*

$\mathcal{F} = 2^X$,
$\mathcal{F} = \{E \subset X : E \text{ open}\}$,
$\mathcal{F} = \{E \subset X : E \text{ closed}\}$.

If $X = \mathbb{R}^n$ (because diam conv A = diam A *for all A), we may add to the list:*

$\mathcal{F} = \{E \subset X : E = \text{convex}\}$.

Proposition IV.1.2 *Let X be a separable metric space. The following are equivalent:*

(a) $\mathcal{H}^s(A) = 0$;
(b) $\mathcal{H}^s_\delta(A) = 0$ *for some* $\delta > 0$;
(c) given $\epsilon > 0$, *there exists a cover* (E_j) *of A such that*

$$\sum_j (\text{diam } E_j)^s < \epsilon.$$

Theorem IV.1.5 *For* $0 \le s < t < +\infty$, $A \subset X$ *we have*

(a) $\mathcal{H}^s(A) < +\infty \Longrightarrow \mathcal{H}^t(A) = 0$.
(b) $\mathcal{H}^t(A) > 0 \Longrightarrow \mathcal{H}^s(A) = +\infty$.

Proof Let (E_j) be a δ-cover of A. Then

$$\mathcal{H}^t_\delta(A) \leq \sum_j (\operatorname{diam} E_j)^t \leq \delta^{t-s} \sum_j (\operatorname{diam} E_j)^s,$$

which implies

$$\mathcal{H}^t_\delta(A) \leq \delta^{t-s} \mathcal{H}^s_\delta(A),$$

which implies the theorem. ■

Definition The *Hausdorff dimension of A in X* is given by

$$\begin{aligned} \dim A &= \sup\{s : \mathcal{H}^s(A) > 0\} \quad = \sup\{s : \mathcal{H}^s(A) = +\infty\} \\ &= \inf\{t : \mathcal{H}^t(A) < +\infty\} = \inf\{t : \mathcal{H}^t(A) = 0\}. \end{aligned}$$

Note that

$$A \subset B \implies \dim A \leq \dim B.$$

Also,

$$\dim \bigcup_{j=1}^{\infty} A_j = \sup_j \dim A_j.$$

IV.1.1 Example. Euclidean Space

Remark IV.1.1 By Proposition IV.1.2, if a Lebesgue-measurable set A in $\mathbb{R}^n$ has Lebesgue measure 0, then $\mathcal{H}^n(A) = 0$.

Remark IV.1.2 If $X = \mathbb{R}^n$, T a Euclidean transformation of $\mathbb{R}^n$, then

$$\mathcal{H}^s(T(A)) = \mathcal{H}^s(A).$$

If $\lambda > 0$ then

$$\mathcal{H}^s(\lambda A) = \lambda^s \mathcal{H}^s(A).$$

Lemma IV.1.1 *Given any bounded open $U \subset \mathbb{R}^n$, and $\delta > 0$, there exists a pairwise disjoint family of closed disks (D_j) all contained in U such that*

$$\operatorname{diam} D_j < \delta \quad \forall\, j, \qquad \mathbf{v}_n(U \setminus \cup_j D_j) = 0.$$

Proof Let $U = \bigcup_k C_k$, where C_k are closed n-cubes with pairwise disjoint interiors, and $\operatorname{diam} C_k < \delta$ for all k. Pick closed disks D_j to satisfy

$$D_j \subset C_j, \qquad \operatorname{diam} D_j > \frac{1}{2} \operatorname{side} C_j \quad \forall\, j.$$

Then the radius $r(D_j)$ of D_j satisfies

$$r(D_j) > \frac{\text{side } C_j}{4},$$

and

$$\mathbf{v}_n(C_j) \geq \mathbf{v}_n(D_j) \geq \frac{\boldsymbol{\omega}_\mathbf{n}}{4^n}\mathbf{v}_n(C_j),$$

which implies

$$\mathbf{v}_n\left(U\Big\backslash\bigcup_j D_j\right) \leq \left(1 - \frac{\boldsymbol{\omega}_\mathbf{n}}{4^n}\right)\mathbf{v}_n(U).$$

Therefore there exists $N_1 > 0$ such that

$$\mathbf{v}_n\left(U\Big\backslash\bigcup_{j=1}^{N_1} D_j\right) \leq \left(1 - \frac{\boldsymbol{\omega}_\mathbf{n}}{2 \times 4^n}\right)\mathbf{v}_n(U).$$

Now set

$$U_1 = U\Big\backslash\bigcup_{j=1}^{N_1} D_j,$$

and find (in the same manner) pairwise disjoint $D_{N_1+1}, \ldots, D_{N_2} \subset U_1$ satisfying

$$\begin{aligned}\mathbf{v}_n\left(U\Big\backslash\bigcup_{j=1}^{N_2} D_j\right) &= \mathbf{v}_n\left(U_1\Big\backslash\bigcup_{j=N_1+1}^{N_2} D_j\right)\\ &\leq \left(1 - \frac{\boldsymbol{\omega}_\mathbf{n}}{2 \times 4^n}\right)\mathbf{v}_n(U_1)\\ &\leq \left(1 - \frac{\boldsymbol{\omega}_\mathbf{n}}{2 \times 4^n}\right)^2\mathbf{v}_n(U).\end{aligned}$$

One continues in the same manner, to obtain the lemma. ∎

Theorem IV.1.6 *For all Lebesgue-measurable $A \subset \mathbb{R}^n$ we have*

$$\mathbf{v}_n(A) = \frac{\boldsymbol{\omega}_\mathbf{n}}{2^n}\mathcal{H}^n(A).$$

Proof Let (E_j) be a δ-cover of A. Then

$$\mathbf{v}_n(A) \leq \sum_j \mathbf{v}_n(E_j) \leq \frac{\boldsymbol{\omega}_\mathbf{n}}{2^n}\sum_j(\text{diam } E_j)^n,$$

by the isodiametric inequality (Theorem III.2.1), which implies

$$\mathbf{v}_n(A) \le \frac{\omega_{\mathbf{n}}}{2^n}\mathcal{H}^n(A).$$

For the opposite inequality: given any $\epsilon > 0$, there exists a sequence of open n-cubes I_j such that

$$\operatorname{diam} I_j < \delta, \qquad A \subset \cup_j I_j, \qquad \mathbf{v}_n(A) \le \sum_j \mathbf{v}_n(I_j) \le \mathbf{v}_n(A) + \epsilon$$

(one first covers A with open U such that $\mathbf{v}_n(U) < \mathbf{v}_n(A) + \epsilon/3$, then fills U with closed n-cubes of diameter $< \delta$, and then thickens them slightly). Now to each I_j we have pairwise disjoint closed disks $(D_{j,k})$ with $\mathbf{v}_n(I_j \setminus \bigcup_k D_{j,k}) = 0$, which implies, by Remark IV.1.1,

$$\mathcal{H}^n\left(I_j \Big\backslash \bigcup_k D_{j,k}\right) = 0.$$

Therefore

$$\begin{aligned}
\mathcal{H}^n_\delta(A) &\le \mathcal{H}^n_\delta\left(\bigcup_j I_j\right) \\
&\le \sum_j \mathcal{H}^n_\delta(I_j) \\
&= \sum_{j,k} \mathcal{H}^n_\delta(D_{j,k}) \\
&\le \sum_{j,k} \frac{2^n}{\omega_{\mathbf{n}}}\mathbf{v}_n(D_{j,k}) \\
&\le \frac{2^n}{\omega_{\mathbf{n}}}\{\mathbf{v}_n(A) + \epsilon\}
\end{aligned}$$

(the next to last inequality follows from the fact that each $D_{j,k}$ covers itself and has diameter $< \delta$), which implies the theorem. ∎

IV.1.2 Example. Riemannian Manifolds

Theorem IV.1.7 *Let X be a separable, locally compact topological space that admits an exhaustion by compact sets, and let (X, d_1) and (X, d_2) be two metric structures on X, whose metric topologies coincide with the original topology, and which satisfy*

$$\lim_{d_1(x,y)\to 0} \frac{d_2(x, y)}{d_1(x, y)} = 1$$

locally uniformly. Then Hausdorff measure is independent of the choice of metric.

Proof Let K be compact in X. Given $\epsilon > 0$, there exists $\rho > 0$ such that

$$d_1(x, y) < \rho \implies 1 - \epsilon < \frac{d_2(x, y)}{d_1(x, y)} < 1 + \epsilon$$

for all $x, y \in K$. Then any δ-cover of K with respect to d_1 is a $(1 + \epsilon)\delta$-cover of K with respect to d_2, from which we imply

$$\mathcal{H}^s_{(1+\epsilon)\delta}(K; d_2) \leq \mathcal{H}^s_\delta(K; d_1)$$

(with obvious notation), which implies

$$\mathcal{H}^s(K; d_2) \leq \mathcal{H}^s(K; d_1)$$

for all compact K. By switching the roles of d_1 and d_2 we obtain the opposite inequality, which implies equality for all compact K. One easily obtains the equality for general subsets of X. ■

Theorem IV.1.8 *Let M be a Riemannian manifold, with associated Riemannian measure dV. Let $d\mathcal{H}^{n,M}$ denote the n-dimensional Hausdorff measure associated with the Riemannian distance in M. Then*

(IV.1.1) $$dV = \frac{\omega_{\mathbf{n}}}{2^n} d\mathcal{H}^{n,M}$$

on all of M.

Proof (For the basic definitions and facts about Riemannian manifolds, see the summary in §V.1 below.) Let M be a Riemannian manifold, $\mathbf{x} : U \to \mathbb{R}^n$ a chart on M. Then the Riemannian measure on M is given in the chart by

$$dV = \sqrt{g}\, dx^1 \cdots dx^n,$$

where the local metric $\langle\, ,\, \rangle$ is given by the positive definite symmetric matrix

$$g_{ij} = \langle \partial/\partial x^i, \partial/dx^j \rangle, \qquad g = \det(g_{ij}).$$

On any compact subset of U, the Riemannian measure dV and the local Euclidean measure $dx = dx^1 \cdots dx^n$ are equivalent, in the sense that each is absolutely continuous with respect to the other.

Fix $x_0 \in U$, and determine Riemann normal coordinates based at x_0. Then $g_{ij}(x) \to \delta_{ij}$, the Kronecker delta, as $x \to x_0$. In particular, $\sqrt{g}(x) \to 1$ as $x \to x_0$ [(V.1.6) below]. Moreover, for any $\rho > 0$, there exists a sufficiently small $r_0 > 0$ such that $B(x_0; r_0) \subset U$ and such that, for any vector $\boldsymbol{\xi}$ in the tangent bundle of $B(x_0; r_0)$, the Riemannian length $|\xi|$ of ξ and the Euclidean length $\|\xi\|$ of ξ induced by Riemann normal coordinates based at x_0 satisfy

$$(1+\rho)^{-1} \leq \frac{|\xi|}{\|\xi\|} \leq (1+\rho).$$

Furthermore, r_0 can be picked sufficiently small so that the geodesic path realizing the distance between two points of $B(x_0; r_0)$ is completely contained in $B(x_0; r_0)$. Therefore,

$$(1+\rho)^{-1} \leq \frac{d_M(x, y)}{d_{\mathbb{R}^n}(x, y)} \leq (1+\rho)$$

(with obvious notation) for all $x, y \in B(x_0; r_0)$, which implies

$$\lim_{x,y \to x_0} \frac{d_M(x, y)}{d_{\mathbb{R}^n}(x, y)} = 1.$$

We first conclude

$$(1+\rho)^{-s}\mathcal{H}^{s,\mathbb{R}^n} \leq \mathcal{H}^{s,M} \leq (1+\rho)^s\mathcal{H}^{s,\mathbb{R}^n}$$

on $B(x_0; r_0)$. Therefore, $d\mathcal{H}^{n,M}$ is equivalent to $d\mathcal{H}^{n,\mathbb{R}^n}$, which is equivalent to dV. Therefore there exists a positive function ϕ on $B(x_0; r_0)$ such that

$$d\mathcal{H}^{n,M} = \phi\, dV.$$

Since $\sqrt{g}(x) \to 1$ as $x \to x_0$, we have

$$\phi(x_0) = \frac{2^n}{\boldsymbol{\omega}_\mathbf{n}}.$$

But x_0 is arbitrary in M, and dV and $d\mathcal{H}^{n,M}$ are independent of the choice of any local coordinate system. We therefore have (IV.1.1) on all of M. ∎

Corollary IV.1.1 *Let X be a k-dimensional submanifold of the n-dimensional Riemannian manifold Y. Then for any measurable $A \subset X$ we have, by an application of Theorems* IV.1.7 *and* IV.1.8,

$$V_k(A) = \frac{\boldsymbol{\omega}_\mathbf{k}}{2^k}\mathcal{H}^k(A \subset X) = \frac{\boldsymbol{\omega}_\mathbf{k}}{2^k}\mathcal{H}^k(A \subset Y)$$

where dV_k denotes the (k-dimensional) Riemannian measure of the Riemannian manifold X, endowed with its induced Riemannian metric.

IV.2 The Area Formula for Lipschitz Maps

Theorem IV.2.1 *Let X be a separable metric space, with family of subsets $\mathcal{F}$ containing all Borel sets, and monotone and countably subadditive $\zeta : \mathcal{F} \to [0, \infty]$. Let ψ be the Hausdorff measure determined by ζ. Then*

$$\psi(A) = \sup_{H} \sum_{B \in H} \zeta(B),$$

for all ψ-measurable A in X, where H varies over all partitions of A by Borel sets.

Also, if $H_1, H_2, \ldots$ is a sequence of partitions of A by Borel sets satisfying

$$\lim_{j \to \infty} \sup_{B \in H_j} \operatorname{diam} B = 0,$$

then

$$\lim_{j \to \infty} \sum_{B \in H_j} \zeta(B) = \psi(A).$$

Proof Let S be a Borel set. If (E_j) is a δ-cover of S by Borel sets, then $\zeta(S) \leq \sum_j \zeta(E_j)$, which implies $\zeta(S) \leq \psi_\delta(S)$ for all $\delta > 0$, which implies

$$\zeta(S) \leq \psi(S) \quad \forall \text{ Borel } S,$$

which implies

$$\psi(A) = \sum_{B \in H} \psi(B) \geq \sum_{B \in H} \zeta(B)$$

for all Borel partitions H of A (the equality is where we use the ψ-measurability of A).

Set

$$\delta_j = \sup_{B \in H_j} \operatorname{diam} B, \qquad \delta_j \to 0.$$

Given $\epsilon > 0$, there exists j_0 such that $j > j_0$ implies

$$\psi(A) \leq \psi_{\delta_j}(A) + \epsilon \leq \sum_{B \in H_j} \zeta(B) + \epsilon,$$

by the very definition of the measure ψ_{δ_j}. Therefore we also have

$$\psi(A) \leq \liminf_{j \to \infty} \sum_{B \in H_j} \zeta(B),$$

which implies the theorem. ∎

Theorem IV.2.2 *Let X be a separable metric space, (Y, μ) a measure space, and assume the map $f : X \to Y$ takes every Borel set B in X to a measurable set in Y. Let $\mathcal{F}$ be the collection of Borel sets in X, define the outer measure*

$$\zeta(B) = \mu(f(B))$$

on $\mathcal{F}$, and let ψ be the induced Hausdorff measure on X. Then for any ψ-measurable set A in X we have

$$\psi(A) = \int_Y \operatorname{card}(A \cap f^{-1}[y])\, d\mu(y).$$

Proof Pick a sequence of Borel partitions $H_1, H_2, \ldots$ of A such that

$$\delta_j = \sup_{B \in H_j} \operatorname{diam} B, \qquad \lim_{j\to\infty} \delta_j = 0,$$

and such that, for every j, each Borel set in H_j is the union of some subfamily of H_{j+1}. Then for any $y \in Y$ we have

$$\sum_{B \in H_j} \mathcal{I}_{f(B)}(y) \uparrow \operatorname{card}(A \cap f^{-1}[y])$$

as $j \to \infty$, which implies

$$\begin{aligned}
\psi(A) &= \lim_{j\to\infty} \sum_{B \in H_j} \zeta(B) \\
&= \lim_{j\to\infty} \sum_{B \in H_j} \mu(f(B)) \\
&= \lim_{j\to\infty} \sum_{B \in H_j} \int \mathcal{I}_{f(B)}(y)\, d\mu(y) \\
&= \int \operatorname{card}(A \cap f^{-1}[y])\, d\mu(y).
\end{aligned}$$

■

Theorem IV.2.3 *Let X be a complete, separable metric space, Y a metric space, and $f : X \to Y$ Lipschitz. Then*

$$\mathcal{H}^m(f(A)) \le \int_Y \operatorname{card}(A \cap f^{-1}[y])\, d\mathcal{H}^m(y) \le (\operatorname{Lip} f)^m \mathcal{H}^m(A)$$

for all Borel sets A in X.

Proof By Proposition I.3.2, the image of every Borel set B in X is $\mathcal{H}^m$-measurable in Y. We may then define, as above,

$$\zeta(B) := \mathcal{H}^m(f(B)) \le (\operatorname{Lip} f)^m \mathcal{H}^m(B)$$

for any Borel set B in X, and let ψ denote the induced Hausdorff measure on X. Then for a sequence of Borel partitions $H_1, H_2, \ldots$ of A such that $\delta_j = \sup_{B \in H_j} \operatorname{diam} B$, $\lim \delta_j = 0$, and for every j, each Borel set in H_j is the union of some subfamily of H_{j+1}, we have

$$\begin{aligned}\psi(A) &= \lim_{j\to\infty} \sum_{B\in H_j} \mathcal{H}^m(f(B)) \\ &\leq (\operatorname{Lip} f)^m \lim_{j\to\infty} \sum_{B\in H_j} \mathcal{H}^m(B) = (\operatorname{Lip} f)^m \mathcal{H}^m(A),\end{aligned}$$

that is,

$$\psi(A) \leq (\operatorname{Lip} f)^m \mathcal{H}^m(A)$$

for all Borel sets A, which implies

$$\begin{aligned}\mathcal{H}^m(f(A)) &= \int_{f(A)} d\mathcal{H}^m(y) \leq \int_Y \operatorname{card}(A \cap f^{-1}[y])\, d\mathcal{H}^m(y) = \psi(A) \\ &\leq (\operatorname{Lip} f)^m \mathcal{H}^m(A)\end{aligned}$$

for any Borel set A in X. ■

Proposition IV.2.1

(a) Assume L and $\mathfrak{s}$ are linear transformations of $\mathbb{R}^m$ such that

$$|L(\boldsymbol{\xi})| \leq |\mathfrak{s}(\boldsymbol{\xi})|$$

for all $\boldsymbol{\xi}$. Then

$$|\det L| \leq |\det \mathfrak{s}|.$$

(b) Any linear transformation $T : \mathbb{R}^m \to \mathbb{R}^n$ may be factored as

$$T = h \circ g, \qquad \text{where } \ g \in \mathrm{GL}(\mathbb{R}^m), \qquad h \in \mathcal{O}(\mathbb{R}^m; \mathbb{R}^n)$$

[where $\mathrm{GL}(\mathbb{R}^n)$ denotes the general linear group of $\mathbb{R}^m$, and $\mathcal{O}(\mathbb{R}^m; \mathbb{R}^n)$ denotes the orthogonal linear mappings of $\mathbb{R}^m$ to $\mathbb{R}^n$].

Lemma IV.2.1 *Given a continuous map $f : \mathbb{R}^m \to \mathbb{R}^n$, $m \leq n$, consider*

$$U_f = \{x : J_f(x) \text{ exists and is univalent}\},$$

where $J_f(x)$ denotes the Jacobian linear transformation of f at x. Then, for any real number $\lambda > 1$, U_f has a countable covering G by Borel sets such that, for every Borel set $E \in \mathsf{G}$,

(i) $f|E$ *is univalent,*
(ii) there exists a linear automorphism $\mathfrak{s}: \mathbb{R}^m \to \mathbb{R}^m$ *such that*

$$\mathrm{Lip}\,((f|E)\circ\mathfrak{s}^{-1}) \le \lambda, \qquad \mathrm{Lip}\,(\mathfrak{s}\circ(f|E)^{-1}) \le \lambda,$$

$$\lambda^{-1}|\mathfrak{s}(\boldsymbol{\xi})| \le |J_f(x)\boldsymbol{\xi}| \le \lambda|\mathfrak{s}(\boldsymbol{\xi})| \quad \forall\, x \in E, \boldsymbol{\xi} \in \mathbb{R}^m,$$

$$\lambda^{-m}|\det \mathfrak{s}| \le |\det J_f(x)| \le \lambda^m |\det \mathfrak{s}|,$$

where $\det J_f(x)$ *is calculated by considering* $J_f(x)$ *as a linear transformation from* $\mathbb{R}^m$ *to its image* $J_f(x)(\mathbb{R}^m)$ *in* $\mathbb{R}^n$.

Proof Pick $\epsilon > 0$ such that

$$\lambda^{-1} + \epsilon < 1 < \lambda - \epsilon,$$

and fix a countable dense subset $\mathcal{S}$ of $\mathrm{GL}(\mathbb{R}^m)$.

With each $\mathfrak{s} \in \mathcal{S}$ and positive integer k associate the subset $Z(\mathfrak{s}, k)$ of $\mathbb{R}^m$ such that $x \in Z(\mathfrak{s}, k)$ if

(i) $$(\lambda^{-1} + \epsilon)|\mathfrak{s}(\boldsymbol{\xi})| \le |J_f(x)(\boldsymbol{\xi})| \le (\lambda - \epsilon)|\mathfrak{s}(\boldsymbol{\xi})| \quad \forall\, \boldsymbol{\xi} \in \mathbb{R}^m,$$

(ii) $$|f(y) - f(x) - J_f(x)(y - x)| \le \epsilon|\mathfrak{s}(y - x)| \quad \forall\, y \in B(x; 1/k).$$

Therefore, if $x, y \in Z(\mathfrak{s}, k)$, $|y - x| < 1/k$, then

$$|f(y) - f(x)| \le |J_f(x)(y - x)| + \epsilon|\mathfrak{s}(y - x)| \le \lambda|\mathfrak{s}(y - x)|,$$
$$|f(y) - f(x)| \ge |J_f(x)(y - x)| - \epsilon|\mathfrak{s}(y - x)| \ge \lambda^{-1}|\mathfrak{s}(y - x)|.$$

Cover $Z(\mathfrak{s}; k)$ with the collection $\{B(x; 1/2k) \cap Z(\mathfrak{s}; k) : x \in \mathbb{R}^m\}$, select a countable subcover G_k, and set $\mathsf{G} = \bigcup_k \mathsf{G}_k$.

It remains to show that any $x \in U_f$ belongs to some $Z(\mathfrak{s}; k)$. Given any $x \in U_f$, factor $J_f(x)$ by

$$J_f(x) = h \circ g, \qquad \text{where} \quad g \in \mathrm{GL}(\mathbb{R}^m), \qquad h \in \mathcal{O}(\mathbb{R}^m; \mathbb{R}^n).$$

Then

$$|J_f(x)(\boldsymbol{\xi})| = |g(\boldsymbol{\xi})| \quad \forall\, \boldsymbol{\xi}.$$

Since $\mathcal{S}$ is dense in $\mathrm{GL}(\mathbb{R}^m)$, we may pick $\mathfrak{s} \in \mathcal{S}$ such that

$$\|\mathfrak{s}\circ g^{-1}\| < \frac{1}{\lambda^{-1} + \epsilon}, \qquad \|g \circ \mathfrak{s}^{-1}\| < \lambda - \epsilon.$$

Then for this choice of $\mathfrak{s}$, x satisfies (i) above. Furthermore, by the existence of

$J_f(x)$, there exists a positive integer k such that

$$|f(y) - f(x) - J_f(x)(y - x)| \leq \frac{\epsilon}{\|\mathfrak{s}^{-1}\|}|y - x|$$

for all $y \in B(x; 1/k)$, which implies

$$|f(y) - f(x) - J_f(x)(y - x)| \leq \epsilon|\mathfrak{s}(y - x)|$$

for all $y \in B(x; 1/k)$, which implies that $x \in Z(\mathfrak{s}; k)$. ■

Theorem IV.2.4 (Area Formula) *Let $f : \mathbb{R}^m \to \mathbb{R}^n$ be Lipschitz, $m \leq n$, and A a Lebesgue measurable subset of $\mathbb{R}^m$. Then*

$$\text{(IV.2.1)} \qquad \int_A |\det J_f(x)|\, d\mathbf{v}_m = \int_{\mathbb{R}^n} \operatorname{card}(A \cap f^{-1}[y])\, d\mathcal{H}^m(y).$$

Proof If $\mathbf{v}_m(A) = 0$ then, certainly, both integrals vanish (the right hand side by Theorem IV.2.3).

First assume that $A \subseteq U_f$, and choose the covering G of A given by the lemma. Pick a Borel partition H of A such that each B in H is contained in some E in G. Then one has a map $B \mapsto \mathfrak{s} \in \mathcal{S}$, for which

$$\begin{aligned}
\lambda^{-m}\mathcal{H}^m(\mathfrak{s}(B)) &= \lambda^{-m}|\det \mathfrak{s}|\mathbf{v}_m(B) \\
&\leq \int_B |\det J_f|\, d\mathbf{v}_m \\
&\leq \lambda^m|\det \mathfrak{s}|\mathbf{v}_m(B) \\
&= \lambda^m\mathcal{H}^m(\mathfrak{s}(B)),
\end{aligned}$$

that is,

$$\lambda^{-m}\mathcal{H}^m(\mathfrak{s}(B)) \leq \int_B |\det J_f|\, d\mathbf{v}_m \leq \lambda^m\mathcal{H}^m(\mathfrak{s}(B)).$$

We also have

$$\lambda^{-m}\mathcal{H}^m(\mathfrak{s}(B)) \leq \mathcal{H}^m(f(B)) \leq \lambda^m\mathcal{H}(\mathfrak{s}(B)),$$

which implies

$$\lambda^{-2m}\mathcal{H}^m(f(B)) \leq \int_B |\det J_f|\, d\mathbf{v}_m \leq \lambda^{2m}\mathcal{H}^m(f(B)).$$

If we sum B over the Borel partition H, then Theorem IV.2.2 implies

$$\begin{aligned}
\lambda^{-2m}\int_{\mathbb{R}^n} \operatorname{card}(A \cap f^{-1}[y])\, d\mathcal{H}^m(y) &\leq \int_A |\det J_f|\, d\mathbf{v}_m \\
&\leq \lambda^{2m}\int_{\mathbb{R}^n} \operatorname{card}(A \cap f^{-1}[y])\, d\mathcal{H}^m(y).
\end{aligned}$$

Since $\lambda > 1$ is arbitrary, we have (IV.2.1) when $A \subseteq U_f$.

Now assume

$$A \subseteq \{x : \dim\ker J_f(x) > 0\} = \{x : \det\ J_f(x) = 0\}.$$

For any $\epsilon > 0$ write

$$f = p \circ g,$$

where

$$g : \mathbb{R}^m \to \mathbb{R}^n \times \mathbb{R}^m, \qquad g(x) = (f(x), \epsilon x),$$
$$p : \mathbb{R}^n \times \mathbb{R}^m \to \mathbb{R}^n, \qquad p(y, z) = y.$$

Then $x \in A$ implies

$$J_g(x)(\boldsymbol{\xi}) = (J_f(x)(\boldsymbol{\xi}), \epsilon\boldsymbol{\xi}),$$

which implies that g and $J_g(x)$ are univalent. Also, we have

$$\|J_g(x)\| \leq \operatorname{Lip} f + \epsilon, \qquad |\det J_g(x)| \leq \epsilon(\operatorname{Lip} f + \epsilon)^{n-1},$$

since $\ker\ J_f(x) \neq \{0\}$. Since $J_g(x)$ is univalent, we have from the first half of the proof that

$$\mathcal{H}^m(f(A)) \leq \mathcal{H}^m(g(A)) = \int_A |\det J_g|\, d\mathbf{v}_m \leq \epsilon(\operatorname{Lip} f + \epsilon)^{n-1}\mathbf{v}_m(A).$$

Let $\epsilon \to 0$. Then

$$\mathcal{H}^m(f(A)) \leq \limsup_{\epsilon\downarrow 0} \int_A |\det J_g|\, d\mathbf{v}_m = 0,$$

which implies the theorem. ∎

Theorem IV.2.5 *Let $\phi : \mathbb{R}^{n-1} \to \mathbb{R}$ be a Lipschitz function on $\mathbb{R}^{n-1}$, and $\Phi : \mathbb{R}^{n-1} \to \mathbb{R}^n$ the Lipschitz hypersurface given by*

$$\Phi(w) = (w, \phi(w)), \qquad w \in \mathbb{R}^{n-1}.$$

Then

$$d\mathcal{H}^{n-1}|\Phi(\mathbb{R}^{n-1}) = \sqrt{1 + |\operatorname{grad}_{n-1}\phi|^2}\ d\mathbf{v}_{n-1}.$$

In particular, if Ω is a Lipschitz domain in $\mathbb{R}^n$, then the perimeter measure $d\delta\Omega$ satisfies

$$|d\delta\Omega| = d\mathcal{H}^{n-1} \quad \text{on } \partial\Omega.$$

Proof One directly calculates J_Φ to be given by

$$|J_\Phi| = \left|\left\{\mathbf{e}_1 + \frac{\partial \phi}{\partial w^1}\mathbf{e}_n\right\} \wedge \cdots \wedge \left\{\mathbf{e}_{n-1} + \frac{\partial \phi}{\partial w^{n-1}}\mathbf{e}_n\right\}\right|$$
$$= \sqrt{1 + |\mathrm{grad}_{n-1}\phi|^2}.$$

The discussion of the perimeter measure of $\partial\Omega$ is found in Example III.4.3. ■

IV.3 Bibliographic Notes

There are many excellent treatments of Hausdorff measure, especially in dynamics. However, our emphasis is toward its employment in geometric measure theory, for which we have already referred the reader (in the previous chapter) to Federer (1969), Morgan (1988), Simon (1984).

§IV.2 Our treatment follows Federer (1969, pp. 241–244).

V
Isoperimetric Constants

In this chapter we change our venue: we move from Euclidean space to the broader collection of Riemannian manifolds of bounded geometry. Thus we skip over the intermediate levels of generalization, namely, the sphere and hyperbolic space of all dimensions, symmetric spaces, and homogeneous spaces. Even in these intermediate spaces, one gives away the abelian translation group of isometries and the homotheties that play such an important role in Euclidean space. Therefore one requires different methods from those we used earlier. In the more general setting of this half of the book, we also have to change our point of view with regard to the questions that we ask.

In a general Riemannian manifold, the chances of finding the domain of minimum boundary area, given the volume of the domain in advance, are essentially nil except maybe in some very special cases. Furthermore, different choices of the prescribed volume may change the whole character of the problem. Thus, from the analytic perspective, we shall not be able to ask for the precise infimum of the functional

$$D \mapsto \frac{A(\partial D)}{V(D)^{(n-1)/n}}$$

(where n is the dimension of the manifold); nor is it obvious that this is the correct analytic functional to study.

Rather, we shall ask the following: Find $\nu \in [1, \infty]$ such that the isoperimetric functional

$$D \mapsto \frac{A(\partial D)}{V(D)^{(\nu-1)/\nu}}$$

is bounded away from 0. [For $\nu = 1$ the functional is $D \mapsto A(\partial D)$, and for $\nu = \infty$ the functional is $D \mapsto A(\partial D)/V(D)$.] And it will suffice to let D vary over relatively compact domains with smooth (that is, C^∞) boundary.

We shall have to adjust even this formulation of the problem to take account of the difference between the local Euclidean character of Riemannian manifolds and their varied global behavior. This distinction will allow us to consider discrete isoperimetric inequalties on graphs and, using the hypothesis of bounded geometry, will allow us to compare global isoperimetric properties of a Riemannian manifold with those of its discretizations (especially, Theorem V.3.1).

In subsequent chapters, we shall continue this theme, employing it in the study of large time heat diffusion in Riemamannian manifolds. But more of that later.

In this chapter, we mostly present just a summary of the basic background, nearly all of which is considered in Chavel (1994). Detailed proofs are given for results not covered there, or for results that are at the heart of the point of view presented here.

V.1 Riemannian Geometric Preliminaries

In this section we describe the universe we inhabit for the rest of the book – Riemannian manifolds. The necessary definitions and results are given that enable us to discuss the type of isoperimetric inequalities in which we are interested.

Whenever we refer to a manifold, unless otherwise noted, we only have in mind the interior. Also, unless otherwise noted, it is C^∞, Hausdorff, with countable base, and connected. The boundary, when it exists, is usually mentioned separately. (We do not assume that the boundary is connected.) Thus a *compact manifold* has no boundary.

We are given an n-dimensional manifold M, $n \geq 1$. For any point $p \in M$, we denote the tangent space to M at p by M_p; and we denote the full tangent bundle by TM, with natural projection $\pi : TM \to M$. If $\mathbf{x} : U \to \mathbb{R}^n$ is a chart on M, we denote the natural frame field associated with $\mathbf{x}$ by

$$\{\partial_1, \ldots, \partial_n\}, \qquad \partial_j = \frac{\partial}{\partial x^j}.$$

At each point p of U, the frame is the *natural basis* of M_p.

A *Riemannian manifold* is a manifold with an inner product on each of its tangent spaces, denoted by $\langle\, ,\, \rangle$; the assignment of inner products, referred to as a *Riemannian metric* on M, is C^∞ in the sense that if X, Y are C^∞ vector fields on M, then the function $\langle X, Y\rangle$ is a C^∞ real-valued function on M. For any $\xi \in TM$ the length of ξ, $|\xi|$, is defined by $|\xi| = \langle \xi, \xi\rangle^{1/2}$.

Given Riemannian manifolds (M, g), (N, h), where g and h denote the respective Riemannian metrics, one naturally considers the *product Riemannian structure* on the product manifold $M \times N$, as follows: For $x \in M$, $y \in N$ the tangent space $(M \times N)_{(x,y)}$ is canonically isomorphic to $M_x \oplus N_y$. For vectors $\xi, \eta \in M_x$, $\zeta, \nu \in N_y$, we define the inner product of $\xi \oplus \zeta$ and $\eta \oplus \nu$ by

$$(g, h)(\xi \oplus \zeta, \eta \oplus \nu) = g(\xi, \eta) + h(\zeta, \nu).$$

Also, given a Riemannian manifold (M, g) and an imbedded submanifold N in M, then N carries a natural Riemannian metric $g|N$ obtained by restricting the inner product on tangent spaces of M to the tangent subspaces of N in M.

Let $\mathbf{x}: U \to \mathbb{R}^n$ be a chart on the fixed Riemannian manifold M. Then for each $p \in U$, the matrix $G(p)$ given by

(V.1.1) $$G(p) = (g_{ij}(p)), \qquad g_{ij}(p) = \langle \partial_{i|p}, \partial_{j|p} \rangle,$$

is positive definite symmetric, and the functions $g_{ij}: U \to \mathbb{R}$, $i, j = 1, \dots, n$, are C^∞ on U.

We also use the notation

(V.1.2) $$G^{-1} = (g^{jk}), \qquad g = \det G.$$

Let D^1 denote the continuous, piecewise C^1 paths in M. For any path $\omega: [\alpha, \beta] \to M \in D^1$, ω' denotes the velocity vector of ω at values of t where ω is C^1, and the length of ω, $\ell(\omega)$, is defined by

$$\ell(\omega) = \int_\alpha^\beta |\omega'(t)|\, dt.$$

For M connected (our usual assumption), $p, q \in M$, we define the *distance between p and q*, $d(p, q)$, by

$$d(p, q) = \inf_\omega \ell(\omega),$$

where ω ranges over all paths $\omega: [\alpha, \beta] \to M \in D^1$ satisfying $\omega(\alpha) = p$, $\omega(\beta) = q$. Then d defines a distance metric on M. A path $\gamma: (\alpha, \beta) \to M \in D^1$, $|\gamma'| = \text{const.}$, is called a *geodesic* if for every $t \in (\alpha, \beta)$ there exists an $\epsilon > 0$ such that

$$\ell(\gamma|[\alpha_0, \beta_0]) = d(\gamma(\alpha_0), \gamma(\beta_0))$$

for all $[\alpha_0, \beta_0] \subset (t - \epsilon, t + \epsilon)$. That is, geodesics locally minimize distance. It is known that any geodesic γ must in fact be C^∞. Furthermore, it is known that given any $\xi \in TM$, there exists one and only one maximal geodesic $\gamma_\xi: (-\alpha_\xi, \beta_\xi) \to M$, $\alpha_\xi, \beta_\xi \in (0, +\infty]$, satisfying $\gamma_\xi(0) = \pi(\xi)$ and $\gamma_\xi{}'(0) = \xi$. Of course, $|\gamma'| = |\xi|$ on all of $(-\alpha_\xi, \beta_\xi)$. We write $I_\xi = (-\alpha_\xi, \beta_\xi)$.

Let $\mathcal{T}M$ denote the subset of TM consisting of those $\xi \in TM$ for which $1 \in I_\xi$. Define the *exponential map*, $\exp : \mathcal{T}M \to M$, by

$$\exp \xi = \gamma_\xi(1).$$

Then for every $t \in \mathbb{R}$, $\xi \in TM$, for which $t\xi \in \mathcal{T}M$, we have

$$\exp t\xi = \gamma_\xi(t).$$

Notation For every $p \in M$ we let $\exp_p = \exp|(\mathcal{T}M \cap M_p)$. For every $p \in M$ and $r > 0$, we let $\mathsf{B}(p;r)$ denote the open n-disk in M_p centered at the origin of radius r. Recall that, for every $p \in M$ and $r > 0$, $B(p;r)$ denotes the metric disk in M centered at p of radius r.

For any $p \in M$ there exists an $\epsilon > 0$ such that $\exp_p$ is defined on $\mathsf{B}(p;\epsilon)$ and is a diffeomorphism of $\mathsf{B}(p;\epsilon)$ onto the metric disk $B(p;\epsilon)$. Furthermore, if for any $r > 0$ we have $\exp_p$ defined on all of $\mathsf{B}(p;r)$, then

$$\exp_p \mathsf{B}(p;r) = B(p;r);$$

one just cannot guarantee that $\exp | \mathsf{B}(p;r)$ is a diffeomorphism. The *injectivity radius of* p, $\operatorname{inj} p$, is defined to be the supremum of all $r > 0$ for which $\exp_p | \mathsf{B}(p;r)$ is a diffeomorphism. By our first remark, $\operatorname{inj} p > 0$ for every p. The *injectivity radius of* M, $\operatorname{inj} M$, is defined to be the infimum of $\operatorname{inj} p$, where p varies over all of M.

The Riemannian metric also has a local convexity property, namely, given any $p \in M$, there exists $r_0 \in (0, \operatorname{inj} p]$ such that for any two points $u, v \in B(p;r_0)$ there is a unique length-minimizing geodesic in M joining u to v, and it is completely contained in $B(p;r_0)$.

Let M be a Riemannian manifold. We say that M is *geodesically complete* if for every $\xi \in TM$, the geodesic γ_ξ is defined on all of $\mathbb{R}$, that is, if exp is defined on all of TM.

Proposition V.1.1 *If M is geodesically complete, then every closed and bounded subset is compact. As a consequence, M is a complete metric space. If M is a complete metric space, then M is geodesically complete. If M is complete, in either of the two meanings, then any two points of M can be joined by a minimal geodesic.*

V.1.1 Connections and Curvature

We must first consider the Levi-Civita connection on M. Of course, differentiation of functions on a manifold is determined by the differentiable structure alone. But the differentiation of vector fields is not uniquely determined. A *connection* is a rule for differentiating vector fields on a manifold. Let $\Gamma^\infty(TM)$ denote C^∞ vector fields on M, that is, the C^∞ sections in TM. Then a *connection on* M is a map $\nabla : TM \times \Gamma^\infty(TM) \to TM$, which we write as $\nabla_\xi Y$ instead of $\nabla(\xi, Y)$, with the following properties: First we require that $\nabla_\xi Y$ be in the same tangent space as ξ, and that for $\alpha, \beta \in \mathbb{R}$, $p \in M$, $\xi, \eta \in M_p$, $Y \in \Gamma^\infty(TM)$,

$$\nabla_{\alpha\xi+\beta\eta} Y = \alpha \nabla_\xi Y + \beta \nabla_\eta Y.$$

Second, we require that for $p \in M, \xi \in M_p, Y, Y_1, Y_2 \in \Gamma^\infty(TM), f \in C^\infty(M)$, we shall have

$$\nabla_\xi (Y_1 + Y_2) = \nabla_\xi Y_1 + \nabla_\xi Y_2, \qquad \nabla_\xi (fY) = (\xi f) Y_{|p} + f(p) \nabla_\xi Y.$$

Finally we require that ∇ be C^∞ in the following sense: if $X, Y \in \Gamma^\infty(TM)$ then $\nabla_X Y \in \Gamma^\infty(TM)$.

A connection is a local operator in the following sense: $\nabla_\xi Y$ is uniquely determined by the values of Y in a neighborhood of $p = \pi(\xi)$; in fact, it is determined by the restriction of Y to a path in a neighborhood of p that passes through p and which has velocity vector ξ at p.

If M is a Riemannian manifold, then there exists a unique connection ∇ (henceforth called the *Levi-Civita connection*) for which

$$\nabla_X Y = \nabla_Y X + [X, Y], \tag{V.1.3}$$

$$X\langle Y, Z\rangle = \langle \nabla_X Y, Z\rangle + \langle Y, \nabla_X Z\rangle \tag{V.1.4}$$

for all differentiable vector fields $X, Y, Z \in \Gamma(TM)$. We shall always work with this connection.

M is a Riemannian manifold with Levi-Civita connection ∇. The *curvature tensor*, R, of M is defined by

$$R(X, Y)Z = \nabla_Y \nabla_X Z - \nabla_X \nabla_Y Z - \nabla_{[Y,X]} Z,$$

where $X, Y, Z, W \in \Gamma(TM)$. One has that $R(X, Y)Z$ at the point p depends only on the values of X, Y, Z at p, and is trilinear on M_p. More generally, if α, β, ϵ are functions on M and X, Y, Z are vector fields on M, then $R(\alpha X, \beta Y)(\epsilon Z) = \alpha\beta\epsilon R(X, Y)Z$. Of course,

$$R(X, Y)Z + R(Y, X)Z = 0;$$

and (V.1.3) implies

$$R(X, Y)Z + R(Z, X)Y + R(Y, Z)X = 0.$$

One also has

$$\langle R(X,Y)Z,W\rangle - \langle R(Z,W)X,Y\rangle = 0,$$
$$\langle R(X,Y)Z,W\rangle + \langle R(X,Y)W,Z\rangle = 0.$$

Since the curvature tensor vanishes identically for $\dim M = 1$, all discussions concerning the curvature tensor will assume $\dim M \geq 2$.

To define the sectional curvature, one defines, for $p \in M$, $\xi, \eta \in M_p$,

$$\mathbf{k}(\xi,\eta) = \langle R(\xi,\eta)\xi,\eta\rangle, \qquad \mathbf{k}_1(\xi,\eta) = \langle R_1(\xi,\eta)\xi,\eta\rangle,$$

where

$$R_1(\xi,\eta)\zeta = \langle \xi,\zeta\rangle\eta - \langle \eta,\zeta\rangle\xi.$$

If ξ, η are linearly independent tangent vectors in M_p, then

$$\mathcal{K}(\xi,\eta) := \frac{\mathbf{k}(\xi,\eta)}{\mathbf{k}_1(\xi,\eta)} = \frac{\langle R(\xi,\eta)\xi,\eta\rangle}{|\xi|^2|\eta|^2 - \langle \xi,\eta\rangle^2}$$

is well defined and only depends on the 2-dimensional subspace determined by ξ and η. We refer to $\mathcal{K}(\xi,\eta)$ as the *sectional curvature of the 2-section* determined by ξ, η. We note that if G_2 is the complete collection of all 2-dimensional spaces tangent to M, then G_2 can be provided a C^∞ structure in a natural manner, and $\mathcal{K} : G_2 \to \mathbb{R}$ will then be C^∞.

For $p \in M$, the *Ricci tensor* $\mathrm{Ric} : M_p \times M_p \to \mathbb{R}$ is defined by

$$\mathrm{Ric}\,(\xi,\eta) = \text{ trace } (\zeta \mapsto R(\xi,\zeta)\eta).$$

In particular, we have for any orthonormal basis of M_p, $\{e_1,\ldots,e_n\}$

$$\mathrm{Ric}\,(\xi,\eta) = \sum_{j=1}^{n} \langle R(\xi,e_j)\eta, e_j\rangle.$$

Thus Ric is a symmetric bilinear form on M_p. To calculate its associated quadratic form (referred to as *Ricci curvature*), pick $\{e_1,\ldots,e_n\}$ so that $e_n = \xi/|\xi|$; then

$$\mathrm{Ric}\,(\xi,\xi) = \left\{\sum_{j=1}^{n-1} \mathcal{K}(e_j,\xi)\right\} |\xi|^2.$$

So, for any unit vector ξ, $\mathrm{Ric}\,(\xi,\xi)/(n-1)$ is the average sectional curvature of all 2-sections containing ξ.

The Riemann and Ricci tensors measure how the Riemannian metric and the Riemannian measure, respectively, of M differ at the infinitesimal level from the Euclidean metric and measure (see §V.1.2 below). It is expressed as follows: Fix $p \in M$ and $\mathsf{U} = \mathsf{B}(p; \text{inj}\, p)$, $U = B(p; \text{inj}\, p)$. Then every choice of orthonormal basis $\{e_1, \ldots, e_n\}$ of M_p determines a chart $\mathbf{n}: U \to \mathbb{R}^n$, referred to as *Riemann normal coordinates*, given by

$$n^j(q) = \langle (\exp |\mathsf{U})^{-1}(q), e_j \rangle$$

for $q \in U$, that is, for $v = \sum_j v^j e_j \in \mathsf{U}$ we have $n^j(\exp v) = v^j$. Then for $v \in \mathsf{U}$ we have

$$g_{jk}(\exp v) = \delta_{jk} - \tfrac{1}{3}\langle R(v, e_j)v, e_k \rangle + O(|v|^3), \tag{V.1.5}$$

$$\det(g_{jk}(\exp v)) = 1 - \tfrac{1}{3}\text{Ric}\,(v, v) + O(|v|^3) \tag{V.1.6}$$

as $v \to 0$.

If M_1 and M_2 are Riemannian manifolds, we say they are *isometric* if there exists a diffeomorphism $\phi: M_1 \to M_2$ such that

$$|\phi_* \xi|_2 = |\xi|_1$$

for all $\xi \in TM_1$ (the subscripts indicate in which Riemannian manifold the length is being evaluated), where ϕ_* denotes the Jacobian map from TM_1 to TM_2. The map ϕ is referred to as an *isometry*.

Let M be a Riemannian manifold of dimension ≥ 2, $\mathcal{K}$ the Riemann sectional curvature of 2-dimensional spaces tangent to M. We say that M *has constant sectional curvature* κ, $\kappa \in \mathbb{R}$, if $\mathcal{K}(\sigma) = \kappa$ for all 2-sections σ. One has the three standard model spaces of simply connected complete Riemannian manifolds of constant sectional curvature. For $\kappa = 0$ one has Euclidean space $\mathbb{R}^n$, with its standard Riemannian metric. Of course, the geodesics are the straight lines in $\mathbb{R}^n$. When $\kappa > 0$, the n-sphere $\mathbb{S}^n(1/\sqrt{\kappa})$ in $\mathbb{R}^{n+1}$ of radius $1/\sqrt{\kappa}$ has constant sectional curvature κ. The geodesics of $\mathbb{S}^n(1/\sqrt{\kappa})$ are given by the "great circles," the intersection with $\mathbb{S}^n(1/\sqrt{\kappa})$ of 2-planes in $\mathbb{R}^{n+1}$ that pass through the center of $\mathbb{S}^n(1/\sqrt{\kappa})$. When $\kappa < 0$, we have hyperbolic space given by the ball model, namely, on the n-disk $\mathbb{B}^n(1/\sqrt{-\kappa})$ we define the Riemannian metric

$$ds^2 = \frac{4|dz|^2}{\{1 + \kappa|z|^2\}^2};$$

then the sectional curvature is constant equal to $\kappa < 0$. One easily sees that, in this model, the geodesics emanating from the origin are given by straight lines emanating from origin, and their length to the boundary $\mathbb{S}^{n-1}(1/\sqrt{-\kappa})$ is infinite. The rest of the geodesics consist of the intersection with $\mathbb{B}^n(1/\sqrt{-\kappa})$ of those circles in $\mathbb{R}^n$ that orthogonally intersect the boundary sphere $\mathbb{S}^{n-1}(1/\sqrt{-\kappa})$.

By the Hopf–Killing theorem, a complete simply connected Riemannian manifold of constant sectional curvature κ is uniquely determined up to isometry. In particular such a space is isometric to the appropriate model among those discussed above. We always refer to the model space as the *model space* $\mathbb{M}_\kappa$ *of constant sectional curvature* κ.

For convenience, we define the function

$$\mathbf{S}_\kappa(r) = \begin{cases} (1/\sqrt{\kappa}) \sin \sqrt{\kappa} r, & \kappa > 0, \\ r, & \kappa = 0, \\ (1/\sqrt{-\kappa}) \sinh \sqrt{-\kappa} r, & \kappa < 0. \end{cases}$$

Of course, $\varphi(r) = \mathbf{S}_\kappa(r)$ satisfies

$$\varphi'' + \kappa\varphi = 0, \qquad \varphi(0) = 0, \qquad \varphi'(0) = 1.$$

Let $\mathbb{M}_\kappa$ be the model space of constant sectional curvature κ. It is standard that for any $x \in M$, unit vector $\xi \in M_x$, and $r > 0$ (that is, when $\kappa \leq 0$ – otherwise, we have to restrict $r \in (0, \pi/\sqrt{\kappa}]$), we have

(V.1.7) $$ds^2(\exp r\xi) = dr^2 + \mathbf{S}_\kappa{}^2(r)|d\xi|^2,$$

where $|d\xi|^2$ denotes the Riemannian metric on the $(n-1)$-dimensional unit tangent sphere (henceforth denoted by) S_x in M_x.

V.1.2 Volumes of Disks and Areas of Spheres

We now consider the Riemannian measure dV on M. Let $\mathbf{x} : U \to \mathbb{R}^n$ be a chart on M, with the Riemannian metric given in local coordinates by (V.1.1) and (V.1.2). Then the local measure

$$dV = \sqrt{g} \circ \mathbf{x}^{-1}\, d\mathbf{v}_n(\mathbf{x}) = \sqrt{g}\, dx^1 \cdots dx^n$$

[where g is given by (V.1.2)] is well defined, that is, the integral

$$I(f; U) = \int_{\mathbf{x}(U)} (f\sqrt{g}) \circ \mathbf{x}^{-1}\, d\mathbf{v}_n(\mathbf{x})$$

depends only on f and U – not on the particular choice of chart $\mathbf{x}$. One then defines a global Riemannian measure on M, also denoted by dV, using a partition of unity.

Definition Given any $x \in M$, we let D_x denote the largest starlike (relative to the origin) neighborhood of the origin of M_x on which exp is a diffeomorphism.

For any $\xi \in \mathsf{S}_x$, we let

$$c(\xi) = \sup\{r > 0 : t\xi \in \mathsf{D}_x\ \forall\, t \in [0, r)\}.$$

So

$$\mathsf{D}_x = \{r\xi : r \in [0, c(\xi))\}.$$

For any $x \in M$, $r > 0$, we define $\mathsf{D}_x(r)$ to be the subset of S_x consisting of those elements ξ for which $r\xi \in \mathsf{D}_x$, which we write as

$$r\mathsf{D}_x(r) = \mathsf{S}(x;r) \cap \mathsf{D}_x.$$

We let $D_x = \exp \mathsf{D}_x$, and refer to D_x as the *domain inside the cut locus of* x. We also refer to $r \in [0, c(\xi))$, $\xi \in \mathsf{S}_x$ as *geodesic spherical coordinates on* D_x.

We write the Riemannian measure on D_x as

(V.1.8) $$dV(\exp r\xi) = \sqrt{\mathbf{g}}(r;\xi)\,dr\,d\mu_x(\xi),$$

where $d\mu_x(\xi)$ denotes the Riemannian measure on S_x induced by the Euclidean Lebesgue measure on M_x. We also let $V(x;r)$ denote the volume of $B(x;r)$.

Given $x \in M$, we let $\mathfrak{A}(x;r)$ denote the *lower area of the metric sphere* $S(x;r)$, with center x and radius r, that is,

$$\mathfrak{A}(x;r) := \int_{\mathsf{D}_x(r)} \sqrt{\mathbf{g}}(r;\xi)\,d\mu_x(\xi),$$

and we let $A(x;r)$ denote the *area of the metric sphere* $S(x;r)$, that is,

$$A(x;r) = \mathcal{H}^{n-1}(S(x;r)),$$

the $(n-1)$-dimensional Hausdorff measure of $S(x;r)$.

Proposition V.1.2 *One has*

$$V(x;r) = \int_0^r \mathfrak{A}(x;t)\,dt$$

and

$$\limsup_{\epsilon \downarrow 0} \frac{V(x;r+\epsilon) - V(x;r)}{\epsilon} \le \mathfrak{A}(x;r) \le A(x;r)$$

for all $r > 0$.

Finally, $A(x;r) = \mathfrak{A}(x;r)$ *for almost all* r.

V.1.3 Curvature and Volume

Recall that $\mathbf{c_{n-1}}$ denotes the $(n-1)$-dimensional Riemannian measure of the $(n-1)$-dimensional unit sphere $\mathbb{S}^{n-1}$ in $\mathbb{R}^n$. For κ a real constant, and $\mathbb{M}_\kappa$ the simply connected complete Riemannian manifold of constant sectional curvature κ described above, let $dV_\kappa = \sqrt{\mathbf{g}_\kappa}(r;\xi)\,dr\,d\mu_x(\xi)$ denote the Riemannian measure on $\mathbb{M}_\kappa$ in geodesic spherical coordinates described above, and $V_\kappa(r)$ the volume of the metric disk in $\mathbb{M}_\kappa$ of radius r. Then

$$\sqrt{\mathbf{g}_\kappa}(r;\xi) = \mathbf{S}_\kappa{}^{n-1}(r), \qquad V_\kappa(r) = \mathbf{c}_{\mathbf{n-1}} \int_0^r \mathbf{S}_\kappa{}^{n-1}(t)\, dt.$$

Bishop's comparison theorem states that if M is an n-dimensional Riemannian manifold, $x \in M$ such that $\mathsf{B}(x;R) \subset \mathcal{T}M$ [that is, the exponential map is defined on all of $\mathsf{B}(x;R)$], and

$$\mathrm{Ric}\,(\xi,\xi) \geq (n-1)\kappa|\xi|^2$$

for all $\xi \in M_q$, $q \in B(x;R)$, then

$$\text{(V.1.9)} \qquad \frac{\partial_r \sqrt{\mathbf{g}}(r;\xi)}{\sqrt{\mathbf{g}}(r;\xi)} \leq \frac{\partial_r \left(\mathbf{S}_\kappa{}^{n-1}\right)}{\mathbf{S}_\kappa{}^{n-1}}(r),$$

which implies

$$\text{(V.1.10)} \qquad \sqrt{\mathbf{g}}(r;\xi) \leq \mathbf{S}_\kappa{}^{n-1}(r),$$

for all $r \in (0, \min\{R, c(\xi)\})$, and

$$\text{(V.1.11)} \qquad V(x;r) \leq V_\kappa(r)$$

for all $r \in [0, R]$. In particular, (V.1.9) implies that the function

$$\text{(V.1.12)} \qquad r \mapsto \sqrt{\mathbf{g}}(r;\xi)/\mathbf{S}_\kappa{}^{n-1}(r) \downarrow,$$

that is, is nonincreasing with respect to r on $(0, \min\{R, c(\xi)\})$. Gromov's refinement of Bishop's comparison theorem states that, under the same hypotheses, the function

$$\text{(V.1.13)} \qquad r \mapsto V(x;r)/V_\kappa(r) \downarrow,$$

that is, is nonincreasing with respect to r on $[0, R]$.

V.1.4 Liouville's Theorem

On the unit tangent bundle of M, $\mathsf{S}M$, one has the *geodesic flow*, Φ_t, defined by

$$\xi \mapsto \Phi_t\xi = \gamma_\xi{}'(t),$$

that is, $\Phi_t\xi$ is the velocity vector of the geodesic $\gamma_\xi(t) = \exp t\xi$ at the point $\gamma_\xi(t)$. Since $\gamma_\xi{}'(t)$ has the same length as ξ, Φ_t indeed maps the unit tangent bundle to itself. The *kinematic density* is the measure $d\mu$ on $\mathsf{S}M$ given by

$$\int_{\mathsf{S}M} F\, d\mu = \int_M dV(x) \int_{\mathsf{S}_x} F(\xi)\, d\mu_x(\xi)$$

for functions F on $\mathsf{S}M$.

Proposition V.1.3 (Liouville's Theorem) *The measure $d\mu$ on $\mathsf{S}M$ is invariant with respect to the geodesic flow.*

V.2 Isoperimetric Constants

In this section we define the basic apparatus of isoperimetric constants that we use in the sequel. We mention some of the most elementary examples that guide the intuition, and we discuss the necessity of introducing a modified apparatus of constants in order to deal with the dichotomy between local and global considerations in the geometry of Riemannian manifolds. Probably the most important result for our point of view here is Theorem V.2.4, which proves the invariance of positivity of modified isoperimetric constants under compact perturbations of a complete Riemannian manifold.

V.2.1 Definitions and Examples

Definition Let M be an n-dimensional Riemannian manifold, $n \geq 2$. Denote by V its n-dimensional Riemannian measure, and by A the Riemannian measure associated with $(n-1)$-dimensional submanifolds of M. For each $\nu > 1$ define the *ν-isoperimetric constant of M*, $\mathfrak{I}_\nu(M)$, to be the infimum

$$\mathfrak{I}_\nu(M) = \inf_\Omega \frac{A(\partial\Omega)}{V(\Omega)^{1-1/\nu}},$$

where Ω varies over open submanifolds of M possessing compact closure and C^∞ boundary.

For $\nu = \infty$ define *Cheeger's constant*, $\mathfrak{I}_\infty(M)$, by

$$\mathfrak{I}_\infty(M) = \inf_\Omega \frac{A(\partial\Omega)}{V(\Omega)},$$

where Ω ranges over open submanifolds of M described above.

Remark V.2.1 If M is compact, then, by considering $M \setminus B(x;\epsilon)$ for small $\epsilon > 0$, one can easily show that $\mathfrak{I}_\nu(M) = 0$ for all ν. Therefore, we shall restrict our interest to the case of noncompact manifolds. One can adjust the definitions of isoperimetric constants for the compact case and obtain a rich theory. See our comments in §VI.2 below. Also, see the remarks in the Bibliographic Notes to Chapter VI (§VI.6).

Definition Let $\phi : M \to N$ be a diffeomorphism between Riemannian manifolds M, N with respective metric tensors g, h. We say that ϕ is a *quasi-isometry* if there exists a constant $c > 1$ such that

$$c^{-1}g(\xi, \xi) \le h(\phi_*\xi, \phi_*\xi) \le cg(\xi, \xi)$$

for all $\xi \in TM$, the tangent bundle of M.

Two metric tensors g, h on a Riemannian manifold M are *quasi-isometric* if the identity map $\mathrm{id}_M : (M, g) \to (M, h)$ is a quasi-isometry, that is, if there exists a constant $c > 1$ such that

$$c^{-1}g(\xi, \xi) \le h(\xi, \xi) \le cg(\xi, \xi)$$

for all $\xi \in TM$.

Remark V.2.2 The property $\mathfrak{I}_\nu(M) > 0$, $\nu \in [1, \infty]$, is invariant under quasi-isometry.

Example V.2.1 The first, $\mathbb{R}^n$, $n \ge 1$, as discussed in previous chapters.

Example V.2.2 The hyperbolic spaces, $\mathbb{M}_\kappa$, $\kappa < 0$. Then the Cheeger constant $\mathfrak{I}_\infty(\mathbb{M}_\kappa) = (n-1)\sqrt{-\kappa}$, where $n = \dim \mathbb{M}_\kappa$. The proof goes as follows:

Let M be arbitrary Riemannian, $x \in M$, and r, ξ geodesic spherical coordinates on D_x. Let ∂_r denote the *radial vector field on* D_x, given by

$$\partial_{r|\exp t\xi} = \gamma_\xi{}'(t).$$

Then it is standard that the Riemannian divergence of ∂_r (see §VII.2.1 below) is given by

$$\operatorname{div} \partial_r(\exp r\xi) = \frac{\partial_r \sqrt{\mathbf{g}}(r;\xi)}{\sqrt{\mathbf{g}}(r;\xi)}.$$

When $M = \mathbb{M}_k$ then

$$\operatorname{div} \partial_r(\exp r\xi) = (n-1)\frac{\mathbf{S}_\kappa{}'}{\mathbf{S}_\kappa}.$$

When $\kappa < 0$, then $c(\xi) = +\infty \quad \forall\ \xi \in \mathsf{S}_x$ – so $D_x = \mathbb{M}_\kappa$. Therefore any relatively compact $\Omega \subset\subset D_x$. Furthermore,

$$\operatorname{div} \partial_r \geq (n-1)\sqrt{-\kappa}.$$

By the Riemannian divergence theorem,

$$A(\partial\Omega) \geq \int_{\partial\Omega} \langle \partial_r, \nu \rangle \, dA = \iint_{\Omega} \operatorname{div} \partial_r \, dV \geq (n-1)\sqrt{-\kappa}\, V(\Omega),$$

where ν denotes the exterior unit normal vector field along $\partial\Omega$. The lower bound

$$\frac{A(\partial\Omega)}{V(\Omega)} \geq (n-1)\sqrt{-\kappa}$$

is sharp, because one has

$$\frac{A(x;r)}{V(x;r)} \to (n-1)\sqrt{-\kappa} \quad \text{as } r \to +\infty.$$

As mentioned in Remark I.1.4, if $M = \mathbb{M}^2_\kappa$, the simply connected Riemannian 2-manifold of constant Gauss curvature equal to κ, then the isoperimetric inequality is (I.1.6):

$$L^2 \geq 4\pi A - \kappa A^2,$$

where A denotes the 2-dimensional measure of the domain, and L the 1-dimensional measure of the boundary, with equality if and only if the domain in question is a metric disk. When $\kappa < 0$, the inequality (I.1.6) is sharp in two senses, because (I.1.6) implies both the inequalities

$$L/A^{1/2} \geq \sqrt{4\pi}, \qquad L/A \geq \sqrt{-\kappa}.$$

The first is sharp for geodesic disks of radius r as $r \downarrow 0$, and the second is sharp for geodesic disks of radius r as $r \uparrow +\infty$.

Thus, one has distinct isoperimetric behavior when discussing the local properties of the manifold versus the global ones. Indeed, one expects all local isoperimetric behavior to be Euclidean in character [see (V.1.5)], and the global behavior to be quite diverse.

Remark V.2.3 The inequality $\mathfrak{I}_\nu(M) > 0$ is only possible for $n \leq \nu \leq \infty$. Indeed, let $\nu < n$, and consider small geodesic disks $B(x;\epsilon)$, with center $x \in M$ and radius $\epsilon > 0$. Then for the isoperimetric quotient of $B(x;\epsilon)$ we have

$$A(x;\epsilon)/V(x;\epsilon)^{1-1/\nu} \sim \text{const.}\epsilon^{n-1-n(1-1/\nu)} = \text{const.}\epsilon^{n/\nu-1}$$

as $\epsilon \downarrow 0$; thus $\mathfrak{I}_\nu(M) = 0$ whenever $\nu < n$. So it seems at first glance that one only has a discussion of isoperimetric constants for $\nu \geq n = \dim M$.

Definition For each $\nu > 1$ and $\rho > 0$ define the *ν-modified isoperimetric constant of M*, $\mathfrak{I}_{\nu,\rho}(M)$, by

$$\mathfrak{I}_{\nu,\rho}(\Omega) = \inf_{\Omega} \frac{A(\partial\Omega)}{V(\Omega)^{1-1/\nu}},$$

where Ω varies over open submanifolds of M possessing compact closure and C^∞ boundary and *inradius greater than* ρ, that is, Ω that contain a closed metric disk of radius ρ.

Definition For $\nu = 1$ and $\rho > 0$ define the 1-*modified isoperimetric constant of M*, $\mathfrak{I}_{1,\rho}(M)$, by

$$\mathfrak{I}_{1,\rho}(\Omega) = \inf_{\Omega} A(\partial\Omega),$$

where Ω varies over open submanifolds of M, described above, with inradius greater than ρ.

For $\nu = \infty$ and $\rho > 0$ define the *modified Cheeger constant*, $\mathfrak{I}_{\infty,\rho}(M)$, by

$$\mathfrak{I}_{\infty,\rho}(M) = \inf_{\Omega} \frac{A(\partial\Omega)}{V(\Omega)},$$

where Ω ranges over open submanifolds of M, described above, with inradius greater than ρ.

Remark V.2.4 Given $\nu \in [1, \infty]$, then the existence of $\rho > 0$ for which $\mathfrak{I}_{\nu,\rho}(M) > 0$ is invariant under quasi-isometry.

Example V.2.3 Consider the Riemannian product $M = M_0 \times \mathbb{R}^k$, where M_0 is an $(n-k)$-dimensional *compact* Riemannian manifold. Then $I_k(M) = 0$, as noted in Remark V.2.3. Yet, for extremely large domains – for example, metric disks of large radius – one expects the volume of these domains and the area of their boundaries to reflect k-dimensional space. We shall see (Example V.3.6 below) that $\mathfrak{I}_{\nu,\rho}(M) > 0$ for any $\rho > 0$ for all $\nu \in [1, k]$.

Example V.2.4 The 2-dimensional jungle gym JG^2 is constructed by (a) considering the integer lattice $\mathbb{Z}^3$ in $\mathbb{R}^3$, (b) connecting points, for which precisely one of their coordinates differ by 1 and the other two coordinates are equal, by a line segment parallel to the coordinate axis, (c) considering the surface consisting of all points with distance from the 1-dimensional network to be precisely equal to ϵ, for some given small $\epsilon > 0$, and finally (d) smoothing out the corners in a bounded periodic (that is, $\mathbb{Z}^3$-invariant) fashion. We shall see (Example V.3.7 below) that $\mathfrak{I}_3(\mathrm{JG}^2) > 0$.

The verification of the positivity of the isoperimetric constants in the last two examples requires discrete isoperimetric inequalities for graphs, which we consider later.

V.2.2 Basic Results

Remark V.2.5 It is an easy consequence of Minkowski's inequality that for domains Ω_j, $j = 1, \ldots, N$, in M, and any $k \geq 1$, we have

$$\sum_j V(\Omega_j)^{1/k} \geq \left\{\sum_j V(\Omega_j)\right\}^{1/k}.$$

From this Yau (1975) has shown that in the definition of $\mathfrak{I}_\nu(M)$, $\nu \in (1, \infty]$, it suffices to let Ω range over open submanifolds of M that are connected.

Definition For each $\nu > 1$, define the *Sobolev constant* $\mathfrak{S}_\nu(M)$ *of* M by

$$\mathfrak{S}_\nu(M) = \inf_f \frac{\|\mathrm{grad}\, f\|_1}{\|f\|_{\nu/(\nu-1)}},$$

where f varies over $C_c^\infty(M)$.

For each $\rho > 0$, let $C_{c,\rho}^\infty(M)$ consist of those compactly supported Lipschitz functions ϕ on M for which (a) there exists an $x \in M$ such that the preimage of $\max|\phi|$, Ω_ϕ, satisfies $\Omega_\phi \supseteq D(x;\rho)$, and (b) $\phi \mid M\backslash\Omega_\phi \in C^\infty$. For each $\nu > 1$ and $\rho > 0$, define the *modified Sobolev constant of* M, $\mathfrak{S}_{\nu,\rho}(M)$, by

(V.2.1) $$\mathfrak{S}_{\nu,\rho}(M) = \inf_f \frac{\|\mathrm{grad}\, f\|_1}{\|f\|_{\nu/(\nu-1)}},$$

where f ranges over $C_{c,\rho}^\infty(M)$.

Remark V.2.6 Even for $\mathfrak{S}_\nu(M)$ we may (by Theorem I.3.3) allow f to vary over compactly supported Lipschitz functions on M.

Theorem V.2.1 (The Federer–Fleming Theorem) *The isoperimetric and Sobolev constants are equal, that is,*

(V.2.2) $$\mathfrak{I}_\nu(M) = \mathfrak{S}_\nu(M),$$

and the modified isoperimetric and modified Sobolev constants are equal, that is,

(V.2.3) $$\mathfrak{I}_{\nu,\rho}(M) = \mathfrak{S}_{\nu,\rho}(M).$$

Proof The proof of (V.2.2) is the same as the proof of Theorem II.2.1, with the substitution of ν for n. The argument for (V.2.3) is a slight modification of the proof of Theorem II.2.1. ∎

Theorem V.2.2 *Suppose, for a given $\nu \in (1, \infty)$, we have $\mathfrak{I}_{\nu,\rho}(M) > 0$. Then for the area and volume of metric spheres and disks we have*

(V.2.4) $$A(x;r) \geq \mathfrak{I}_{\nu,\rho}(M)V(x;r)^{1-1/\nu}$$

for all $x \in M$ *and* $r > \rho$.

For the volume of metric disks we have

(V.2.5) $$V'(x;r) \geq \mathfrak{I}_{\nu,\rho}(M)V(x;r)^{1-1/\nu}$$

(where the prime denotes differentiation with respect to r) for almost all $r > \rho$.

In particular,

(V.2.6) $$\liminf_{r\uparrow\infty} V(x;r)r^{-\nu} > 0.$$

Proof The point, of course, is even when $r > \operatorname{inj} x$.

Assume $r > \rho$. For $\epsilon > 0$ define the function $\tau_\epsilon : [0, \infty) \to [0, 1]$ by (a) $\tau_\epsilon(s) = 1$ when $s \in [0, r]$, (b) $\tau_\epsilon(s) = (r + \epsilon - s)/\epsilon$ when $s \in [r, r + \epsilon]$, and (c) $\tau_\epsilon(s) = 0$ when $s > r + \epsilon$. Also define the Lipschitz function $f_\epsilon : M \to \mathbb{R}$ by

$$f_\epsilon(y) = \tau_\epsilon(d(x, y)).$$

Then

$$\begin{aligned} V(x;r)^{1-1/\nu} &\leq \|f_\epsilon\|_{\nu/(\nu-1)} \\ &\leq \mathfrak{I}_{\nu,\rho}(M)^{-1}\|\operatorname{grad} f_\epsilon\|_1 \\ &\leq \mathfrak{I}_{\nu,\rho}(M)^{-1}\frac{V(x;r+\epsilon) - V(x;r)}{\epsilon}. \end{aligned}$$

Now let $\epsilon \downarrow 0$. Then

$$V(x;r)^{1-1/\nu} \leq \mathfrak{I}_{\nu,\rho}(M)^{-1}\mathfrak{A}(x;r) \leq \mathfrak{I}_{\nu,\rho}(M)^{-1}A(x;r),$$

which implies the first claim.

The second claim follows from the fact that $A(x;r) = V'(x;r)$ for almost all r. Now integrate (V.2.5) to obtain (V.2.6). ∎

A similar argument, using the corresponding Federer–Fleming theorem for $\nu = \infty$, shows

Theorem V.2.3 *Suppose $\mathfrak{I}_\infty(M) > 0$. Then for the volume of metric disks we have for all $x \in M$*

(V.2.7) $$V'(x;r) \geq \mathfrak{I}_\infty(M)V(x;r)$$

for almost all $r > 0$.

In particular,

(V.2.8) $$\liminf_{r\uparrow\infty} V(x;r)e^{-\mathfrak{I}_\infty(M)r} > 0.$$

Proposition V.2.1 *If $\mathfrak{I}_{\nu,\rho}(M) > 0$ for some $\nu \in [1,\infty)$, $\rho > 0$, then $V(x;\rho+\epsilon)$ is uniformly bounded from below for all x with $d(x,\partial M) > \rho+\epsilon$, for any $\epsilon > 0$. Also, $I_{\mu,\rho+\epsilon}(M) > 0$ for all μ in $[1,\nu)$.*

Proposition V.2.2 *Let Ω be a domain in M with compact closure and C^∞ boundary. Then $\mathfrak{I}_\infty(\Omega) > 0$.*

Theorem V.2.4 *Let D be a relatively compact domain in the Riemannian complete M with C^∞ boundary Γ, and D' an n-dimensional Riemannian manifold with compact closure and C^∞ boundary Γ, such that M' given by $M' = \{M\backslash D\} \cup D'$ is C^∞ Riemannian. If $\mathfrak{I}_{\nu,\rho}(M) > 0$ for given $\nu \in [1,\infty)$ and $\rho > 0$, then there exists $\rho' > 0$ such that $\mathfrak{I}_{\nu,\rho'}(M') > 0$.*

Proof (See Figure V.2.1.) Suppose we are given any $\alpha \in (0,\infty)$. The number α will be fixed throughout the argument, although its estimated value will be determined as we go along.

Since $\mathfrak{I}_{\nu,\rho}(M) > 0$, there exists $R > \rho$ such that

$$V(x;R)^{1-1/\nu} \geq \alpha A(\Gamma)$$

for all $x \in M$. Let δ' denote the diameter of D'. We pick

$$\rho' = 2\rho + R + \delta'.$$

Let $E = M \setminus D$, and suppose we are given $\Omega' \subseteq M'$, with compact closure and C^∞ boundary, and $\Omega' \supseteq B(y';\rho')$.

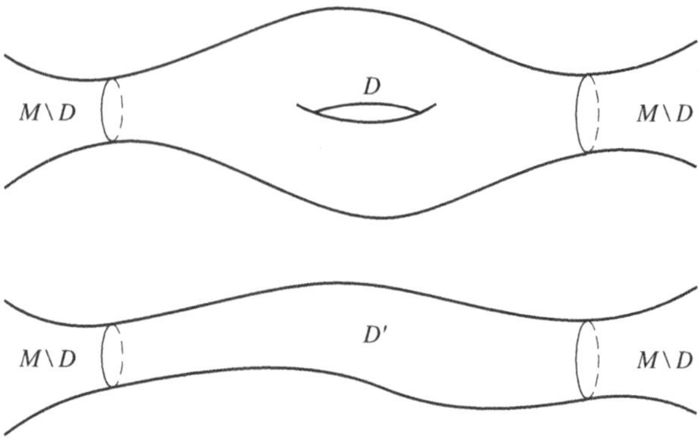

Figure V.2.1: The compact perturbation.

(i) We first assume that $d(y', D') < R$. Then $B(y'; \rho') \supseteq D'$, which implies $\partial\Omega' \subseteq E$. Set $\Omega = (\Omega' \setminus D') \cup D$. Then $\partial\Omega = \partial\Omega'$, and Ω contains a disk in M with inradius ρ, which implies

$$A(\partial\Omega') \geq \mathfrak{I}_{\nu,\rho}(M)V(\Omega)^{1-1/\nu}.$$

We obtain

$$\begin{aligned} A(\partial\Omega') &\geq \mathfrak{I}_{\nu,\rho}(M)\{V(D) + V(\Omega' \cap E)\}^{1-1/\nu} \\ &\geq \min\{1, V(D)/V(D')\}^{1-1/\nu}\mathfrak{I}_{\nu,\rho}(M)\{V(D') + V(\Omega' \cap E)\}^{1-1/\nu} \\ &= \min\{1, V(D)/V(D')\}^{1-1/\nu}\mathfrak{I}_{\nu,\rho}(M)V(\Omega')^{1-1/\nu}, \end{aligned}$$

that is,

$$\text{(V.2.9)} \quad A(\partial\Omega') \geq \min\{1, V(D)/V(D')\}^{1-1/\nu}\mathfrak{I}_{\nu,\rho}(M)V(\Omega')^{1-1/\nu}.$$

(ii) We now assume $d(y', D') \geq R$. Then $B(y'; R) \subseteq E$, which implies

$$\text{(V.2.10)} \qquad V(\Omega')^{1-1/\nu} \geq V(y'; R)^{1-1/\nu} \geq \alpha A(\Gamma).$$

(ii-a) Assume first that $V(\Omega' \cap E) \geq \frac{1}{2}V(\Omega')$. Then

$$\begin{aligned} A(\partial\Omega') &\geq A(\partial\Omega' \cap E) \\ &\geq A(\partial(\Omega' \cap E)) - A(\Gamma) \\ &\geq \mathfrak{I}_{\nu,\rho}(M)V(\Omega' \cap E)^{1-1/\nu} - \frac{1}{\alpha}V(\Omega')^{1-1/\nu}, \end{aligned}$$

because $R \geq \rho$. So

$$A(\partial\Omega') \geq \left\{\frac{\mathfrak{I}_{\nu,\rho}(M)}{2^{1-1/\nu}} - \frac{1}{\alpha}\right\} V(\Omega')^{1-1/\nu}.$$

Therefore we pick α at the very outset to also be greater than or equal to $\mathfrak{I}_{\nu,\rho}(M)/2^{2-1/\nu}$. We obtain

$$A(\partial\Omega') \geq (\mathfrak{I}_{\nu,\rho}(M)/2^{2-1/\nu})V(\Omega')^{1-1/\nu}. \tag{V.2.11}$$

(ii-b) The final situation to consider is, therefore, $d(y', D') \geq R$ and $V(\Omega' \cap D') \geq \frac{1}{2}V(\Omega')$. We still have (V.2.10); then

$$\begin{aligned} A(\partial\Omega') &\geq A(\partial\Omega' \cap D') \\ &\geq A(\partial(\Omega' \cap D')) - A(\Gamma) \\ &\geq \mathfrak{I}_\infty(D')V(\Omega' \cap D') - \frac{1}{\alpha}V(\Omega')^{1-1/\nu} \\ &\geq \mathfrak{I}_\infty(D')\frac{V(\Omega')}{2} - \frac{1}{\alpha}V(\Omega')^{1-1/\nu} \\ &\geq \left\{\frac{\mathfrak{I}_\infty(D')V(\Omega')^{1/\nu}}{2} - \frac{1}{\alpha}\right\}V(\Omega')^{1-1/\nu} \\ &\geq \left\{\frac{\mathfrak{I}_\infty(D')\{\alpha A(\Gamma)\}^{1/(\nu-1)}}{2} - \frac{1}{\alpha}\right\}V(\Omega')^{1-1/\nu} \end{aligned}$$

(the third inequality uses Proposition V.2.2). In addition to the above, pick, at the outset, α sufficiently large so that

$$\frac{\mathfrak{I}_\infty(D')\{\alpha A(\Gamma)\}^{1/(\nu-1)}}{2} - \frac{1}{\alpha} \geq \frac{\mathfrak{I}_\infty(D')A(\Gamma)^{1/(\nu-1)}}{2}. \tag{V.2.12}$$

Then (V.2.9)–(V.2.12) will imply the theorem. ■

Example V.2.5 The proofs of Proposition V.2.1 and Theorem V.2.4 break down when $\nu = \infty$ – unless we *postulate* the existence of $\rho_0 > 0$ for which

$$V(x; \rho_0) \geq \text{const.} > 0 \tag{V.2.13}$$

for all x (the constant independent of x). Moreover, one has a counterexample for Proposition V.2.1. Consider the Riemannian metric on $M = \mathbb{R}^1 \times \mathbb{S}^1$ given by

$$ds^2 = dr^2 + e^{2r}\,d\theta^2.$$

Certainly there is no $\rho_0 > 0$ for which (V.2.13) is satisfied for all x. At the same time, one easily shows

$$\mathfrak{I}_{\infty,\rho}(M) = \mathfrak{I}_\infty(M) = 1, \qquad \mathfrak{I}_{\nu,\rho}(M) = 0$$

for all $\nu \in [1, \infty)$, $\rho > 0$.

V.2.3 Bounded Geometry

Proposition V.2.3

(a) (Croke's inequality) *Let M be an arbitrary Riemannian manifold. Given any $o \in M$, $\rho > 0$, such that $\exp_o$ is defined on $D(o;\rho)$, then for*

$$r < \frac{1}{2}\min\left\{\inf_{x\in B(o;\rho)} \operatorname{inj} x,\ \rho\right\}$$

we have

$$A(\partial\Omega) \geq \text{const.}_n V(\Omega)^{n/(n-1)} \tag{V.2.14}$$

(where const._n *is a constant depending only on $n = \dim M$) for all $\Omega \subseteq B(o;r)$, which implies*

$$V(o;r) \geq \text{const.}_n r^n \tag{V.2.15}$$

for all $r < \frac{1}{2}\min\{\inf_{x\in B(o;\rho)} \operatorname{inj} x,\ \rho\}$.

(b) (Buser's inequality) *Let M be a complete n-dimensional Riemannian manifold, with*

$$\operatorname{Ric}(\xi,\xi) \geq (n-1)\kappa|\xi|^2, \qquad \kappa \leq 0, \tag{V.2.16}$$

for all $\xi \in TM$. Then there exists a positive constant $c(n,\kappa,r)$ depending on n, κ, and r, such that for any given $x \in M$, $r > 0$, and a dividing C^∞ hypersurface Γ in $B(x;r)$ with $\overline{\Gamma}$ imbedded in $D(x;r)$ and $B(x;r)\setminus\Gamma = \Omega_1 \cup \Omega_2$, where Ω_1, Ω_2 are open in $B(x;r)$, we have

$$\min\{V(\Omega_1), V(\Omega_2)\} \leq c(n,\kappa,r)A(\Gamma). \tag{V.2.17}$$

Moreover, for any fixed $r_0 > 0$ we have

$$c(n,\kappa,r) < c(n,\kappa,r_0)r \qquad \forall\, r \in (0,r_0]. \tag{V.2.18}$$

Remark V.2.7 Croke's inequalities (V.2.14) and (V.2.15) are statements that the local isoperimetric behavior of an arbitrary Riemannian manifold is indeed Euclidean.

Definition A Riemannian manifold M has *bounded geometry* if the Ricci curvature of M is bounded uniformly from below [as in (V.2.16)], and if the injectivity radius of M is bounded uniformly away from 0 on all of M.

Theorem V.2.5 *If M is Riemannian complete with bounded geometry, then $\mathfrak{I}_{1,\rho}(M) > 0$ for every $\rho > 0$.*

Proof Set

$$\delta = \min\{\rho, \operatorname{inj} M\}.$$

Assume we are given Ω as above, containing $B(x;\rho)$ for some $x \in M$. One easily has the existence of $z \in M$ for which $\partial\Omega \cap B(z;\delta/2)$ divides $B(z;\delta/2)$ into two open subsets for which the smaller volume is greater than or equal to $V(z;\delta/2)/3$. Then Buser's inequality (V.2.17) implies

$$A(\partial\Omega \cap B(z;\delta/2)) \geq \text{const.}\frac{V(z;\delta/2)}{3},$$

and Croke's inequality (V.2.15) implies

$$V(z;\delta/2) \geq c_n \delta^n,$$

which bounds $A(\partial\Omega)$ away from 0. ∎

Here is an alternate characterization of modified isoperimetric constants.

Theorem V.2.6 *Let M be a complete Riemannian manifold with bounded geometry. Then $\mathfrak{I}_{\nu,\rho}(M) > 0$ for some $\nu \geq 1$, $\rho > 0$ if and only if there exists $v_0 > 0$ such that*

$$A(\partial\Omega) \geq \text{const.}\begin{cases} V(\Omega)^{1-1/n}, & V(\Omega) \leq v_0, \\ V(\Omega)^{1-1/\nu}, & V(\Omega) \geq v_0, \end{cases} \tag{V.2.19}$$

for all domains Ω with compact closure and C^∞ boundary.

Proof Assume that (V.2.19) is valid for all Ω. Since M has bounded geometry, then for any $r_0 > 0$ there exists $v_0 > 0$ such that $V(x;r_0) \geq v_0$ for all x in M. Then $\mathfrak{I}_{\nu,r_0}(M) > 0$. ∎

To prove the converse, we first prove

Lemma V.2.1 *Let M be Riemannian complete with bounded geometry, and set $R = \frac{1}{4}\operatorname{inj} M$. Then there exists a positive constant c such that for any domain Ω in M with compact closure and C^∞ boundary, satisfying*

$$V(\Omega \cap B(x;R)) < \frac{1}{2}V(x;R)$$

for all x in Ω (so Ω is uniformly thin), we have

$$A(\partial\Omega) \geq cV(\Omega)^{1-1/n}.$$

Proof Given such an Ω, then for each $x \in \Omega$ there exists $r(x) \in (0, R)$ such that

$$V(\Omega \cap B(x; r(x))) = \frac{1}{2} V(x; r(x)).$$

Pick a finite collection $\{x_\iota, r_\iota\}$, $r_\iota = r(x_\iota)$, such that $\{B(x_\iota; r_\iota)\}$ are pairwise disjoint and $\{B(x_\iota; 3r_\iota)\}$ cover Ω. Then Buser's inequality (V.2.18) and Croke's inequality (V.2.15) imply

$$\begin{aligned} A(\partial\Omega \cap B(x_\iota; r_\iota)) &\geq \frac{\text{const.}}{r_\iota} V(\Omega \cap B(x_\iota; r_\iota)) \\ &= \frac{2\text{const.}}{r_\iota} V(x_\iota; r_\iota) \\ &\geq \text{const.} r_\iota{}^{n-1}, \end{aligned}$$

which implies

$$\begin{aligned} A(\partial\Omega) &\geq \text{const.} \sum_\iota r_\iota{}^{n-1} \\ &\geq \text{const.} \sum_\iota V(x_\iota; r_\iota)^{1-1/n} \\ &\geq \text{const.} \left\{ \sum_\iota V(x_\iota; r_\iota) \right\}^{1-1/n} \\ &\geq \text{const.} \left\{ \sum_\iota V(x_\iota; 3r_\iota) \right\}^{1-1/n} \\ &\geq \text{const.} V(\Omega)^{1-1/n} \end{aligned}$$

[the second inequality is (V.1.11); the third line in Minkowski's inequality; and the fourth line follows from (V.1.13)]. ■

Lemma V.2.2 *Let M be Riemannian complete with bounded geometry, and assume $\mathfrak{I}_{\nu,\rho}(M) > 0$ for some $\nu \geq 1$, $\rho > 0$. Let $R = \frac{1}{4}\operatorname{inj} M$. There exists a positive constant c' such that, if Ω is a domain in M with compact closure and C^∞ boundary for which there exists $x' \in \Omega$ satisfying*

$$V(\Omega \cap B(x'; R)) \geq \frac{1}{2} V(x'; R)$$

(so Ω is not uniformly thin), then

$$A(\Omega) \geq c' V(\Omega)^{1-1/\nu}.$$

Proof There exists x_0 such that

$$V(\Omega \cap B(x_0; R)) = \frac{1}{2} V(x_0; R).$$

Let $\Omega' = \Omega \cup B(x_0; R)$. Then Ω' has inradius greater than or equal to R, which implies

$$A(\partial\Omega') \geq \mathfrak{I}_{\nu,R} V(\Omega')^{1-1/\nu} \geq \mathfrak{I}_{\nu,R} V(\Omega)^{1-1/\nu}.$$

Also,

$$A(\partial\Omega') \leq A(\partial\Omega) + A(x_0; R),$$

and

$$A(x_0; R) \leq \text{const.} R^{n-1} \leq \text{const.} A(\partial\Omega \cap B(x_0; R)) \leq \text{const.} A(\partial\Omega)$$

(the second inequality follows from the same one in the proof of the previous lemma), which implies

$$A(\partial\Omega') \leq \text{const.} A(\partial\Omega),$$

which implies the lemma. ∎

Conclusion of the Proof of Theorem V.2.6 We are assuming that $\mathfrak{I}_{\nu,\rho}(M) > 0$ for some $\nu \geq 1$ and $\rho > 0$. Pick $v_0 = \frac{1}{3}\inf_{x\in M} V(x; R)$, $R = \frac{1}{4}\operatorname{inj} M$.

Then $V(\Omega) \leq v_0$ implies the hypotheses of Lemma V.2.1, which implies

$$A(\partial\Omega) \geq \text{const.} V(\Omega)^{1-1/n}.$$

Assume $V(\Omega) \geq v_0$. If $\nu \geq n$, then, by Theorem V.3.2 below,

$$A(\partial\Omega) \geq \text{const.} V(\Omega)^{1-1/\nu}$$

for *all* Ω, which implies the theorem in this case. Therefore assume $\nu < n$. If Ω satisfies the hypotheses of Lemma V.2.2, then we are done. Otherwise we are in the case of Lemma V.2.1, that Ω is "uniformly thin" but has large volume. Then

$$A(\partial\Omega) \geq \text{const.} V(\Omega)^{1-1/n} \geq \text{const.} V(\Omega)^{1-1/\nu} {v_0}^{1/\nu-1/n} = \text{const.} V(\Omega)^{1-1/\nu},$$

which implies the theorem. ∎

V.3 Discretizations and Isoperimetric Inequalities

Here we introduce discrete approximations to Riemannian manifolds. There is no pretense here to fine approximation; rather, we highlight what one might refer to as the coarse geometry of the manifold. First we discuss graph structures in their own right, and an apparatus of measures and isoperimetric constants associated with graphs. And then we discuss them as discretizations of smooth Riemannian manifolds. The most important result is Theorem V.3.1, which allows one to verify the positivity of an isoperimetric constant in one category, by checking in the other.

V.3.1 Volume Growth and Graphs

Definition Let X and Y be metric spaces with a bijection $\phi : X \to Y$. We say that ϕ is a *quasi-isometry* if there exists a constant $c \geq 1$ such that

$$c^{-1}d(x_1, x_2) \leq d(\phi(x_1), \phi(x_2)) \leq cd(x_1, x_2)$$

for all x_1, x_2 in X.

Let $\mathcal{G}$ be a countable set such that for each $\xi \in \mathcal{G}$ we have a finite nonempty subset $\mathsf{N}(\xi) \subseteq \mathcal{G} \setminus \{\xi\}$, of cardinality $m(\xi)$, each element of which is referred to as a *neighbor of* ξ, with the property that $\eta \in \mathsf{N}(\xi)$ if and only if $\xi \in \mathsf{N}(\eta)$ for all $\xi, \eta \in \mathcal{G}$. Then one determines a graph structure $\mathbf{G}$ by postulating the existence of precisely one oriented edge from any ξ to each of its neighbors, the elements of $\mathsf{N}(\xi)$. We refer to $m(\xi)$ as the *valence of* $\mathbf{G}$ at ξ. We say that the graph $\mathbf{G}$ has *bounded geometry* if the valence function $m(\xi)$ is bounded uniformly from above on all of $\mathcal{G}$. In such a case we let $\mathbf{m} = \max\{m(\xi) : \xi \in \mathcal{G}\}$.

A sequence of points $(\xi_0, \ldots, \xi_k)$ is a *combinatorial path of length k* if $\xi_j \in \mathsf{N}(\xi_{j-1})$ for all $j = 1, \ldots, k$. The graph $\mathbf{G}$ is called *connected* if any two points are connected by a combinatorial path. Note that $m(\xi) \geq 1$ for all ξ if $\mathbf{G}$ is connected. For any two vertices ξ and η in the connected graph $\mathbf{G}$, one defines their *distance* $\mathsf{d}(\xi, \eta)$ to be the infimum of the length of all combinatorial paths connecting ξ to η. We also refer to d as the *combinatorial metric*. We set the notation for the respective metric "disks":

$$\beta(\xi; k) = \{\eta \in \mathcal{G} : \mathsf{d}(\eta, \xi) \leq k\}$$

for any $\xi \in \mathcal{G}$.

Example V.3.1 Let Γ be a finitely generated group, with generator set $\mathcal{A}$. Then consider every element of Γ to be written as a word of minimum length in the generators in $\mathcal{A}$ and their inverses. For the graph structure, given any $\gamma \in \Gamma$, we let $\mathsf{N}(\gamma) = \gamma \cdot \{\mathcal{A} \cup \mathcal{A}^{-1}\}$ be the neighbors of γ. Then the combinatorial metric of the graph structure may be realized by the word metric as follows: Let $\mathcal{A} := \{\gamma_1, \ldots, \gamma_k\}$ be a given set of generators of Γ. With every $\gamma \in \Gamma$ we associate the *word norm of* γ, $|\gamma|_{\mathcal{A}}$, defined to be the minimum length of γ as a word in the given set of generators $\mathcal{A}$ and their inverses. Note that

$$|\gamma|_{\mathcal{A}} \geq 0, \quad \text{with} \quad |\gamma|_{\mathcal{A}} = 0 \Leftrightarrow \gamma = \text{id},$$
$$|\beta\gamma|_{\mathcal{A}} \leq |\beta|_{\mathcal{A}} + |\gamma|_{\mathcal{A}}, \qquad |\gamma^{-1}|_{\mathcal{A}} = |\gamma|_{\mathcal{A}},$$

for all β, γ in Γ. The *word metric on* Γ is then given by

$$\delta_{\mathcal{A}}(\beta, \gamma) = |\beta^{-1}\gamma|_{\mathcal{A}}.$$

Thus the group theoretic word metric is equal to the graph theoretic combinatorial metric. It is common to refer to this graph as the *Cayley graph of* Γ.

Note that if we use a different set of generators $\mathcal{B} := \{\gamma_1^*, \ldots, \gamma_l^*\}$, then we have the metrics induced by $\mathcal{A}$ and $\mathcal{B}$ quasi-isometric to each other, namely,

$$N := \max\{|\gamma_r^*|_{\mathcal{A}} : r = 1, \ldots, l\}$$
$$\Rightarrow N^{-1}|\gamma|_{\mathcal{A}} \leq |\gamma|_{\mathcal{B}} \leq N|\gamma|_{\mathcal{A}} \qquad \forall\, \gamma \in \Gamma.$$

Example V.3.2 Let M be a complete Riemannian manifold, and let Γ be any finitely generated subgroup of isometries M acting freely and properly discontinuously on M, such that M/Γ is compact. For each $x \in M$ let $\|\cdot\|_x$ denote the *displacement norm on* Γ, given by

$$\|\gamma\|_x = d(x, \gamma\cdot x)$$

for all $\gamma \in \Gamma$, where d denotes distance in M. Then again we have

$$\|\gamma\|_x \geq 0, \quad \text{with} \quad \|\gamma\|_x = 0 \Leftrightarrow \gamma = \text{id},$$
$$\|\beta\gamma\|_x \leq \|\beta\|_x + \|\gamma\|_x, \qquad \|\gamma^{-1}\|_x = \|\gamma\|_x,$$

for all β, γ in Γ. Given a set of generators $\mathcal{A}$ of Γ, one has the existence of a constant $a \geq 1$ such that

$$a^{-1}|\gamma|_{\mathcal{A}} \leq \|\gamma\|_x \leq a|\gamma|_{\mathcal{A}}$$

for all $\gamma \in \Gamma$, which implies that the induced metrics are quasi-isometric. Furthermore, the map $\phi : \Gamma \to M$, given by

$$\phi(\gamma) = \gamma\cdot x,$$

satisfies

$$a^{-1}\delta_{\mathcal{A}}(\beta, \gamma) \leq d(\phi(\beta), \phi(\gamma)) \leq a\delta_{\mathcal{A}}(\beta, \gamma)$$

for all β, γ in Γ – so ϕ is a quasi-isometry (at the metric level) of Γ to its image in M.

Definition Let X and Y be metric spaces with map $\phi : X \to Y$. We say that ϕ is a *rough isometry* if there exist constants $a \geq 1$, $b > 0$, and $\epsilon > 0$ such that

$$a^{-1}d(x_1, x_2) - b \leq d(\phi(x_1), \phi(x_2)) \leq ad(x_1, x_2) + b$$

for all x_1, x_2 in X, and ϕ is *ϵ-full*, that is,

$$\bigcup_{x \in X} B(\phi(x); \epsilon) = Y.$$

Note that the definition of rough isometry does not require that ϕ be a bijection, just a map; in fact, the map ϕ need not be continuous.

Proposition V.3.1 *If $\phi : X_1 \to X_2$ and $\psi : X_2 \to X_3$ are rough isometries, then so is $\psi \circ \phi$. If $\phi : X \to Y$ is a rough isometry, then there exists $\phi^- : Y \to X$ a rough isometry, for which both $d(\phi^- \circ \phi(x), x)$ and $d(\phi \circ \phi^-(y), y)$ are uniformly bounded on X and Y, respectively. Any two spaces of finite diameter are roughly isometric. If X and Y are roughly isometric, then X and $Y \times K$ are roughly isometric, for any compact metric space K.*

Return to the graph **G**. On the collection of vertices $\mathcal{G}$ we have two natural measures. The first is simply the *counting measure $d\iota$*; thus, for any subset $\mathcal{K}$ of $\mathcal{G}$ we have

$$\iota(\mathcal{K}) = \operatorname{card} \mathcal{K}.$$

The second is what we call the *volume measure $d\mathsf{V}$* on $\mathcal{G}$, defined by

$$d\mathsf{V}(\xi) = m(\xi)\, d\iota(\xi).$$

Of course, when **G** has bounded geometry, the two measures are commensurate in the sense that the Radon–Nikodym derivative of $d\mathsf{V}$ with respect to $d\iota$ is uniformly bounded away from 0 and $+\infty$. Because in what follows we generally discuss graphs of bounded geometry, and we are only interested in qualitative estimates on volumes, we shall work with the counting measure $d\iota$ – even when we announce the results in terms of the volume measure $d\mathsf{V}$.

Now denote the collection of oriented edges of the connected graph **G** by $\mathcal{G}_e$. The oriented edge from ξ to η will be denoted by $[\xi, \eta]$; and when we wish to consider the unoriented edge connecting ξ and η, we denote it by $[\xi \sim \eta]$.

Any finite subset $\mathcal{K}$ in $\mathcal{G}$ determines a finite subgraph **K** of **G**, for which one can describe a variety of suitable definitions for its boundary. Our definition will be that the *boundary of* **K**, ∂**K**, will be the subset of $\mathcal{G}_e$ consisting of those oriented edges that connect points of $\mathcal{K}$ to the complement of $\mathcal{K}$ in $\mathcal{G}$. We define the *area measure $d\mathsf{A}$ on $\mathcal{G}_e$* to be the counting measure for the oriented edges.

Thus, for any finite subset of vertices, the area of its boundary will be equal to the number of edges in the boundary.

Another definition of the boundary of a finite subgraph **K** of **G** is given by

$$\partial\mathbf{K} = \{\xi \in \mathcal{G} : \mathsf{d}(\xi, \mathbf{K}) = 1\}.$$

Thus, by this definition, $\partial\mathbf{K}$ is a subset of vertices in the complement of **K**; its area is defined to be its cardinality. When **G** has bounded geometry, the two choices of area functions

$$\mathbf{K} \mapsto \mathsf{A}(\partial\mathbf{K}),$$

as functions on the collection of subgraphs **K** of **G**, are commensurate each with respect to the other in the sense that the quotient of the two functions is bounded uniformly away from 0 and ∞.

Therefore, when **G** has bounded geometry, we will work with the counting measure for the volume of **K**, and the second definition of $\partial\mathbf{K}$ as a subset of **G** with counting measure for its area – despite the fact that the theorems are formulated with respect to the original notions of volume, boundary, and area.

Definition We say that the Riemannian manifold M *has exponential volume growth* if $\ln\, V(x;r) \geq \text{const.}r$ for sufficiently large r; otherwise we say that M *has subexponential volume growth*. Also, we say that M *has polynomial volume growth* if there exists $k > 0$ such that $V(x;r) \leq \text{const.}r^k$ for sufficiently large $r > 0$.

Similarly, we say that the graph **G** *has exponential volume growth* if $\ln\, \mathsf{V}(\xi;r) \geq \text{const.}r$ for sufficiently large r; otherwise we say that **G** *has subexponential volume growth*. Also, we say that **G** *has polynomial volume growth* if there exists $k > 0$ such that $\mathsf{V}(\xi;r) \leq \text{const.}r^k$ for sufficiently large $r > 0$.

Example V.3.3 (Example V.3.1 continued.) When one is given a finitely generated group Γ with κ generators $\mathcal{A} = \{\gamma_1, \ldots \gamma_\kappa\}$, then for the counting function

$$n_{\mathcal{A}}(\lambda) := \text{card}\,\{\gamma : |\gamma|_{\mathcal{A}} \leq \lambda\}$$

one has

$$n_{\mathcal{A}}(\lambda) = \mathsf{V}(\gamma;\lambda)/2\kappa,$$

where γ is any element of Γ (because all elements of Γ have 2κ elements in each neighborhood). Moreover (Milnor, 1968), the limit of $n_{\mathcal{A}}(\lambda)^{1/\lambda}$, as $\lambda \to \infty$, exists.

Furthermore, when M covers a compact Riemannian manifold with deck transformation group Γ, then there exist constants $\mathsf{a} \geq 1$, $\mathsf{b} \geq 0$, and $\mathsf{c} \geq 1$ such that $\mathsf{c}^{-1} n_{\mathcal{A}}(\mathsf{a}^{-1}\lambda - \mathsf{b}) \leq V(x;\lambda) \leq \mathsf{c} n_{\mathcal{A}}(\mathsf{a}\lambda + \mathsf{b})$ for every $x \in M$.

Proposition V.3.2 *Let* **G**, **F** *be connected, roughly isometric graphs, both with bounded geometry. Then* **G** *has polynomial (has exponential) volume growth if and only if* **F** *has polynomial (has exponential) volume growth.*

Definition For any $\nu \geq 1$ define the *ν-isoperimetric constant* of **G** by

$$\mathsf{I}_\nu(\mathbf{G}) = \inf_{\mathbf{K}} \frac{\mathsf{A}(\partial \mathbf{K})}{\mathsf{V}(\mathbf{K})^{1-1/\nu}},$$

where **K** varies over finite subgraphs of **G**.

Proposition V.3.3 *If* **G**, **F** *are roughly isometric graphs, both with bounded geometry, then* $\mathsf{I}_\nu(\mathbf{G}) > 0$ *if and only if* $\mathsf{I}_\nu(\mathbf{F}) > 0$.

Example V.3.4 (Example V.3.3 continued.) Consider a finitely generated group Γ, with fixed collection of generators $\mathcal{A}$. We (henceforth) denote the combinatorial disk centered at the identity e of radius k by $\beta(k)$. On the graph associated with Γ we work with the counting measure (and in this example we denote it by V); so when discussing the volume of combinatorial disks, we are working with the counting function. We define its "inverse" $\Phi(\lambda)$, $\lambda > 0$, by

$$\Phi(\lambda) = \min\{k \in \mathbb{Z}^+ : \mathsf{V}(k) > \lambda\}.$$

That is, $\mathsf{V}(\Phi(\lambda) - 1) \leq \lambda < \mathsf{V}(\Phi(\lambda))$. Then for any finite subgraph **K** of Γ,

$$\mathsf{A}(\partial \mathbf{K}) \geq \frac{1}{2} \frac{\mathsf{V}(\mathbf{K})}{\Phi(2\mathsf{V}(\mathbf{K}))}. \tag{V.3.1}$$

The proof goes as follows:

Definition In any graph, not necessarily a group, given any function f on the vertices, one defines the *differential of* f, $\mathfrak{D}f$, on the edges by

$$\mathfrak{D}f([x,y]) = f(y) - f(x),$$

and one integrates $\mathfrak{D}f$ relative to $d\mathsf{A}$ (see §VI.5 below).

For any function $\phi : \Gamma \to \mathbb{R}$ on the group Γ of finite support, and nonnegative integer k, define the mean value function ϕ_k on Γ by

$$\phi_k(x) = \frac{1}{\mathsf{V}(k)} \sum_{y \in \beta(k)} \phi(xy).$$

So $\phi_k(x)$ is the mean value of ϕ on the combinatorial disk centered at x with radius k.

For $k = 1$, y varies over $\mathcal{A} \cup \mathcal{A}^{-1}$, which implies

$$\phi_1(x) - \phi(x) = \frac{1}{2\,\text{card}\,\mathcal{A} + 1} \sum_{y \in \mathcal{A} \cup \mathcal{A}^{-1}} \mathfrak{D}\phi([x, xy]),$$

which implies

$$\begin{aligned} \|\phi_1 - \phi\|_1 &\leq \frac{1}{2\,\text{card}\,\mathcal{A}} \sum_x \sum_{y \in \mathcal{A} \cup \mathcal{A}^{-1}} |\mathfrak{D}\phi([x, xy])| \\ &= \frac{1}{2\,\text{card}\,\mathcal{A}} \sum_{y \in \mathcal{A} \cup \mathcal{A}^{-1}} \sum_x |\mathfrak{D}\phi([x, xy])| \\ &= \|\mathfrak{D}\phi\|_1. \end{aligned}$$

For any $k > 1$, let $y = z_1 \cdots z_\ell$, $\ell \leq k$, where z_j varies over $\mathcal{A} \cup \mathcal{A}^{-1}$. One has (set $z_0 = e$)

$$|\phi(xy) - \phi(x)| \leq \sum_{j=1}^{\ell} |\mathfrak{D}\phi([xz_0 \cdots z_{j-1}, xz_1 \cdots z_j])|,$$

which implies

$$\|\phi_k - \phi\|_1 \leq k\|\mathfrak{D}\phi\|_1.$$

Next, for every k, the triangle inequality implies

$$\{|\phi| \geq \lambda\} \subseteq \{|\phi - \phi_k| \geq \lambda/2\} \cup \{|\phi_k| \geq \lambda/2\},$$

which implies

$$\mathsf{V}(\{|\phi| \geq \lambda\}) \leq \mathsf{V}(\{|\phi - \phi_k| \geq \lambda/2\}) + \mathsf{V}(\{|\phi_k| \geq \lambda/2\}).$$

Also, one certainly has

$$\|\phi_k\|_\infty \leq \mathsf{V}(k)^{-1}\|\phi\|_1;$$

so if we pick k_0 to be the smallest k for which $\mathsf{V}(k)^{-1}\|\phi\|_1 < \lambda/2$, that is, $k_0 = \Phi(2\|\phi\|_1/\lambda)$, then

$$\begin{aligned} \mathsf{V}(\{|\phi| \geq \lambda\}) &\leq \mathsf{V}(\{|\phi - \phi_{k_0}| \geq \lambda/2\}) \\ &\leq \frac{2}{\lambda}\|\phi - \phi_{k_0}\|_1 \\ &\leq k_0 \frac{2}{\lambda}\|\mathfrak{D}\phi\|_1 \\ &= \frac{2}{\lambda}\Phi(2\|\phi\|_1/\lambda)\|\mathfrak{D}\phi\|_1, \end{aligned}$$

that is,

$$\mathsf{V}(\{|\phi| \geq \lambda\}) \leq \frac{2}{\lambda}\Phi(2\|\phi\|_1/\lambda)\|\mathfrak{D}\phi\|_1.$$

Pick $\phi = \mathcal{I}_{\mathbf{K}}$, and $\lambda = 1$. Then one obtains (V.3.1).

In particular,

$$\text{(V.3.2)} \quad \mathsf{V}(k) \geq \text{const.}k^{\nu} \quad \forall\, k \quad \Rightarrow \quad \mathsf{A}(\partial\mathbf{K}) \geq \text{const.}\mathsf{V}(\mathbf{K})^{1-1/\nu}$$

for all $\mathbf{K}$ of finite cardinality. So in the case of a group, a lower bound for volume growth alone is equivalent to the corresponding isoperimetric inequality.

For $\nu = \infty$ the argument yields a slightly weaker result, namely,

$$\mathsf{V}(k) \geq \text{const.}e^{\text{const.}k}\ \forall\, k \quad \Rightarrow \quad \mathsf{A}(\partial\mathbf{K}) \geq \frac{\text{const.}\mathsf{V}(\mathbf{K})}{\text{const.}\log\, \mathsf{V}(k) + \text{const.}}$$

for all $\mathbf{K}$ of finite cardinality.

Example V.3.5 If G is a Lie group with left-invariant Riemannian metric and bi-invariant Riemannian (Haar) measure, then one can formulate and prove, in similar manner, a corresponding result, that volume growth uniformly bounded from below implies an isoperimetric inequality.

V.3.2 Discretizations

Definition Let M be a Riemannian manifold. A subset $\mathcal{G}$ of M is said to be *ϵ-separated*, $\epsilon > 0$, if the distance between any two distinct points of $\mathcal{G}$ is greater than or equal to ϵ.

Remark V.3.1 If $\mathcal{G}$ is ϵ-separated, then one always has only a finite number of elements of $\mathcal{G}$ in $B(x;r)$ when the exponential map is defined on $\overline{\mathsf{B}(x;r+\epsilon/2)}$ [where, as above, $\mathsf{B}(x;\rho)$ denotes the disk of radius ρ centered at the origin of the tangent space at x, M_x]. Indeed,

$$B(x;r+\epsilon/2) \supseteq \bigcup_{\xi\in\mathcal{G}\cap B(x;r)} B(\xi;\epsilon/2),$$

where the union on the right hand side is disjoint union. Therefore

$$\begin{aligned} V(x;r+\epsilon/2) &\geq \sum_{\xi\in\mathcal{G}\cap B(x;r)} V(\xi;\epsilon/2) \\ &\geq \text{card}\,\{\mathcal{G}\cap B(x;r)\} \inf_{\eta\in\mathcal{G}\cap B(x;r)} V(\eta;\epsilon/2). \end{aligned}$$

But the compactness of $\overline{B(x;r)}$ implies $\inf_{\eta\in\mathcal{G}\cap B(x;r)} V(\eta;\epsilon/2) > 0$, which implies an upper bound on card $\{\mathcal{G}\cap B(x;r)\}$. So the real question is to obtain a uniform upper bound for card $\mathcal{G}\cap B(x;r)$ independent of x and $\mathcal{G}$.

Lemma V.3.1 *Let M be complete, with Ricci curvature bounded from below as in* (V.2.16) *on all of TM, and $\mathcal{G}$ an ϵ-separated subset of M. Then*

$$\operatorname{card}\{\mathcal{G} \cap B(x;r)\} \leq \frac{V_\kappa(2r + \epsilon/2)}{V_\kappa(\epsilon/2)}$$

for all $x \in M$ and $r > 0$.

Proof Since card $\mathcal{G} \cap B(x;r)$ is finite, there exists $\xi \in \mathcal{G} \cap B(x;r)$ such that

$$V(\xi;\epsilon/2) = \inf_{\eta \in \mathcal{G} \cap B(x;r)} V(\eta;\epsilon/2).$$

Therefore, the Bishop–Gromov volume comparison theorem (V.1.13) implies

$$\operatorname{card}\{\mathcal{G} \cap B(x;r)\} \leq \frac{V(x;r+\epsilon/2)}{V(\xi;\epsilon/2)} \leq \frac{V(\xi;2r+\epsilon/2)}{V(\xi;\epsilon/2)} \leq \frac{V_\kappa(2r+\epsilon/2)}{V_\kappa(\epsilon/2)}.$$

■

Definition Let M be a Riemannian manifold. A *discretization of M* is a graph $\mathbf{G}$, determined by an ϵ-separated subset $\mathcal{G}$ of M, for which there exists $\rho > 0$ such that $M = \bigcup_{\xi \in \mathcal{G}} B(\xi;\rho)$. Then ϵ is called the *separation*, and ρ the *covering, radius of the discretization*. The graph structure $\mathbf{G}$ is determined by the collection of neighbors of ξ, $\mathsf{N}(\xi) := \{\mathcal{G} \cap B(\xi;3\rho)\} \setminus \{\xi\}$, for each $\xi \in \mathcal{G}$.

Remark V.3.2 Note that card $\mathsf{N}(\xi) \geq 1$, so $\mathbf{G}$ is connected.

Remark V.3.3 Lemma V.3.1 implies that when the Ricci curvature is bounded from below as in (V.2.16), then for the graph $\mathbf{G}$ we have $1 + m(\xi) \leq V_\kappa(6\rho + \epsilon/2)/V_\kappa(\epsilon/2) := \mathbf{M}_{\epsilon,\rho}$ for all $\xi \in \mathbf{G}$ – so $\mathbf{G}$ has bounded geometry.

Proposition V.3.4 *Let M be Riemannian and $\mathbf{G}$ a discretization of M. Then there exist $a \geq 1$ and $b > 0$ for which*

$$a^{-1}d(\xi_1,\xi_2) \leq \mathsf{d}(\xi_1,\xi_2) \leq a d(\xi_1,\xi_2) + b \qquad \text{(V.3.3)}$$

for all ξ_1, ξ_2 in $\mathcal{G}$. Thus, M is roughly isometric to any of its discretizations, and any two of its discretizations are roughly isometric.

Lemma V.3.2 *Let M be a complete Riemannian manifold, with Ricci curvature bounded from below as in* (V.2.16)*, and assume there exist positive constants r_0 and V_0 such that*

$$V(x;r_0) \geq V_0$$

for all $x \in M$. *Then for any* $r > 0$ *one has a positive constant* const._r *such that*

$$V(x;r) \geq \text{const.}_r$$

for all $x \in M$.

Proof If $r > r_0$, then simply use V_0. If $r < r_0$, then simply note that the Bishop–Gromov theorem (V.1.13) implies

$$V(x;r) \geq \frac{V_\kappa(r)}{V_\kappa(r_0)} V(x;r_0) \geq \frac{V_\kappa(r)}{V_\kappa(r_0)} V_0,$$

which implies the claim. ∎

Proposition V.3.5 *Let* M *be a complete Riemannian manifold, with Ricci curvature bounded from below as in* (V.2.16). *Then for any discretization* **G** *of* M, **G** *has polynomial (has exponential) volume growth only if (if)* M *has polynomial (has exponential) volume growth.*

If, on the other hand, there exist positive constants r_0 *and* V_0 *such that*

$$V(x;r_0) \geq V_0$$

for all $x \in M$, *then for any discretization* **G** *of* M, **G** *has polynomial (has exponential) volume growth if (only if)* M *has polynomial (has exponential) volume growth.*

V.3.3 Isoperimetry and Discretizations

This is the main theorem of the chapter.

Theorem V.3.1 *Let* M *be a complete Riemannian manifold with bounded geometry. Then for any* $\nu \geq 1$ *we have* $\mathfrak{I}_{\nu,\rho}(M) > 0$ *if and only if* $\mathsf{I}_\nu(\mathbf{G}) > 0$ *for any discretization* **G** *of* M.

Proof First, given $\mathfrak{I}_{\nu,\rho}(M) > 0$. By Proposition V.3.4, we may work with *any* discretization. Therefore we consider a discretization **G** of M with separation constant $\epsilon > 0$ and covering radius $R = \rho$. To show that $\mathsf{I}_\nu(\mathbf{G}) > 0$, it suffices to prove the existence of positive constants such that given any $\mathcal{K} \subseteq \mathcal{G}$ we may find $\Omega \subseteq M$ of inradius $\geq \rho$ for which

$$A(\partial\Omega) \leq \text{const.}\ \text{card}\ \partial\mathcal{K}, \tag{V.3.4}$$

and

$$V(\Omega) \geq \text{const.}\ \text{card}\ \mathcal{K}. \tag{V.3.5}$$

We proceed as follows: Given a finite subset $\mathcal{K}$, set

$$\Omega := \bigcup_{\xi \in \mathcal{K}} B(\xi; R).$$

Then

$$\sum_{\xi \in \mathcal{K}} V(\xi; R) \leq \mathbf{M}_{\epsilon,R} V\left(\bigcup_{\xi \in \mathcal{K}} B(\xi; R)\right) = \mathbf{M}_{\epsilon,R} V(\Omega),$$

where $\mathbf{M}_{\epsilon,R}$ is an upper bound (depending on ϵ, R, and the lower bound of the Ricci curvature) of the maximum number of ϵ-separated points in a disk of radius R (see Remark V.3.3). So for

$$V_R := \inf_{x \in M} V(x; R) > 0$$

[the positivity of V_R follows from Croke's inequality (V.2.15)], we have

$$V_R \operatorname{card} \mathcal{K} \leq \mathbf{M}_{\epsilon,R} V(\Omega),$$

which implies (V.3.5). For the upper bound of $A(\partial\Omega)$ we note that

$$\partial\Omega \subseteq \bigcup_{\xi \in \partial(\mathcal{G} \setminus \mathcal{K})} S(\xi; R).$$

Indeed, if $x \in \partial\Omega$, then $\mathsf{d}(x, \xi) \geq R$ for all $\xi \in \mathcal{K}$, and there exists $\xi_0 \in \mathcal{K}$ such that $x \in S(\xi_0; R)$. But there must exist $\xi' \in \mathcal{G}$ such that $\mathsf{d}(x, \xi') < R$, which implies $\xi' \notin \mathcal{K}$. Then $\mathsf{d}(\xi_0, \xi') < 2R$, which implies $\xi_0 \in \mathsf{N}(\xi')$. So $\xi_0 \in \partial(\mathcal{G} \setminus \mathcal{K})$, which is the claim. Therefore,

$$A(\partial\Omega) \leq A_\kappa(R) \operatorname{card} \partial(\mathcal{G} \setminus \mathcal{K}) \leq \mathbf{m} A_\kappa(R) \operatorname{card} \partial\mathcal{K},$$

which implies (V.3.4). So we have the "only if" claim of the theorem.

For the "if" claim, we again note that we may work with any discretization. Therefore assume that we are given the graph $\mathbf{G}$, for which $\mathsf{I}_\nu(\mathbf{G}) > 0$, with covering radius $R = \rho < \operatorname{inj} M/2$.

Suppose we are given Ω, with compact closure, C^∞ boundary, and inradius greater than ρ. Set

$$\mathcal{K}_0 := \{\xi \in \mathcal{G} : V(\Omega \cap B(\xi; \rho)) > V(\xi; \rho)/2\},$$
$$\mathcal{K}_1 := \{\xi \in \mathcal{G} : 0 < V(\Omega \cap B(\xi; \rho)) \leq V(\xi; \rho)/2\}.$$

So both $\mathcal{K}_0$ and $\mathcal{K}_1$ are contained in $[\Omega]_\rho$ (the set of points with distance from

Ω less than or equal to ρ). Then for at least one of $j = 0, 1$ we have

(V.3.6) $$\frac{V(\Omega)}{2} \leq V\left(\Omega \cap \bigcup_{\xi \in \mathcal{K}_j} B(\xi;\rho)\right).$$

If (V.3.6) is valid for $j = 1$, then we have directly from Buser's inequality (V.2.18)

$$\begin{aligned}\frac{V(\Omega)}{2} &\leq \sum_{\xi \in \mathcal{K}_1} V(\Omega \cap B(\xi;\rho)) \\ &\leq \text{const.} \sum_{\xi \in \mathcal{K}_1} A(\partial\Omega \cap B(\xi;\rho)) \leq \text{const.}\mathbf{M}_{\epsilon,\rho} A(\partial\Omega)\end{aligned}$$

[without any hypothesis on $\mathsf{I}_\nu(\mathbf{G})$], which implies

$$A(\partial\Omega) \geq \text{const.} V(\Omega) = \text{const.} V(\Omega)^{1/\nu} V(\Omega)^{1-1/\nu} \geq \text{const.} V(\Omega)^{1-1/\nu},$$

since Ω contains a disk of radius ρ, which, by Croke's inequality (V.2.15), has volume uniformly bounded from below. So we must consider the case when (V.3.6) is valid only for $j = 0$.

First,

$$\frac{V(\Omega)}{2} \leq \sum_{\eta \in \mathcal{K}_0} V(\Omega \cap B(\eta;\rho)) \leq V_\kappa(\rho)\, \text{card}\, \mathcal{K}_0.$$

Therefore it suffices to give a lower bound of $A(\partial\Omega)$ by a multiple of card $\partial\mathcal{K}_0$ – the multiple independent of the choice of $\mathcal{K}_0$. To this end, define $H \subset [\Omega]_\rho$ by

$$H := \{x \in M : V(x;\rho)/2 = V(\Omega \cap B(x;\rho))\}.$$

For each $\xi \in \partial\mathcal{K}_0$ there exists $\eta \in \mathsf{N}(\xi)$, $\eta \in \mathcal{K}_0$; we have, of course,

$$d(\xi, \eta) < 3\rho.$$

By definition,

$$V(\Omega \cap B(\eta;\rho)) > V(\eta;\rho)/2, \qquad V(\Omega \cap B(\xi;\rho)) \leq V(\xi;\rho)/2,$$

which implies the minimizing geodesic connecting ξ to η contains an element $\zeta \in H$, which implies $\partial\mathcal{K}_0 \subseteq [H]_{3\rho}$, which implies

$$\bigcup_{\xi \in \partial\mathcal{K}_0} B(\xi;\rho) \subseteq [H]_{4\rho}.$$

Let Q be a maximal 2ρ-separated subset of H. Thus

$$\bigcup_{\xi\in\partial\mathcal{K}_0} B(\xi;\rho) \subseteq [Q]_{6\rho},$$

which implies

$$\begin{aligned}
V_\rho \text{card}\, \partial\mathcal{K}_0 &\leq \sum_{\xi\in\partial\mathcal{K}_0} V(\xi;\rho) \\
&\leq \mathbf{M}_{\epsilon,\rho} \sum_{\zeta\in Q} V(\zeta;6\rho) \\
&\leq \mathbf{M}_{\epsilon,\rho} \text{const.} \sum_{\zeta\in Q} V(\zeta;\rho) \\
&= 2\mathbf{M}_{\epsilon,\rho} \text{const.} \sum_{\zeta\in Q} V(\Omega\cap B(\zeta;\rho)) \\
&\leq 2\mathbf{M}_{\epsilon,\rho} \text{const.} \sum_{\zeta\in Q} A(\partial\Omega\cap B(\zeta;\rho)) \\
&\leq 2{\mathbf{M}_{\epsilon,\rho}}^2 \text{const.} A(\partial\Omega)
\end{aligned}$$

– the third inequality uses the Bishop–Gromov comparison theorem (V.1.13); the following equality follows from the definition of $H \supseteq Q$; and the fourth inequality uses Buser's inequality (V.2.18). ■

Example V.3.6 (Example V.2.3 continued.) Consider the Riemannian product $M = M_0 \times \mathbb{R}^k$, where M_0 is an $(n-k)$-dimensional *compact* Riemannian manifold. Then Proposition V.2.1 and Theorem V.3.1 imply that $\mathfrak{I}_{\nu,\rho}(M) > 0$ for any $\rho > 0$ for all $\nu \in [1,k]$. Indeed, both M and $\mathbb{R}^k$ are discretized by $\mathbb{Z}^k$. Since $V(x;r) \leq \text{const.} r^k$ for all $r > 0$, then k is the maximum of all ν for which $\mathfrak{I}_{\nu,\rho}(M) > 0$ for some $\rho > 0$.

Theorem V.3.2 *Let M have bounded geometry. Then, for any $\nu \geq n$, we have $\mathfrak{I}_\nu(M) > 0$ if and only if $\mathsf{I}_\nu(\mathbf{G}) > 0$ for any discretization $\mathbf{G}$ of M.*

Proof The "only if" is precisely as above.

So we assume that $\mathsf{I}_\nu(\mathbf{G}) > 0$. Suppose we are given Ω, with compact closure, C^∞ boundary, and no assumption on the inradius. As above, we set

$$\mathcal{K}_0 := \{\xi \in \mathcal{G} : V(\Omega\cap B(\xi;\rho)) > V(\xi;\rho)/2\},$$
$$\mathcal{K}_1 := \{\xi \in \mathcal{G} : 0 < V(\Omega\cap B(\xi;\rho)) \leq V(\xi;\rho)/2\}.$$

Again, both $\mathcal{K}_0$ and $\mathcal{K}_1$ are contained in $[\Omega]_\rho$, and for at least one of $j = 0, 1$

we have (V.3.6):

$$\frac{V(\Omega)}{2} \leq V\left(\Omega \cap \bigcup_{\xi \in \mathcal{K}_j} B(\xi;\rho)\right).$$

If (V.3.6) is valid for $j = 0$, then we may argue as above; for the only place we invoked the hypothesis of the inradius uniformly bounded away from 0 was when (V.3.6) is valid only for $j = 1$. We therefore adjust the argument for (V.3.6) valid only for $j = 1$. ■

Lemma V.3.3 *There exists a constant $j_\nu > 0$ such that*

$$A(\partial\Omega \cap B(\xi;\rho)) \geq j_\nu V(\Omega \cap B(\xi;\rho))^{1-1/\nu}, \qquad \rho < \frac{\operatorname{inj} M}{2}$$

for all $\xi \in \mathcal{K}_1$.

Conclusion of the Proof of Theorem V.3.2. Assume the lemma is valid. Then (I.1.5) implies

$$\sum_{\xi \in \mathcal{K}_1} V(\Omega \cap B(\xi;\rho)) \leq \left\{\sum_{\xi \in \mathcal{K}_1} V(\Omega \cap B(\xi;\rho))^{1-1/\nu}\right\}^{\nu/(\nu-1)}.$$

Therefore, (V.3.6) and the lemma imply

$$\begin{aligned}\frac{V(\Omega)}{2} &\leq V\left(\Omega \cap \bigcup_{\xi \in \mathcal{K}_1} B(\xi;\rho)\right)\\ &\leq \sum_{\xi \in \mathcal{K}_1} V(\Omega \cap B(\xi;\rho))\\ &\leq \text{const.}\, A(\partial\Omega)^{\nu/(\nu-1)},\end{aligned}$$

which implies the theorem. ■

So it remains to prove the lemma.

Proof of Lemma V.3.3. Let $D = \Omega \cap B(\xi;\rho)$; then

$$V(D) \leq \frac{V(\xi;\rho)}{2} \leq \frac{V_\kappa(\rho)}{2}.$$

Therefore $\nu \geq n$ implies

$$V(D)^{1-1/\nu} \leq \left\{\frac{V_\kappa(\rho)}{2}\right\}^{1/n-1/\nu} V(D)^{1-1/n};$$

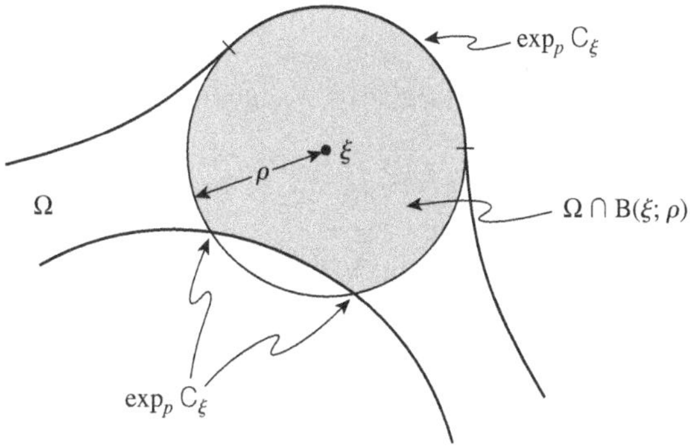

Figure V.3.1: C_ξ.

so it suffices to prove

$$V(D)^{1-1/n} \leq \text{const.} A(\partial D \cap B(\xi;\rho)).$$

Since $\rho < \operatorname{inj} M/2$ we have, by Croke's inequality (V.2.14),

$$\begin{aligned} V(D)^{1-1/n} &\leq \text{const.} A(\partial D) \\ &= \text{const.}\{A(\partial D \cap B(\xi;\rho)) + A(\partial D \cap S(\xi;\rho))\}. \end{aligned}$$

So we want to show

$$A(\partial D \cap S(\xi;\rho)) \leq \text{const.} A(\partial D \cap B(\xi;\rho)).$$

Consider geodesic spherical coordinates centered at ξ. Let C_ξ denote the subset of S_ξ for which

$$\exp \rho C_\xi = \partial D \cap S(\xi;\rho).$$

(See Figure V.3.1.) For each $\theta \in C_\xi$, let

$$\sigma(\theta) = \sup\{t \geq 0 \colon \exp t\theta \in \partial D \cap B(\xi;\rho)\}.$$

Note that if $\sigma(\theta) < \rho$, then the geodesic segment from $\exp \sigma(\theta)\theta$ to $\exp \rho\theta$ is contained in D. The Bishop–Gromov theorem (V.1.13) implies

$$\begin{aligned} \int_{\sigma(\theta)}^{\rho} \sqrt{\mathbf{g}}(s;\theta)\,ds &\geq \frac{V_\kappa(\rho) - V_\kappa(\sigma(\theta))}{A_\kappa(\rho)} \sqrt{\mathbf{g}}(\rho;\theta) \\ &= \frac{V_\kappa(\rho)}{A_\kappa(\rho)} \sqrt{\mathbf{g}}(\rho;\theta) - \frac{V_\kappa(\sigma(\theta))}{A_\kappa(\rho)} \sqrt{\mathbf{g}}(\rho;\theta) \\ &\geq \frac{V_\kappa(\rho)}{A_\kappa(\rho)} \sqrt{\mathbf{g}}(\rho;\theta) - \frac{V_\kappa(\sigma(\theta))}{A_\kappa(\sigma(\theta))} \sqrt{\mathbf{g}}(\sigma(\theta);\theta) \\ &\geq \frac{V_\kappa(\rho)}{A_\kappa(\rho)} \{\sqrt{\mathbf{g}}(\rho;\theta) - \sqrt{\mathbf{g}}(\sigma(\theta);\theta)\}, \end{aligned}$$

which implies

$$\begin{aligned} V(D) &\geq \frac{V_\kappa(\rho)}{A_\kappa(\rho)} \int_{\mathsf{C}_\xi} \{\sqrt{\mathbf{g}}(\rho;\theta) - \sqrt{\mathbf{g}}(\sigma(\theta);\theta)\}\, d\boldsymbol{\mu}_\xi(\theta) \\ &\geq \frac{V_\kappa(\rho)}{A_\kappa(\rho)} \{A(\partial D \cap S(\xi;\rho)) - A(\partial D \cap B(\xi;\rho))\}, \end{aligned}$$

which implies, by Buser's inequality (V.2.18),

$$\begin{aligned} A(\partial D \cap S(\xi;\rho)) &\leq \frac{A_\kappa(\rho)}{V_\kappa(\rho)} V(D) + A(\partial D \cap B(\xi;\rho)) \\ &\leq \text{const.} A(\partial D \cap B(\xi;\rho)) + A(\partial D \cap B(\xi;\rho)) \\ &= \text{const.} A(\partial D \cap B(\xi;\rho)), \end{aligned}$$

which implies the claim. ∎

Example V.3.7 (Example V.2.4 continued.) Let JG^2 denote the 2-dimensional jungle gym in $\mathbb{R}^3$. Then Theorem V.3.2 implies that $\mathfrak{I}_3(\mathrm{JG}^2) > 0$. Indeed, JG^2 and $\mathbb{R}^3$ are simultaneously discretized by $\mathbb{Z}^3$.

Example V.3.8 Given a noncompact M that covers of a compact Riemannian manifold M_0, the deck transformation group Γ of the covering is a discretization (see Example V.3.2) of M. Then, by the proof of Proposition V.3.5 (see Kanai, 1985), a volume growth of order $\geq r^k$ in M implies a volume growth in Γ of order $\geq k$, which, by Example V.3.4, implies a k-isoperimetric inequality in Γ, which implies a (modified) k-isoperimetric inequality in M.

V.3.4 Isoperimetric Inequalities on Products

For future reference we note:

Given the Riemannian manifolds M_j, $j = 1, 2$, we consider the respective volume and area measures associated with the standard product metric on $M_1 \times M_2$.

Given the graphs $\mathbf{G}^{(j)}$, $j = 1, 2$, we consider the graph $\mathbf{G}^{(1)} \times \mathbf{G}^{(2)}$ with vertices

$$(\mathbf{G}^{(1)} \times \mathbf{G}^{(2)})_0 = (\mathbf{G}^{(1)})_0 \times (\mathbf{G}^{(2)})_0,$$

and oriented edges $[(\xi_1, \xi_2), (\eta_1, \eta_2)]$ where either (a) $\xi_1 \sim \xi_2$ and $\eta_1 = \eta_2$, or (b) $\xi_1 = \xi_2$ and $\eta_1 \sim \eta_2$, with corresponding volume and area measures on the vertices and edges.

Given the Riemannian manifolds M_j, $j = 1, 2$, of bounded geometry, with respective discretizations $\mathbf{G}^{(j)}$, then the inclusion of the discretization $(\mathbf{G}^{(1)} \times \mathbf{G}^{(2)})_0$ in $M_1 \times M_2$ is a rough isometry.

Proposition V.3.6

(a) Let $\mathbf{G}^{(j)}$, $j = 1, 2$, *be two graphs, both with uniformly bounded valence functions* $m^{(j)}(\xi)$, *and assume* $\mathsf{I}_{\nu_j}(\mathbf{G}^{(j)}) > 0$, $\nu_j \geq 1$, $j = 1, 2$. *Then*

$$\mathsf{I}_{\nu_1+\nu_2}(\mathbf{G}^{(1)} \times \mathbf{G}^{(2)}) > 0.$$

(b) Let M_j, $j = 1, 2$, *be Riemannian manifolds of respective dimension* n_j. *Let* $\nu_j \geq n_j$, $j = 1, 2$, *and assume that*

$$I_{\nu_j}(M_j) > 0, \qquad j = 1, 2.$$

Then

$$I_{\nu_1+\nu_2}(M_1 \times M_2) > 0.$$

(c) Let M_j, $j = 1, 2$, *be Riemannian manifolds as in (b). Let* $\nu_j \geq 1$ *and* $\rho > 0$, *and assume that*

$$I_{\nu_j,\rho}(M_j) > 0, \qquad j = 1, 2.$$

Then

$$I_{\nu_1+\nu_2,\sqrt{2}\rho}(M_1 \times M_2) > 0. \tag{V.3.7}$$

If, in addition, both M_1 *and* M_2 *have bounded geometry, and*

$$\nu_1 + \nu_2 \geq n_1 + n_2,$$

then

$$I_{\nu_1+\nu_2}(M_1 \times M_2) > 0. \tag{V.3.8}$$

V.4 Bibliographic Notes

Nearly all the material of this chapter is detailed, with references, in Chavel (1994). We chose proofs from among the material there to emphasize our major themes, isoperimetric inequalities and discretizations. An very nice introductory survey is Burstall's article in the 1998 Edinburgh Lectures (Davies and Safarov, 1999, pp. 1–29).

§**V.1** Proposition V.1.1 is known as the Hopf–Rinow theorem; see Hopf and Rinow (1931).

§**V.2.3** Propositions V.2.2 and V.2.3(**a**) were first proved in Croke (1980), and Proposition V.2.3(**b**) in Buser (1982). Theorem V.2.6 was communicated to me by A. Grigor'yan; see also his Grigor'yan (1994b).

§**V.3.1** The proof in Example V.3.4 is from Coulhon and Saloff-Coste (1993).

§V.3.2 Proposition V.3.4 was first proved in Kanai (1985), with the hypothesis of Ricci curvature bounded from below to obtain the upper bound. Subsequently, this hypothesis was shown in Holopainen (1994) to be unnecessary. Theorem V.3.2 was first proved in Kanai (1985).

§V.3.4 An elegant proof of the product theorem, Proposition V.3.6, for the discrete case is given in §3 of Varopoulos (1985). (It is surely valid in the continuous case.) Also, see Grigor'yan (1985).

VI
Analytic Isoperimetric Inequalities

In this chapter we explore, in greater detail, the applications of isoperimetric inequalities to Sobolev inequalities, a phenomenon only hinted at in the Federer–Fleming theorem (Theorem II.2.1) and the Faber–Krahn inequality (Theorem III.3.1). The application of isoperimetric inequalities to analysis is a rich subject in its own right, although we only develop here what we need for our particular study of the heat equation in later chapters. We also present the analogue of these arguments in the discrete case, and discuss the equivalence of Sobolev inequalities on Riemannian manifolds of bounded geometry, and the corresponding Sobolev inequalities on their discretizations.

VI.1 L^2 Sobolev Inequalities

Let M be an n-dimensional Riemannian manifold.

Definition Let f be a C^1 function on M. We define the *gradient of* f to be the vector field on M satisfying

$$\langle \operatorname{grad} f, \xi \rangle = df(\xi) = \xi f \quad \forall\, \xi \in TM,$$

where df denotes the differential of f on M.

For C^1 functions f, h on M we have

$$\begin{aligned} \operatorname{grad}(f + h) &= \operatorname{grad} f + \operatorname{grad} h, \\ \operatorname{grad} fh &= f \operatorname{grad} h + h \operatorname{grad} f. \end{aligned}$$

If $\mathbf{x} : U \to \mathbb{R}^n$ is a chart on M, then, with the usual notation

$$g_{ij} = \langle \partial_i, \partial_j \rangle, \qquad G = (g_{ij}), \qquad G^{-1} = (g^{ij}),$$

we have

$$\operatorname{grad} f = \sum_{j,k} \frac{\partial(f \circ \mathbf{x}^{-1})}{\partial x^j} g^{jk} \frac{\partial}{\partial x^k}.$$

Note that the Riemannian metric on M naturally induces a metric on fibers of TM^*, namely, if α and β are covectors in the same fiber, then $\langle \alpha, \beta \rangle = \langle \theta^{-1}(\alpha), \theta^{-1}(\beta) \rangle$, where $\theta : TM \to TM^*$ is the bundle isomorphism defined by $\{\theta(\xi)\}(\eta) = \langle \xi, \eta \rangle$ for any $\xi, \eta \in TM$ that belong to the same fiber. One verifies, rather easily, that for any differentiable function on M we have

$$|df| = |\operatorname{grad} f|$$

on all of M.

Lemma VI.1.1 *If the isoperimetric constant $\mathfrak{I}_\nu(M)$ is positive for some given $\nu > 2$, then*

$$\|\operatorname{grad} \phi\|_2 \geq \frac{\nu - 2}{2(\nu - 1)} \mathfrak{I}_\nu(M) \|\phi\|_{2\nu/(\nu-2)}. \tag{VI.1.1}$$

for any function ϕ in C_c^∞.

Proof Of course, we have $\mathfrak{I}_\nu = \mathfrak{S}_\nu > 0$, by the Federer–Fleming theorem. Therefore,

$$\|\operatorname{grad} f\|_1 \geq \mathfrak{I}_\nu \|f\|_{\nu/(\nu-1)}$$

for all $f \in C_c^\infty$. Given any $\phi \in C_c^\infty$, let $f = |\phi|^p$, $p = 2(\nu - 1)/(\nu - 2)$. Then

$$|\operatorname{grad} f| = p|\phi|^{p-1} |\operatorname{grad} |\phi|\,| = p|\phi|^{p-1} |\operatorname{grad} \phi|$$

(the last equality valid a.e.-dV – see Lemma II.2.1), which implies, by the Cauchy–Schwarz inequality, that

$$\int_M |\operatorname{grad} f| \, dV \leq p \|\operatorname{grad} \phi\|_2 \|\phi\|_{2(p-1)}{}^{p-1},$$

which easily implies the claim. ■

Remark VI.1.1 We refer to (VI.1.1) as the *Nirenberg–Sobolev inequality.*

Lemma VI.1.2 *If $\mathfrak{I}_\nu(M)$ is positive for some given $\nu \geq 2$, then there exists a positive* const.$_\nu$ *such that*

$$\|\phi\|_2{}^{2+4/\nu} \leq \text{const.}_\nu \|\operatorname{grad} \phi\|_2{}^2 \|\phi\|_1{}^{4/\nu} \tag{VI.1.2}$$

for all $\phi \in C_c^\infty$.

Proof Start with the previous lemma for $\nu > 2$. The first step is to show that $\mathfrak{S}_\nu > 0$ for some $\nu \geq 2$ implies

$$\int \phi^{2+4/\nu}\, dV \leq \text{const.} \|\operatorname{grad} \phi\|_2{}^2 \|\phi\|_2{}^{4/\nu}. \tag{VI.1.3}$$

Indeed, if $\nu > 2$, set $f = |\phi|^2$, $p = \nu/(\nu - 2)$, and $g = |\phi|^{4/\nu}$, $q = \nu/2$, and use Hölder's inequality to obtain (VI.1.3) from (VI.1.1). On the other hand, if $\nu = 2$, one has directly

$$\|\phi^2\|_2 \leq \text{const.} \|\operatorname{grad}(\phi^2)\|_1 \leq \text{const.} \|\phi\|_2 \|\operatorname{grad} \phi\|_2.$$

Square both sides to obtain (VI.1.3). To obtain (VI.1.2), simply apply Hölder's inequality to $f = |\phi|^{4/(\nu+4)}$, $p = (\nu + 4)/4$, and $g = |\phi|^{(2\nu+4)/(\nu+4)}$, $q = (\nu + 4)/\nu$. ∎

Remark VI.1.2 We refer to (VI.1.2) as the *Nash–Sobolev inequality.*

Theorem VI.1.1 *The Nash–Sobolev inequality is equivalent to the Nirenberg–Sobolev inequality.*

Proof The proof of Lemma VI.1.2 shows that the Nirenberg–Sobolev inequality implies the Nash–Sobolev inequality. So it remains to consider the converse.

Suppose we are given a positive constant c such that

$$\|\phi\|_2{}^{1+2/\nu} \leq \mathsf{c} \|\operatorname{grad} \phi\|_2 \|\phi\|_1{}^{2/\nu}$$

for all $\phi \in C_c^\infty$, for some $\nu > 2$. Set

$$q = \frac{2\nu}{\nu - 2}, \qquad \theta = \frac{\nu}{\nu + 2}.$$

Then

$$2(1 - \theta) = \frac{4}{\nu + 2}.$$

Given any $f \in C_c^\infty$ define, for every $k \in \mathbb{Z}$,

$$f_k = \min\{(f - 2^k)^+, 2^k\} = \begin{cases} 0, & f < 2^k, \\ f - 2^k, & 2^k \leq f < 2^{k+1}, \\ 2^k, & 2^{k+1} \leq f \end{cases}$$

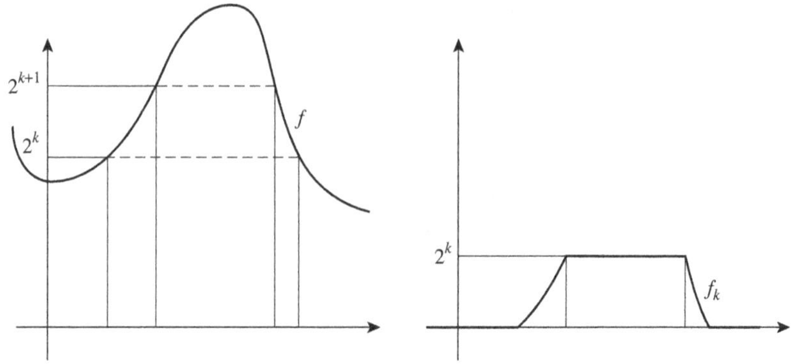

Figure VI.1.1: f_k.

(see Figure VI.1.1) and

$$B_k = \{2^k \le f < 2^{k+1}\}.$$

Then

$$\left\{\int {f_k}^2\right\}^{1+2/\nu} \le \mathsf{c}^2 \left\{\int |\text{grad } f_k|^2\right\} \left\{\int f_k\right\}^{4/\nu},$$

which implies

$$\begin{aligned}\{2^{2k} V(f \ge 2^{k+1})\}^{1+2/\nu} &= \left[\int_{\{f \ge 2^{k+1}\}} {f_k}^2\right]^{1+2/\nu} \\ &\le \left[\int {f_k}^2\right]^{1+2/\nu} \\ &\le \mathsf{c}^2\{2^k V(f \ge 2^k)\}^{4/\nu} \int_{B_k} |\text{grad } f|^2,\end{aligned}$$

that is,

$$\{2^{2k} V(f \ge 2^{k+1})\}^{1+2/\nu} \le \mathsf{c}^2\{2^k V(f \ge 2^k)\}^{4/\nu} \int_{B_k} |\text{grad } f|^2. \tag{VI.1.4}$$

Let

$$a_k = 2^{kq} V(f \ge 2^k), \qquad b_k = \int_{B_k} |\text{grad } f|^2;$$

then (VI.1.4) implies

$$a_{k+1} \le 2^q \mathsf{c}^{2\theta} {b_k}^\theta {a_k}^{2(1-\theta)},$$

which implies

$$\sum_{k\in\mathbb{Z}} a_k = \sum_{k\in\mathbb{Z}} a_{k+1}$$
$$\leq 2^q \mathsf{c}^{2\theta}\left\{\sum_k b_k\right\}^\theta \left\{\sum_k {a_k}^2\right\}^{1-\theta}$$
$$\leq 2^q \mathsf{c}^{2\theta}\left\{\sum_k b_k\right\}^\theta \left\{\sum_k a_k\right\}^{2(1-\theta)},$$

which implies the theorem. ■

Theorem VI.1.2 (Cheeger's Inequality) *If $\mathfrak{I}_\infty(M) > 0$, then for any function ϕ in C_c^∞ we have*

$$\|\operatorname{grad}\phi\|_2 \geq \frac{1}{2}\mathfrak{I}_\infty(M)\|\phi\|_2.$$

In particular, $\lambda(M) = \inf \operatorname{spec} -\Delta$ *(see* §VII.1 *and* §VII.2.2) *satisfies*

$$\lambda(M) \geq \frac{1}{4}\mathfrak{I}_\infty(M)^2.$$

Proof The inequality for grad f is simply an adjustment of the argument for (VI.1.1) to the case $\nu = \infty$. To obtain the Cheeger inequality for $\lambda(M)$ consider the case $\phi = f^2$, for any $f \in C_c^\infty$, and then use the Cauchy–Schwarz inequality, followed by the characterization of eigenvalues (VII.1.2) and (VII.2.4). ■

Remark VI.1.3 When our manifold M is a relatively compact domain in some larger Riemannian manifold, and M has C^∞ boundary, then our Sobolev constant $\mathfrak{S}_\nu$, and the inequalities that follow from it, correspond to vanishing Dirichlet data on the boundary ∂M. Therefore, in the Cheeger inequality in Theorem VI.1.2, $\lambda(M)$ is the lowest *Dirichlet* eigenvalue of the Laplacian on M. For vanishing Neumann boundary data, we use the definitions of the Sobolev constants $\mathbf{s}_\nu$ of the compact case (below), in which case $\lambda_1(M)$ is the lowest *nonvanishing Neumann* eigenvalue of the Laplacian on M.

VI.2 The Compact Case

Here *compact* means compact without boundary, what was once called a *closed manifold.*

Let M be compact Riemannian, $n \geq 1$ the dimension of M. Then, as mentioned in Remark V.2.1, all isoperimetric constants vanish. Alternatively, by considering the function $f \equiv 1$ on M one has that all the Sobolev constants of M vanish. Nevertheless, one can adjust the definitions as follows (here one only needs, for the geometry, the isoperimetric dimensions n and ∞):

Definition Define the *isoperimetric constant* $\mathbf{i}_n(M)$ by

$$\mathbf{i}_n(M) = \inf_{\Gamma} \frac{A(\Gamma)}{[\min\{V(\Omega_1), V(\Omega_2)\}]^{1-1/n}},$$

where Γ varies over compact $(n-1)$-dimensional C^∞ submanifolds of M that divide M into two disjoint open submanifolds Ω_1, Ω_2 of M.

Define *Cheeger's constant* $\mathbf{i}_\infty(M)$ by

$$\mathbf{i}_\infty(M) = \inf_{\Gamma} \frac{A(\Gamma)}{\min\{V(\Omega_1), V(\Omega_2)\}},$$

where Γ varies over compact $(n-1)$-submanifolds of M described above.

Remark VI.2.1 One has here, as in Remark V.2.5, that it suffices to assume that the open submanifolds Ω_1 and Ω_2 are connected.

Definition For $\nu \in (1, \infty]$ define the *Sobolev constant of* M, $\mathbf{s}_\nu(M)$, by

$$\mathbf{s}_\nu(M) = \inf_f \frac{\|\text{grad}\, f\|_1}{\inf_\alpha \|f - \alpha\|_{\nu/(\nu-1)}}$$

(with the obvious interpretation for $\nu = \infty$), where α varies over $\mathbb{R}$, and f over C^∞. The analogue of the Federer–Fleming theorem (Theorem II.2.1) reads as follows:

(VI.2.1) $$\mathbf{i}_n(M) \leq \mathbf{s_n}(M) \leq 2\mathbf{i}_n(M);$$

and Cheeger's inequality reads as

$$\lambda_1(M) \geq \frac{\mathbf{i}_\infty(M)^2}{4}$$

where $\lambda_1(M)$ denotes the lowest nonzero eigenvalue of M.

Definition Given any function on M, we define its *mean value* f_M by

$$f_M := \frac{1}{V(M)} \int_M f\, dV.$$

Remark VI.2.2 For any $s > 0$, the α_0 that realizes $\inf_\alpha \|f - \alpha\|_{s+1}$ is characterized by

$$\int_M \{\text{sgn}\,(f - \alpha_0)\}|f - \alpha_0|^s = 0.$$

For $\nu = 2$, we have $\inf_\alpha \|f - \alpha\|_2$ realized by the mean value f_M of f over M. For $\nu = \infty$, we have $\inf_\alpha \|f - \alpha\|_1$ realized by that α_0 for which

$$V(\{f > \alpha_0\}) = V(\{f < \alpha_0\}).$$

Definition For M compact, we define the *alternate Sobolev constant* $\mathbf{s}'_\nu(M)$ by

$$\mathbf{s}'_\nu(M) = \inf_f \frac{\|\text{grad } f\|_1}{\|f - f_M\|_{\nu/(\nu-1)}}, \qquad \nu \in [1, \infty],$$

where f varies over $C^\infty(M)$.

Remark VI.2.3 We comment on the relation between $\mathbf{i}_\infty$, $\mathbf{s}_\infty$, $\mathbf{s}'_\infty$. Clearly,

$$\mathbf{s}'_\infty \le \mathbf{s}_\infty.$$

Suppose we are given a disconnection of $M = \Omega_1 \cup \Omega_2 \cup \Gamma$, where Ω_1, Ω_2 are domains in M and Γ is a compact $(n-1)$-submanifold of M. Let Γ_ϵ, $\epsilon > 0$, denote all points in M whose distance from Γ is less than ϵ, and assume $V(\Omega_1) \le V(\Omega_2)$. Let f_ϵ be the function on M which is equal to 1 on $\Omega_1 \setminus \Gamma_\epsilon$, equal to -1 on $\Omega_2 \setminus \Gamma_\epsilon$, and linear across Γ on Γ_ϵ. Then a standard argument shows that

$$\lim_{\epsilon \downarrow 0} \int_M |\text{grad } f_\epsilon|\, dV = 2A(\Gamma),$$

and

$$\begin{aligned}\int_M |f_\epsilon - \alpha|\, dV &\ge |1 - \alpha|\{V(\Omega_1) - \text{const.}\delta_\epsilon\} + |1 + \alpha|\{V(\Omega_2) - \text{const.}\delta_\epsilon\} \\ &\ge \{|1 - \alpha| + |1 + \alpha|\}\{V(\Omega_1) - \text{const.}\delta_\epsilon\}, \\ &\ge 2V(\Omega_1) - \text{const.}\delta_\epsilon,\end{aligned}$$

where $\delta_\epsilon \to 0$ as $\epsilon \downarrow 0$, which implies

$$\inf_\alpha \int_M |f_\epsilon - \alpha|\, dV \ge 2V(\Omega_1) - \text{const.}\delta_\epsilon,$$

which implies $\mathbf{s}_\infty \le A(\Gamma)/V(\Omega_1)$; so

$$\mathbf{s}_\infty \le \mathbf{i}_\infty.$$

Now given *any* function $f \in C^\infty(M)$, pick the constant β so that, for

$$\Omega_1 = \{f > \beta\}, \qquad \Omega_2 = \{f < \beta\},$$

we have

$$V(\Omega_1) = V(\Omega_2)$$

(see Remark VI.2.2 above); and let

$$D_t = \{q \in \Omega_1 : f(q) > \beta + t\}.$$

Then $V(D_t) \leq V(M)/2$ for all $t > 0$, which implies, by the co-area formula (I.3.7) and Cavalieri's principle (I.3.3),

$$\begin{aligned} \int_{\Omega_1} |\mathrm{grad}\,(f-\beta)|\,dV &= \int_0^\infty A(\partial D_t)\,dt \\ &\geq \mathbf{i}_\infty \int_0^\infty V(D_t)\,dt = \mathbf{i}_\infty \int_{\Omega_1} |f-\beta|\,dV. \end{aligned}$$

The same argument yields

$$\int_{\Omega_2} |\mathrm{grad}\,(f-\beta)|\,dV \geq \mathbf{i}_\infty \int_{\Omega_2} |f-\beta|\,dV,$$

which implies

$$\begin{aligned} \int_{\Omega} |\mathrm{grad}\,f|\,dV &= \int_{\Omega} |\mathrm{grad}\,(f-\beta)|\,dV \\ &\geq \mathbf{i}_\infty \int_{\Omega} |f-\beta|\,dV \geq \inf_\alpha \mathbf{i}_\infty \int_{\Omega} |f-\alpha|\,dV, \end{aligned}$$

which implies $\mathbf{s}_\infty \geq \mathbf{i}_\infty$; so

$$\mathbf{s}_\infty = \mathbf{i}_\infty.$$

On the other hand, if we work with f_M instead of β, then we are only guaranteed that at least one of $\{f - f_M > 0\}$, $\{f - f_M < 0\}$ has volume $\leq V(M)/2$. The same argument then yields

$$\frac{\mathbf{i}_\infty}{2} \leq \mathbf{s}'_\infty.$$

In sum, we have

$$\frac{\mathbf{i}_\infty}{2} \leq \mathbf{s}'_\infty \leq \mathbf{s}_\infty = \mathbf{i}_\infty. \tag{VI.2.2}$$

Remark VI.2.4 If M is not closed, but has compact closure and C^∞ boundary, then the above apparatus of isoperimetric and Sobolev constants is still well defined, with the above inequalities. We only have to note that $\lambda_1(M)$ in Cheeger's inequality is now replaced by $\mu_1(M)$, the first nonzero Neumann eigenvalue of M.

VI.3 Faber–Krahn Inequalities

One may easily check that, for $M = \mathbb{R}^n$, $n \geq 1$, one may express the Faber–Krahn inequality, Theorem III.3.1, as the analytic inequality

$$\lambda(\Omega) \geq \mathsf{C}_n V(\Omega)^{-2/n}$$

for relatively compact Ω in $\mathbb{R}^n$ with C^∞ boundary, where C_n depends only on the dimension n, with equality if and only if Ω is an n-disk in $\mathbb{R}^n$ [in particular, $\mathsf{C}_n = \lambda(B)V(B)^{2/n}$, where B denotes an n-disk in $\mathbb{R}^n$]. The bottom of the spectrum, $\lambda(\Omega)$, is discussed in VII.1 and VII.2.2. For general Riemannian manifolds, we do not even hope for precise geometric or eigenvalue inequalities. Nonetheless, we can consider qualitative versions of the Faber–Krahn inequality, in the spirit of the geometric isoperimetric inequality.

Lemma VI.3.1 *Let $\nu > 2$. Then*

(a) The Nirenberg–Sobolev inequality

$$\|\operatorname{grad} u\|_2 \geq \alpha \|u\|_{2\nu/(\nu-2)},$$

for all $u \in C_c^\infty$, implies the Faber–Krahn inequality

$$\lambda(\Omega) \geq \alpha^2 V(\Omega)^{-2/\nu}$$

for relatively compact domains Ω in M with C^∞ boundary.

(b) The Nash–Sobolev inequality

$$\|\operatorname{grad} u\|_2 \geq \beta \|u\|_2{}^{1+2/\nu} \|u\|_1{}^{-2/\nu},$$

for all $u \in C_c^\infty$, implies the Faber–Krahn inequality

$$\lambda(\Omega) \geq \beta^2 V(\Omega)^{-2/\nu}$$

for relatively compact domains Ω in M with C^∞ boundary.

Proof Let $u > 0$ be an eigenfunction of $\lambda(\Omega)$, $\|u\| = 1$ (see Proposition VII.4.1). Then

$$\lambda = \|\operatorname{grad} u\|_2^2 \geq \alpha^2 \|u\|_{2\nu/(\nu-2)}{}^2.$$

Also,

$$1 = \int u^2 \leq \left\{\int u^{2\nu/(\nu-2)}\right\}^{(\nu-2)/\nu} \left\{\int 1^{\nu/2}\right\}^{2/\nu},$$

which implies

$$\|u\|_{2\nu/(\nu-2)}{}^2 \geq V(\Omega)^{-2/\nu},$$

which implies the claim.

Again, $u > 0$ be an eigenfunction of $\lambda(\Omega)$, $\|u\| = 1$. Then $\lambda \geq \beta^2 \|u\|_1{}^{-4/\nu}$, and $\int u \leq \|u\|_2 V(\Omega)^{1/2} = V(\Omega)^{1/2}$, which implies the claim. ∎

Remark VI.3.1 We refer to the Faber–Krahn inequality of this lemma as a *Faber–Krahn inequality of ν-Euclidean type.*

Theorem VI.3.1 *The Riemannian manifold M satisfies a Faber–Krahn inequality of ν-Euclidean type, ν > 2, if and only if it satisfies the corresponding Nirenberg– and Nash–Sobolev inequalities.*

Proof We already know the "if" part of the theorem, so we only consider "only if." Let

$$q = \frac{2\nu}{\nu - 2},$$

and, for any relatively compact Ω in M with smooth boundary, set

$$\alpha_\nu(\Omega) = \lambda(\Omega)V(\Omega)^{2/\nu}, \qquad \mu_\nu(\Omega) = \inf_{\phi \in C_c^\infty(\Omega)} \|\operatorname{grad}\phi\|^2/\|\phi\|_q{}^2.$$

It suffices to show that there exists const. > 0 such that

(VI.3.1) $$\alpha_\nu(\Omega) \leq \text{const.}\mu_\nu(\Omega)$$

for all Ω. ∎

Proposition VI.3.1 *There exists a bounded* $u \in C^\infty(\Omega) \cap \mathfrak{H}_c(\Omega)$ *such that:* (i) $u > 0$, (ii) $\int_\Omega u^q \, dV = 1$, (iii) $\Delta u = -\mu_\nu(\Omega)u^{q-1}$.

Remark VI.3.2 For the definition of $\mathfrak{H}_c$ see Remark I.3.3 and the discussion in §VII.2.2.

Lemma VI.3.2 *Assume M satisfies the Faber–Krahn inequality of ν-Euclidean type. Let u satisfy Proposition* VI.3.1. *Then there exists* const.$_\nu > 0$ *such that*

$$V(\{u > \|u\|_\infty - t\}) \geq \text{const.}_\nu \left\{ \frac{\alpha_\nu}{\mu_\nu} \frac{t}{\|u\|_\infty{}^{q-1}} \right\}^{\nu/2} \quad \forall t \in [0, \|u\|_\infty] \cap \text{regval}_u,$$

where regval_u *denotes the set of regular values of u,* $\alpha_\nu = \alpha_\nu(\Omega)$, $\mu_\nu = \mu_\nu(\Omega)$.

Proof Set

$$L = \|u\|_\infty, \qquad \Omega_t = \{u > L - t\}.$$

Then

$$\lambda(\Omega_t) \geq \alpha_\nu V(\Omega_t)^{-2/\nu},$$

and

$$\begin{aligned}\lambda(\Omega_t) &\leq \frac{\int_{\Omega_t} |\operatorname{grad} u|^2}{\int_{\Omega_t} (u - L + t)^2} \\ &= -\frac{\int_{\Omega_t} (u - L + t)\Delta u}{\int_{\Omega_t} (u - L + t)^2} \\ &\leq \mu_\nu(\Omega) L^{q-1} \frac{\|u - L + t\|_{1,\Omega_t}}{\|u - L + t\|_{2,\Omega_t}{}^2}.\end{aligned}$$

Now

$$\int_{\Omega_t} \{u - L + t\}^2 \geq \int_{\Omega_{t/2}} \{u - L + t\}^2 \geq (t/2)^2 V(\Omega_{t/2}),$$

which implies

$$\alpha_\nu V(\Omega_t)^{-2/\nu} \leq \mu_\nu L^{q-1} \frac{2}{t} \left\{ \frac{V(\Omega_t)}{V(\Omega_{t/2})} \right\}^{1/2},$$

which implies

$$\frac{\alpha_\nu}{\mu_\nu L^{q-1}} \frac{t}{2} V(\Omega_{t/2})^{1/2} \leq V(\Omega_t)^{1/2+2/\nu} = V(\Omega_t)^{(\nu+4)/2\nu}.$$

Therefore,

$$\left\{ \frac{\alpha_\nu}{\mu_\nu L^{q-1}} \right\}^{\frac{2\nu}{\nu+4}} \left(\frac{t}{2} \right)^{\frac{2\nu}{\nu+4}} V(\Omega_{t/2})^{\frac{\nu}{\nu+4}} \leq V(\Omega_t),$$

which implies

$$\begin{aligned}V(\Omega_t) &\geq \left\{ \frac{\alpha_\nu}{\mu_\nu L^{q-1}} \right\}^{\frac{2\nu}{\nu+4}} \left(\frac{t}{2} \right)^{\frac{2\nu}{\nu+4}} V(\Omega_{t/2})^{\frac{\nu}{\nu+4}} \\ &\geq \left\{ \frac{\alpha_\nu}{\mu_\nu L^{q-1}} \right\}^{\frac{2\nu}{\nu+4}} \left(\frac{t}{2} \right)^{\frac{2\nu}{\nu+4}} \left\{ \frac{\alpha_\nu}{\mu_\nu L^{q-1}} \right\}^{\frac{2\nu}{\nu+4}\frac{\nu}{\nu+4}} \left(\frac{t}{2^2} \right)^{\frac{2\nu}{\nu+4}\frac{\nu}{\nu+4}} V(\Omega_{t/2^2})^{\left(\frac{\nu}{\nu+4}\right)^2} \\ &\geq \left\{ \frac{\alpha_\nu t}{\mu_\nu L^{q-1}} \right\}^{2\sum_{j=1}^{\ell} (\nu/(\nu+4))^j} 2^{-2\sum_{j=1}^{\ell} j(\nu/(\nu+4))^j} \left\{ V(\Omega_{t/2^\ell}) \right\}^{\left(\frac{\nu}{\nu+4}\right)^\ell}.\end{aligned}$$

Also,

$$\sum_{j=1}^{\infty} \left(\frac{\nu}{\nu+4} \right)^j = \frac{\nu}{4}, \qquad \sum_{j=1}^{\infty} j \left(\frac{\nu}{\nu+4} \right)^j = \text{const.}_\nu.$$

Next, fix $o \in \Omega$ that realizes $\|\operatorname{grad} u\|_\infty = \max |\operatorname{grad} u| := \delta$. Then

$$\|u\|_\infty - u(x) \leq \delta d(x, o);$$

therefore,

$$\delta d(x, o) \leq \frac{t}{2^\ell} \quad \Rightarrow \quad x \in \Omega_{t/2^\ell},$$

which implies

$$V\left(\Omega_{t/2^\ell}\right) \geq V(o; t/\delta 2^\ell) \sim \boldsymbol{\omega}_{\mathbf{n}} \left(\frac{t}{\delta 2^\ell}\right)^n$$

as $\ell \to \infty$. Thus,

$$\lim_{\ell \to \infty} V\left(\Omega_{t/2^\ell}\right)^{(\nu/(\nu+4))^\ell} = 1,$$

which implies the claim of the lemma. ■

Conclusion of the Proof of Theorem VI.3.1 Given u of Proposition VI.3.1, then, by Cavalieri's principle (Proposition I.3.3),

$$\begin{aligned}
1 &= \int u^q \, dV \\
&= \int_0^L qV(u > t) t^{q-1} \, dt \\
&= \int_0^L qV(u > L - t)(L - t)^{q-1} \, dt \\
&\geq \text{const.}_\nu \left(\frac{\alpha_\nu}{\mu_\nu}\right)^{\nu/2} \frac{1}{L^{(q-1)\nu/2}} \int_0^L t^{\nu/2} (L - t)^{q-1} \, dt \\
&= \text{const.}_\nu \left(\frac{\alpha_\nu}{\mu_\nu}\right)^{\nu/2} L^{-(q-1)\nu/2 + \nu/2 + q} \\
&= \text{const.}_\nu \left(\frac{\alpha_\nu}{\mu_\nu}\right)^{\nu/2},
\end{aligned}$$

since $-(q-1)\nu/2 + \nu/2 + q = 0$.

Thus Faber–Krahn implies Nirenberg–Sobolev, which implies Nash–Sobolev. ■

All the analytic isoperimetric inequalities considered thus far require $\nu \geq n$ (check!), but a simple way to deal with $\nu < n$ goes as follows:

Definition Let M be an arbitrary Riemannian manifold. Given a positive increasing function $g(\nu)$, $\nu \geq 0$. We say that a domain Ω in M satisfies a *geometric* *g-isoperimetric inequality* if

$$A(\partial D) \geq g(V(D))$$

for all domains $D \subset\subset \Omega$.

Given a positive decreasing function $\Lambda(v)$, $v \geq 0$. We say that a domain Ω in M satisfies an *eigenvalue* Λ*-isoperimetric inequality* if

$$\lambda(D) \geq \Lambda(V(D))$$

for all $D \subset\subset \Omega$.

Remark VI.3.3 Thus, Λ-isoperimetric inequalities are generalizations of Faber–Krahn inequalities of Euclidean type.

Theorem VI.3.2 *Suppose Ω satisfies a geometric g-isoperimetric inequality, with $g(v)/v$ a decreasing function of v. Then Ω satisfies an eigenvalue Λ-isoperimetric inequality with*

$$\Lambda(v) = \frac{1}{4}\left(\frac{g(v)}{v}\right)^2.$$

Proof This is the usual proof of Cheeger's inequality. For any $u \in C_c^\infty(D)$ we apply the co-area formula to u^2. We define

$$D(\tau) = \{x \in D : |u(x)|^2 > \tau\},$$

and

$$V(\tau) = V(D(\tau)), \qquad A(\tau) = A(\partial D(\tau)).$$

Then

$$\begin{aligned}
2\|u\|_2\|\text{grad}\, u\|_2 &\geq \int_D |\text{grad}\,(u^2)|\, dV \\
&= \int_0^\infty A(\tau)\, d\tau \\
&\geq \int_0^\infty g(V(\tau))\, d\tau \\
&\geq \int_0^\infty \frac{g(V(\iota))}{V(\tau)} V(\tau)\, d\tau \\
&\geq \frac{g(V(D))}{V(D)} \int_0^\infty V(\tau)\, d\tau \\
&= \frac{g(V(D))}{V(D)} \|u\|_2{}^2,
\end{aligned}$$

which implies, by (VII.1.2) and (VII.2.4),

$$\lambda(D) \geq \frac{1}{4}\left(\frac{g(V(D))}{V(D)}\right)^2,$$

which implies the theorem. ∎

Example VI.3.1 Let M be a complete Riemannian manifold with bounded geometry, and assume $\mathfrak{I}_{\nu,\rho}(M) > 0$ for some $\nu \geq 1$, $\rho > 0$. Then, by Theorem V.2.6, M satisfies a geometric g-isoperimetric inequality, with $g(v)$ given by

$$g(v) = \text{const.}\begin{cases} v^{1-1/n}, & v \leq v_0, \\ \text{const.}_{v_0} v^{1-1/\nu} & v \geq v_0, \end{cases}$$

Then Theorem VI.3.2 implies that M satisfies a eigenvalue Λ-isoperimetric inequality, where

$$\Lambda(v) = \text{const.}\begin{cases} v^{-2/n}, & v \leq v_0, \\ \text{const.}_{v_0} v^{-2/\nu}, & v \geq v_0. \end{cases}$$

Example VI.3.2 Similarly, let M be a complete Riemannian manifold with bounded geometry and positive Cheeger constant, that is, $\mathfrak{I}_\infty(M) > 0$. Then, by the argument of Theorem V.2.6, M satisfies a geometric g-isoperimetric inequality, with $g(v)$ given by

$$g(v) = \text{const.}\begin{cases} v^{1-1/n}, & v \leq v_0, \\ {v_0}^{1-1/n}, & v \geq v_0. \end{cases}$$

Then Theorem VI.3.2 implies that M satisfies a eigenvalue Λ-isoperimetric inequality, where

$$\Lambda(v) = \text{const.}\begin{cases} v^{-2/n}, & v \leq v_0, \\ {v_0}^{-2/n}, & v \geq v_0. \end{cases}$$

We shall consider a variant of the Nirenberg–Sobolev inequality in §VIII.4.

VI.4 The Federer–Fleming Theorem: The Discrete Case

We now formulate and prove the discrete Federer–Fleming theorem. But first,

Definition (Recall.) To every function f on a graph **G** we associate its *differential*, defined on $\mathcal{G}_e$ the oriented edges of **G**, by

$$\mathfrak{D}f([\xi,\eta]) = f(\eta) - f(\xi).$$

Theorem VI.4.1 (Discrete Co-area Formula) *Let $f : \mathcal{G} \to [0, +\infty)$ have finite support. Denote the collection of its values by $f(\mathcal{G}) := \{0 = \beta_0 < \beta_1 < \cdots < \beta_N\}$. To each $i \in 0, \ldots, N$ associate $\mathbf{K}_i$ the subgraph of $\mathbf{G}$ determined by the vertices*

$$\mathcal{K}_i := \{\xi \in \mathcal{G} : f(\xi) \geq \beta_i\}.$$

Then

$$\int_{\mathcal{G}_e} |\mathfrak{D} f| \, d\mathsf{A} = 2 \sum_{i=1}^{N} \mathsf{A}(\partial \mathbf{K}_i)(\beta_i - \beta_{i-1}).$$

Proof First note that $\partial \mathbf{K}_i = \{[\xi, \eta] : f(\xi) \geq \beta_i > f(\eta)\}$; so

$$f(\xi) \geq \beta_i, \qquad \text{and} \qquad f(\eta) = \beta_{i-k}$$

for some $k \in \{1, \ldots, i\}$. Therefore, for any such $[\xi, \eta] \in \partial \mathbf{K}_i$ we have

$$[\xi, \eta] \in \partial \mathbf{K}_i \cap \partial \mathbf{K}_{i-1} \cap \cdots \cap \partial \mathbf{K}_{i-(k-1)};$$

which implies

$$\begin{aligned}
\int_{\mathcal{G}_e} |\mathfrak{D} f| \, d\mathsf{A} &= \sum_{[\xi,\eta]} |f(\eta) - f(\xi)| \\
&= 2 \sum_{i=1}^{N} \sum_{\xi \in f^{-1}[\beta_i]} \sum_{\eta \in \mathsf{N}(\xi) : f(\eta) < \beta_i} \{f(\xi) - f(\eta)\} \\
&= 2 \sum_{i=1}^{N} \sum_{\xi \in f^{-1}[\beta_i]} \sum_{\eta \in \mathsf{N}(\xi) : f(\eta) < \beta_i} \sum_{j=1}^{k(\eta)} (\beta_{i-(j-1)} - \beta_{i-j}) \\
&= 2 \sum_{i=1}^{N} \sum_{[\xi,\eta] \in \partial \mathbf{K}_i} (\beta_i - \beta_{i-1}) \\
&= 2 \sum_{i=1}^{N} \mathsf{A}(\partial \mathbf{K}_i)(\beta_i - \beta_{i-1}),
\end{aligned}$$

which is the claim. ∎

Lemma VI.4.1 *Suppose we are given the real number $p > 1$. Then there exist positive constants c_1 and c_2 such that for all $0 < y < x$ we have*

$$c_1 \leq \frac{x^p - y^p}{(x - y)(x + y)^{p-1}} \leq c_2.$$

Proof First,

$$\frac{x^p - y^p}{(x-y)(x+y)^{p-1}} \le \frac{px^{p-1}}{(x+y)^{p-1}} \le \frac{px^{p-1}}{x^{p-1}} = p;$$

so we may choose $c_2 = p$. Next, fix any $\alpha > 1$. If $y < x \le \alpha y$, then

$$\frac{x^p - y^p}{(x-y)(x+y)^{p-1}} \ge \frac{py^{p-1}}{(x+y)^{p-1}} \ge \frac{p(x/\alpha)^{p-1}}{(2x)^{p-1}} = \frac{p}{(2\alpha)^{p-1}};$$

and if $\alpha y \le x$, then

$$x^p - y^p \ge (1-\alpha^{-p})x^p, \qquad x - y \le x, \qquad x + y \le (1+\alpha^{-1})x,$$

which implies

$$\frac{x^p - y^p}{(x-y)(x+y)^{p-1}} \ge \frac{1-\alpha^{-p}}{1+\alpha^{-1}}.$$

So we may pick

$$c_1 = \min\left\{\frac{p}{(2\alpha)^{p-1}}, \frac{1-\alpha^{-p}}{1+\alpha^{-1}}\right\}.$$

This proves the lemma. ■

Lemma VI.4.2 *Let α_j be a decreasing sequence of nonnegative numbers, and β_j an increasing sequence of nonnegative numbers, with $\beta_0 = 0$. Then for any $p > 1$ we have*

$$\text{(VI.4.1)} \qquad \left\{\sum_{j=1}^{N} \alpha_j{}^{1/p}(\beta_j - \beta_{j-1})\right\}^p \ge \text{const.}_p \sum_{j=1}^{N} \alpha_j\left(\beta_j{}^p - \beta_{j-1}{}^p\right)$$

for all $N = 1, 2, \ldots$.

Proof If $N = 1$, then both the right and left hand sides of (VI.4.1) are equal to $\alpha_1\beta_1{}^p$. So (VI.4.1) is valid for $N = 1$. For the induction step we have

$$\left\{\sum_{j=1}^{N+1} \alpha_j{}^{1/p}(\beta_j - \beta_{j-1})\right\}^p - \left\{\sum_{j=1}^{N} \alpha_j{}^{1/p}(\beta_j - \beta_{j-1})\right\}^p$$

$$\ge c_1\alpha_{N+1}{}^{1/p}(\beta_{N+1} - \beta_N)$$

$$\cdot\left\{\sum_{j=1}^{N+1} \alpha_j{}^{1/p}(\beta_j - \beta_{j-1}) + \sum_{j=1}^{N} \alpha_j{}^{1/p}(\beta_j - \beta_{j-1})\right\}^{p-1}$$

$$\ge c_1\alpha_{N+1}{}^{(p-1)/p}\alpha_{N+1}{}^{1/p}(\beta_{N+1} - \beta_N)$$

$$\cdot\left\{\sum_{j=1}^{N+1}(\beta_j - \beta_{j-1}) + \sum_{j=1}^{N}(\beta_j - \beta_{j-1})\right\}^{p-1}$$

$$= c_1\{\beta_{N+1} + \beta_N\}^{p-1}\alpha_{N+1}(\beta_{N+1} - \beta_N)$$
$$\geq \frac{c_1}{c_2}\alpha_{N+1}\left(\beta_{N+1}{}^p - \beta_N{}^p\right)$$
$$= \frac{c_1}{c_2}\left\{\sum_{j=1}^{N+1}\alpha_j\left(\beta_j{}^p - \beta_{j-1}{}^p\right) - \sum_{j=1}^{N}\alpha_j\left(\beta_j{}^p - \beta_{j-1}{}^p\right)\right\}.$$

Therefore, if for any N we have

$$\left\{\sum_{j=1}^{N}\alpha_j{}^{1/p}(\beta_j - \beta_{j-1})\right\}^p \geq \frac{c_1}{c_2}\sum_{j=1}^{N}\alpha_j\left(\beta_j{}^p - \beta_{j-1}{}^p\right),$$

then we also have

$$\left\{\sum_{j=1}^{N+1}\alpha_j{}^{1/p}(\beta_j - \beta_{j-1})\right\}^p$$
$$\geq \left\{\sum_{j=1}^{N}\alpha_j{}^{1/p}(\beta_j - \beta_{j-1})\right\}^p$$
$$+ \frac{c_1}{c_2}\left\{\sum_{j=1}^{N+1}\alpha_j\left(\beta_j{}^p - \beta_{j-1}{}^p\right) - \sum_{j=1}^{N}\alpha_j\left(\beta_j{}^p - \beta_{j-1}{}^p\right)\right\}$$
$$\geq \frac{c_1}{c_2}\sum_{j=1}^{N+1}\alpha_j\left(\beta_j{}^p - \beta_{j-1}{}^p\right),$$

which is the claim. ∎

Definition Given a graph **G**. Then, for any $\nu > 1$, the *Sobolev constant* $\mathsf{S}_\nu(\mathbf{G})$ is defined by

$$\mathsf{S}_\nu(\mathbf{G}) = \inf_f \|\mathfrak{D} f\|_1/\|f\|_{\nu/(\nu-1)},$$

where f ranges over functions on $\mathcal{G}$ with finite support, and the L^p norms are taken with respect to $d\mathsf{V}$ and $d\mathsf{A}$ on $\mathcal{G}$ and $\mathcal{G}_e$, respectively. (We have already defined $\mathfrak{D} f$ above.)

Theorem VI.4.2 (Discrete Federer–Fleming Theorem) *For any $\nu > 1$, we have $\mathsf{I}_\nu(\mathbf{G}) > 0$ if and only if $\mathsf{S}_\nu(\mathbf{G}) > 0$.*

Proof Given a finite subgraph **K** of **G**, consider the indicator function of $\mathcal{K}$, $\mathcal{I}_\mathcal{K}$. Then

$$\|\mathfrak{D}\mathcal{I}_\mathcal{K}\|_1 = 2\mathsf{A}(\partial\mathbf{K}), \qquad \|\mathcal{I}_\mathcal{K}\|_{\nu/(\nu-1)} = \{\mathsf{V}(\mathbf{K})\}^{(\nu-1)/\nu},$$

which implies $\mathsf{S}_\nu(\mathbf{G}) \leq 2\mathsf{A}(\partial\mathbf{K})/\{\mathsf{V}(\mathbf{K})\}^{(\nu-1)/\nu}$ for all such choices of **K**, which implies $\mathsf{S}_\nu(\mathbf{G}) \leq 2\mathsf{I}_\nu(\mathbf{G})$. Therefore, if $\mathsf{S}_\nu(\mathbf{G}) > 0$, then $\mathsf{I}_\nu(\mathbf{G}) > 0$. To show the opposite direction, we first note that

$$\|\mathfrak{D} f\|_1 \geq \|\mathfrak{D}|f|\|_1,$$

since $|a-b| \geq ||a|-|b||$ for all real a and b; therefore it suffices to consider the case where f is nonnegative. Assume that $\mathsf{I}_\nu(\mathbf{G}) > 0$. Then for f nonnegative we have

$$\begin{aligned}
\int_{\mathcal{G}_e} |\mathfrak{D} f|\, d\mathsf{A} &= 2\sum_{i=1}^{N} \mathsf{A}(\partial \mathbf{K}_i)(\beta_i - \beta_{i-1}) \\
&\geq 2\mathsf{I}_\nu(\mathbf{G}) \sum_{i=1}^{N} \{\mathsf{V}(\mathbf{K}_i)\}^{(\nu-1)/\nu}(\beta_i - \beta_{i-1}) \\
&\geq \text{const.}\mathsf{I}_\nu(\mathbf{G}) \left\{ \sum_{i=1}^{N} \mathsf{V}(\mathbf{K}_i)\left\{\beta_i{}^{\nu/(\nu-1)} - \beta_{i-1}{}^{\nu/(\nu-1)}\right\}\right\}^{(\nu-1)/\nu} \\
&= \text{const.}\mathsf{I}_\nu(\mathbf{G})\|f\|_{\nu/(\nu-1)}
\end{aligned}$$

– the third line follows from the argument of Lemma VI.4.2. Therefore, $\mathsf{I}_\nu(\mathbf{G}) > 0$ implies $\mathsf{S}_\nu(\mathbf{G}) > 0$. ■

Remark VI.4.1 The results of §VI.1 above all follow from the positivity of the L^1 Sobolev constant, and integral inequalities. Therefore, since the Federer–Fleming theorem holds for graphs (as far as positivity of the isoperimetric and Sobolev constants are concerned), we have the corresponding versions of the above results for graphs.

VI.5 Sobolev Inequalities and Discretizations

Given a graph $\mathbf{G}$ with bounded geometry, let $\mathcal{K}$ be a finite subset of $\mathcal{G}$. Then for any function f on $\mathcal{G}$, and any $s > 0$, we have

$$\sum_{[\xi,\eta]\in\mathcal{K}_e\cup\partial\mathcal{K}} |\mathfrak{D} f([\xi,\eta])|^s \asymp \sum_{\xi\in\mathcal{K}} \sum_{\eta\in\mathsf{N}(\xi)} |\mathfrak{D} f([\xi,\eta])|^s;$$

that is, the two expressions are equivalent in that their quotient is contained in a compact subset of $(0, +\infty)$, this subset independent of the choice of $\mathcal{K}$ and f. For convenience define

$$_{\mathsf{V}}|\mathfrak{D} f|_s{}^s(\xi) := \sum_{\eta\in\mathsf{N}(\xi)} |\mathfrak{D} f([\xi,\eta])|^s.$$

In what follows, we will work with $s = 2$, so the notation will not get out of hand. And whenever we write $|\mathfrak{D} f|(\xi)$ we mean $_{\mathsf{V}}|\mathfrak{D} f|_2(\xi)$. One can then work out the details for any other fixed value of s, for example $s = 1$.

Let X be a graph with bounded geometry, Y a metric space, and $\phi : X \to Y$ a rough isometry with rough isometry constants $a > 1, b > 0$. Then card $\phi^{-1}[y]$ is uniformly bounded as y varies over Y. Indeed, if $\phi(x_1) = \phi(x_2)$ then $a^{-1}\mathsf{d}(x_1, x_2) - b \le 0$, which implies that $\mathsf{d}(x_1, x_2)$ is uniformly bounded, independent of y. The bounded geometry hypothesis now implies the claim.

Similarly, if X is a graph with bounded geometry, Y a metric space, and $\phi : X \to Y$ a rough isometry with isometry constants $a > 1, b > 0$, then for every $R > 0, y \in Y$, the number of disks $B(\phi(x); R)$ that cover y is bounded uniformly from above independent of y. Indeed, if $y \in B(\phi(x_1); R) \cap B(\phi(x_2); R)$ then $\mathsf{d}(\phi(x_1), \phi(x_2)) < 2R$, which again implies the claim by the rough isometry hypothesis on ϕ and the bounded geometry hypothesis on X.

Lemma VI.5.1 *Let X be a graph with bounded geometry, Y either a connected graph or a complete Riemannian manifold, and $\phi : X \to Y$ a rough isometry. Let $f : X \to \mathbb{R}$ and $g : Y \to \mathbb{R}$ be nonnegative functions satisfying*

$$f(x) \le \text{const.} \int_{\mathsf{B}(\phi(x);\rho)} g \tag{VI.5.1}$$

for some $\rho > 0$, for all $x \in X$, where B *stands for β or B depending on whether Y is a graph or a Riemannian manifold. Then*

$$\int_{\beta(x:R)} f \le \text{const.} \int_{\mathsf{B}(\phi(x);aR+b+\rho)} g. \tag{VI.5.2}$$

Proof Recall that the constants a and b are from the definition of the rough isometry.

Because $\phi(\beta(x; R)) \subseteq \mathsf{B}(\phi(x); aR + b)$, we only have to count the number of times any y in $\mathsf{B}(\phi(x); aR + b + \rho)$ contributes to the integral on the right in (VI.5.2). Well, all multiple counting of $\phi(x)$ comes from $\phi^{-1}[\phi(x)]$, which has cardinality uniformly bounded from above. For the multiple contribution from y within ρ of $\phi(X)$, we only have to note that $\mathsf{B}(y; \rho) \cap \phi(X)$ has cardinality uniformly bounded from above for all $y \in Y$. ∎

Notation When we have a map from $\phi : X \to Y$, and a function $f : Y \to \mathbb{R}$, we denote by $\phi^* f$ the function defined on X by the usual $\phi^* f(x) = f(\phi(x))$.

Lemma VI.5.2 *Suppose X and Y are graphs, X with bounded geometry, $\phi : X \to Y$ a rough isometry. Let $f : Y \to \mathbb{R}$. Then*

$$\int_{\beta(x;R)} |\phi^* f|^2 \le \text{const.} \int_{\beta(\phi(x);\text{const.}\,R)} |f|^2,$$

$$\int_{\beta(x;R)} |\mathfrak{D}(\phi^* f)|^2 \le \text{const.} \int_{\beta(\phi(x);\text{const.}\,R)} |\mathfrak{D} f|^2$$

for sufficiently large R, independent of x.

Proof We only prove the second claim. We want to estimate $|\mathfrak{D}(\phi^* f)|^2$ from above, for each $x \in X$. Namely,

$$|\mathfrak{D}(\phi^* f)|^2(x) = \sum_{x' \in \mathsf{N}(x)} |(\phi^* f)(x') - (\phi^* f)(x)|^2.$$

Set $a + b := L$. Then $\mathsf{d}(x, x') = 1 \ \Rightarrow\ \mathsf{d}(\phi(x), \phi(x')) \le L$, which implies

$$|\mathfrak{D}(\phi^* f)|^2(x) \le \text{const.} \int_{\beta(\phi(x);L)} |\mathfrak{D} f|^2.$$

Now use Lemma VI.5.1. ∎

Lemma VI.5.3 *Let X be a graph with bounded geometry, and $\Phi : X \to X$ a rough isometry for which $\mathsf{d}(x, \Phi(x))$ is uniformly bounded from above. Then*

$$\int_{\beta(x;R)} |f - \Phi^* f|^2 \le \text{const.} \int_{\beta(x;\text{const. } R)} |\mathfrak{D} f|^2$$

for sufficiently large $R > 0$, independent of x.

Proof Because $\mathsf{d}(x, \Phi(x)) \le K$ for some $K > 0$, for all x, we have

$$|f(x) - f(\Phi(x))|^2 \le \text{const.} \int_{B(x;K)} |\mathfrak{D} f|^2,$$

which implies the claim. ∎

Lemma VI.5.4 *Let $\phi : X \to Y$ be a rough isometry between graphs of bounded geometry, and $f : Y \to \mathbb{R}$ a function on Y. Then, for sufficiently large R,*

$$\int_{\beta(y;R)} |f|^2 \le \text{const.} \int_{\beta(y;\text{const.} R)} |\mathfrak{D} f|^2 + \text{const.} \int_{\beta(\psi(y);\text{const.} R)} |\phi^*(f)|^2,$$

where ψ denotes a rough inverse of ϕ.

Proof Recall that every rough isometry ϕ has a "rough" inverse, that is, there exists a rough isometry $\psi : Y \to X$ such that

$$\mathsf{d}(x, \psi \circ \phi(x)), \qquad \mathsf{d}(y, \phi \circ \psi(y))$$

are uniformly bounded on X and Y respectively. Now for any $y \in Y$, we have

$$\begin{aligned} |f|^2(y) &\le \text{const.}\{|f - (\phi \circ \psi)^*(f)|^2 + |(\phi \circ \psi)^*(f)|^2\}(y) \\ &\le \text{const.} \int_{\beta(y;K)} |\mathfrak{D} f|^2 + \text{const.} |(\phi \circ \psi)^*(f)|^2(y), \end{aligned}$$

which implies

$$\int_{\beta(y;R)} |f|^2 \le \text{const.} \int_{\beta(y;\text{const.}R)} |\mathfrak{D} f|^2 + \text{const.} \int_{\beta(y;R)} |(\psi^* \circ \phi^*)(f)|^2$$
$$\le \text{const.} \int_{\beta(y;\text{const.}R)} |\mathfrak{D} f|^2 + \text{const.} \int_{\beta(\psi(y);\text{const.}R)} |\phi^*(f)|^2,$$

which implies the proposition. ■

Lemma VI.5.5 (Kanai, 1985.) *Let M be a complete Riemannian manifold with Ricci curvature bounded uniformly from below. Then for any $r > 0$ there exists a constant depending on r such that, for any $u \in C^\infty(D(o;r))$,*

$$\int_{B(o;r)} |\text{grad}\, u|\, dV \ge \text{const.}_r \int_{B(o;r)} |u - u_{B(o;r)}|\, dV$$

for all $o \in M$, where $u_{B(o;r)}$ denotes the mean value of u over $B(o;r)$.

Proof Write B for $B(o;r)$. Assume $u_B = 0$ and

$$V(\{x \in B : u(x) > 0\}) \le V(B)/2.$$

For $t > 0$ set $D_t := \{x \in B : u > t\}$. Then, by Buser's inequality (Proposition V.2.6(b)), we have for regular values of u

$$A(\partial D_t) \ge \text{const.}_r V(D_t),$$

which implies, by the co-area formula,

$$\begin{aligned} \int_B |\text{grad}\, u|\, dV &\ge \int_{D_0} |\text{grad}\, u|\, dV \\ &= \int_0^\infty A(\partial D_t)\, dt \\ &\ge \text{const.}_r \int_0^\infty V(D_t)\, dt \\ &= \text{const.}_r \int_{D_0} u\, dV \\ &= \frac{\text{const.}_r}{2} \int_D |u|\, dV, \end{aligned}$$

which is the claim. ■

Given the complete Riemannian manifold M and a discretization $\mathbf{G}$ of M, with bounded geometry, for which the separation radius is ϵ and covering radius is ρ, with $\epsilon \le \rho$. Recall that $\eta \in \mathsf{N}(\xi)$ if $d(\xi, \eta) < 3\rho$. We now investigate the simultaneous validity of the various Sobolev inequalities on M and $\mathbf{G}$. Of course, we need maps between the function spaces on M and $\mathbf{G}$.

VI.5.1 The Discretization $\mathcal{D}$ of Functions on M

Definition Given a smooth function $F : M \to \mathbb{R}$, its *discretization* $f = \mathcal{D}F : \mathcal{G} \to \mathbb{R}$ is defined by

$$f(\xi) = \frac{1}{V(\xi; 3\rho)} \int_{B(\xi;3\rho)} F\, dV, \qquad \xi \in \mathcal{G}.$$

($\mathcal{D}$:i) Assume $V(x; 3\rho) \geq$ const. > 0 for all $x \in M$. Then for any $p \in [1, \infty)$ we have

$$|f|^p(\xi) \leq \frac{1}{V(\xi; 3\rho)} \int_{B(\xi;3\rho)} |F|^p\, dV \leq \text{const.} \int_{B(\xi;3\rho)} |F|^p\, dV,$$

which implies (by Lemma VI.5.1)

$$\int_{\beta(\xi;R)} |\mathcal{D}F|^p\, d\mathsf{V} \leq \text{const.} \int_{B(\xi;\text{const.}R+\text{const.})} |F|^p\, dV.$$

So one has $\|\mathcal{D}\|_{p\to p} \leq$ const. for all $p \in [1, \infty]$.

($\mathcal{D}$:ii) Similarly, one has

$$|f|(\xi) \leq \text{const.} \int_{B(\xi;3\rho)} |F|\, dV \leq \text{const.} \int_M |F|\, dV;$$

so $\|\mathcal{D}\|_{1\to\infty} \leq$ const.

($\mathcal{D}$:iii) Assume M has Ricci curvature bounded from below, and assume $V(x; 3\rho) \geq$ const. > 0 for all $x \in M$. Then for $\eta \in \mathsf{N}(\xi)$ we have

$$(\mathcal{D}F)(\xi) - (\mathcal{D}F)(\eta) = \int_{B(\xi;3\rho)\cap B(\eta;3\rho)} \frac{(\mathcal{D}F)(\xi) - (\mathcal{D}F)(\eta)}{V(B(\xi; 3\rho) \cap B(\eta; 3\rho))}\, dV(x).$$

Since $B(\xi; 3\rho) \cap B(\eta; 3\rho) \supseteq B(\text{midpt}_{\xi,\eta}; \rho)$, which implies $V(B(\xi; 3\rho) \cap B(\eta; 3\rho)) \geq$ const., we have

$$\begin{aligned}
&|(\mathcal{D}F)(\xi) - (\mathcal{D}F)(\eta)| \\
&\quad\leq \text{const.} \left\{ \int_{B(\xi;3\rho)\cap B(\eta;3\rho)} |F(x) - (\mathcal{D}F)(\xi)|\, dV(x) \right. \\
&\qquad \left. + \int_{B(\xi;3\rho)\cap B(\eta;3\rho)} |F(x) - (\mathcal{D}F)(\eta)|\, dV(x) \right\} \\
&\quad\leq \text{const.} \left\{ \int_{B(\xi;3\rho)} |F(x) - (\mathcal{D}F)(\xi)|\, dV(x) \right. \\
&\qquad \left. + \int_{B(\eta;3\rho)} |F(x) - (\mathcal{D}F)(\eta)|\, dV(x) \right\}
\end{aligned}$$

$$\leq \text{const.} \left\{ \int_{B(\xi;3\rho)} |\text{grad } F|\, dV + \int_{B(\eta;3\rho)} |\text{grad } F|\, dV \right\}$$

$$\leq \text{const.} \int_{B(\xi;6\rho)} |\text{grad } F|\, dV,$$

(we used Lemma VI.5.5 in the next to last inequality), that is,

$$|(\mathcal{D}F)(\xi) - (\mathcal{D}F)(\eta)| \leq \text{const.} \int_{B(\xi;6\rho)} |\text{grad } F|\, dV$$

for all $\eta \in \mathsf{N}(\xi)$, for all $\xi \in \mathcal{G}$, which implies

$$|\mathfrak{D}(\mathcal{D}F)|^2(\xi) \leq \text{const.} \int_{B(\xi;6\rho)} |\text{grad } F|^2\, dV$$

for all $\xi \in \mathcal{G}$. Then Lemma VI.5.1 implies

$$\int_{\beta(\xi;R)} |\mathfrak{D}(\mathcal{D}F)|^2\, d\mathsf{V} \leq \text{const.} \int_{B(\xi;\text{const.}R+\text{const.})} |\text{grad } F|^2\, dV$$

for all $\xi \in \mathcal{G}$; letting $R \to +\infty$, we have

$$\int_{\mathbf{G}} |\mathfrak{D}(\mathcal{D}F)|^2\, d\mathsf{V} \leq \text{const.} \int_M |\text{grad } F|^2\, dV.$$

VI.5.2 The Smoothing $\mathcal{S}$ of Functions on $\mathcal{G}$

Definition Fix a function $\psi : [0, +\infty) \to [0, 1] \in C_c^\infty([0, 2\rho))$ such that $\psi|[0, \rho] = 1$. With $\xi \in \mathcal{G}$ associate the function $\psi_\xi : M \to [0, 1]$ defined by

$$\psi_\xi(x) = \psi(d(x, \xi)),$$

and define

$$\phi_\xi(x) = \frac{\psi_\xi(x)}{\sum_{\eta \in \mathcal{G}} \psi_\eta(x)}.$$

So the collection $\{\phi_\xi\}_{\xi \in \mathcal{G}}$ is a partition of unity on M subordinate to the locally finite cover $\{B(\xi; \rho)\}_{\xi \in \mathcal{G}}$ of M. Then for each $f : \mathcal{G} \to \mathbb{R}$ define its *smoothing* $F = \mathcal{S}f : M \to \mathbb{R}$ by

$$(\mathcal{S}f)(x) = \sum_{\xi \in \mathcal{G}} \phi_\xi(x) f(\xi).$$

($\mathcal{S}$:i) Then

$$|\mathcal{S}f|(x) \leq \sum_{\xi \in \mathcal{G}} |f(\xi)| \phi_\xi(x) \leq \begin{cases} \|f\|_\infty \sum_\xi \phi_\xi(x) = \|f\|_\infty, \\ \sum_\xi |f(\xi)| \qquad\quad = \|f\|_1. \end{cases}$$

So $\|\mathcal{S}\|_{1\to\infty} \leq 1$ and $\|\mathcal{S}\|_{\infty\to\infty} \leq 1$.

Also, assume the Ricci curvature is bounded from below. Then one has from the Hölder inequality that

$$|\mathcal{S}f|^p(x) \le \sum_{\xi\in\mathcal{G}} |f|^p(\xi)\phi_\xi(x) = \sum_{\xi\in B(x;\rho)} |f|^p(\xi)\phi_\xi(x) \le \sum_{\xi\in B(x;\rho)} |f|^p(\xi),$$

which implies

$$\begin{aligned}\int_{B(o;R)} |\mathcal{S}f|^p\,dV &\le \int_{B(o;R)} \left\{\sum_{\xi\in B(x;2\rho)} |f|^p(\xi)\right\} dV(x)\\ &= \sum_{\xi\in B(o;R+\rho)} |f|^p(\xi)V(\xi;2\rho)\\ &\le \text{const.} \sum_{\xi\in B(o;R+\rho)} |f|^p(\xi)\\ &\le \text{const.} \int_{\beta(\eta_o;\text{const.}R+\text{const.})} |f|^p\,d\mathsf{V}\end{aligned}$$

[the third line follows from the Bishop comparison theorem (V.1.11)], that is,

$$\int_{B(o;R)} |\mathcal{S}f|^p\,dV \le \text{const.} \int_{\beta(\eta_o;\text{const.}R+\text{const.})} |f|^p\,d\mathsf{V},$$

where η_o is a vertex in $\mathcal{G}$ within Riemannian distance ρ of o. In particular, we also have

$$\int_M |\mathcal{S}f|^p\,dV \le \text{const.} \int_{\mathbf{G}} |f|^p\,d\mathsf{V},$$

that is, $\|\mathcal{S}\|_{p\to p} \le \text{const.}$ for all $p \ge 1$.

($\mathcal{S}$:ii) Furthermore, assume $V(x;\rho) \ge \text{const.} > 0$ for all $x \in M$. If $f \ge 0$ then

$$\begin{aligned}\int_M (\mathcal{S}f)^p\,dV &\ge \sum_\xi \int_{B(\xi;\epsilon/2)} (\mathcal{S}f)^p\,dV\\ &\ge \sum_\xi \int_{B(\xi;\epsilon/2)} f^p(\xi){\phi_\xi}^p(x)\,dV(x)\\ &= \sum_\xi \int_{B(\xi;\epsilon/2)} f^p(\xi)\,dV \ge \text{const.} \sum_\xi f^p(\xi)\end{aligned}$$

[by the Bishop–Gromov theorem (V.1.13)], that is, $\|\mathcal{S}f\|_p \ge \text{const.}\|f\|_p$ for all $f \ge 0$.

($\mathcal{S}$:iii) Now consider

$$(\operatorname{grad}\mathcal{S}f)_{|x} = \sum_\xi f(\xi)\operatorname{grad}\phi_{\xi\,|x} = \sum_{\xi\in B(x;2\rho)} f(\xi)\operatorname{grad}\phi_{\xi\,|x}.$$

Given x, there exists $\eta_x \in \mathcal{G} \cap B(x;\rho)$, which implies

$$(\operatorname{grad} \mathcal{S} f)_{|x} = \sum_{\xi \in B(\eta_x;3\rho)} f(\xi) \operatorname{grad} \phi_{\xi\,|x}$$
$$= \sum_{\xi \in B(\eta_x;3\rho)} \{f(\xi) - f(\eta_x)\} \operatorname{grad} \phi_{\xi\,|x}.$$

Therefore

$$|(\operatorname{grad} \mathcal{S} f)|(x) \le \text{const.} \sum_{\xi \in B(\eta_x;3\rho)} |f(\xi) - f(\eta_x)|,$$

which implies for any $x \in B(\eta;\rho)$, $\eta \in \mathcal{G}$,

$$|(\operatorname{grad} \mathcal{S} f)|^2(x) \le \text{const.} \sum_{\xi \in B(\eta;3\rho)} |f(\xi) - f(\eta)|^2 = \text{const.} |\mathfrak{D} f|^2(\eta).$$

One now obtains, by the argument of Lemma VI.5.1,

$$\int_{B(o;R)} |\operatorname{grad} \mathcal{S} f|^2 \, dV \le \text{const.} \int_{\beta(\eta_o;R+1)} |\mathfrak{D} f|^2 \, d\mathsf{V},$$
$$\|\operatorname{grad} \mathcal{S} f\|_2 \le \text{const.} \|\mathfrak{D} f\|_2.$$

VI.5.3 First Smooth, Then Discretize: $\mathcal{DS}$

($\mathcal{DS}$:i) No assumption on M and $\mathbf{G}$. We have

$$(\mathcal{DS} f)(\xi) = \frac{1}{V(\xi;3\rho)} \int_{B(\xi;3\rho)} \mathcal{S} f \, dV$$
$$= \frac{1}{V(\xi;3\rho)} \int_{B(\xi;3\rho)} \sum_{\eta \in B(\xi;3\rho)} f(\eta) \phi_\eta(x) \, dV(x)$$
$$= \sum_{\eta \in B(\xi;3\rho)} f(\eta) \frac{1}{V(\xi;3\rho)} \int_{B(\xi;3\rho)} \phi_\eta(x) \, dV(x)$$
$$\le \sum_{\eta \in B(\xi;3\rho)} f(\eta)$$
$$= \int_{\beta(\xi;1)} f \, d\mathsf{V}.$$

($\mathcal{DS}$:ii) Assume M has Ricci curvature bounded from below, and $V(x;\rho) \ge$ const. for all $x \in M$. Let f be a nonnegative function on $\mathbf{G}$; then

$$(\mathcal{DS} f)(\xi) = \frac{1}{V(\xi;3\rho)} \int_{B(\xi;3\rho)} \mathcal{S} f \, dV$$
$$\ge \text{const.} \int_{B(\xi;\epsilon/2)} \mathcal{S} f \, dV \ge \text{const.} f(\xi),$$

which implies $\|f\|_p \le \text{const.}_p \|\mathcal{DS} f\|_p$ for all $p \ge 1$.

VI.5.4 How Smoothing Followed by Discretization Differs from the Identity: $id_{\mathcal{G}} - \mathcal{DS}$

Note that

$$(f - \mathcal{DS}f)(\xi) = \frac{1}{V(\xi;3\rho)} \int_{B(\xi;3\rho)} \sum_{\eta\in\mathcal{G}} \phi_\eta(x)\{f(\xi) - f(\eta)\}\, dV(x)$$

$$= \frac{1}{V(\xi;3\rho)} \int_{B(\xi;3\rho)} \sum_{\eta\in B(\xi;3\rho)\cap\mathcal{G}} \phi_\eta(x)\{f(\xi) - f(\eta)\}\, dV(x),$$

which implies

$$|f - \mathcal{DS}f|(\xi) \le \sum_{\eta\in B(\xi;3\rho)\cap\mathcal{G}} |f(\xi) - f(\eta)|.$$

Since **G** has bounded geometry, we have

$$|f - \mathcal{DS}f|^2(\xi) \le \text{const.}|\mathfrak{D}f|^2(\xi).$$

Therefore,

$$\int_{\beta(\xi;R)} |f - \mathcal{DS}f|^2\, d\mathsf{V} \le \text{const.} \int_{\beta(\xi;R)} |\mathfrak{D}f|^2\, d\mathsf{V},$$

$$\|f - \mathcal{DS}f\|_2 \le \text{const.}\|\mathfrak{D}f\|_2.$$

VI.5.5 How Discretization Followed by Smoothing Differs from the Identity: $id_M - \mathcal{SD}$

Again, we require both Ricci curvature bounded uniformly from below, and $V(x;\rho) \ge$ const. for all $x \in M$. For a smooth function $F : M \to \mathbb{R}$ we have

$$(F - \mathcal{SD}F)(x) = \sum_{\xi\in B(x;2\rho)\cap\mathcal{G}} \frac{\phi_\xi(x)}{V(\xi;3\rho)} \int_{B(\xi;3\rho)} \{F(x) - F(y)\}\, dV(y),$$

which implies

$$|F - \mathcal{SD}F|(x) \le \text{const.} \int_{B(x;5\rho)} |F(x) - F(y)|\, dV(y).$$

The Bishop comparison theorem (V.1.10) implies

$$|F - \mathcal{SD}F|(x) \le \text{const.} \int_{\mathsf{S}_x} d\mu_x(v) \int_0^{5\rho} |\text{grad}\, F|(\exp sv)\, ds,$$

which implies

$$\int_{B(o;R)} |F - \mathcal{SD}F|^2\, dV \le \text{const.} \int_{\mathsf{S}B(o;R)} d\mu(v) \int_0^{5\rho} |\text{grad}\, F|^2(\exp sv)\, ds,$$

where $\mathsf{S}B(o;R)$ denotes the unit tangent bundle over $B(o;R)$ and $d\mu$ the Liouville measure on $\mathsf{S}B(o;R)$ (see V.1.4). Therefore,

$$\begin{aligned}\int_{B(o;R)} |F - \mathcal{SD}F|^2\, dV &\le \text{const.} \int_{\mathsf{S}B(o;R)} d\mu(v) \int_0^{5\rho} |\text{grad}\, F|^2(\exp sv)\, ds \\ &= \text{const.} \int_0^{5\rho} ds \int_{\mathsf{S}B(o;R)} |\text{grad}\, F|^2(\pi \circ \Phi_s(v))\, d\mu(v),\end{aligned}$$

where $\pi : \mathsf{S}M \to M$ denotes the natural projection, and Φ_s the geodesic flow. By Liouville's theorem on the invariance of the Liouville measure under the action of the geodesic flow (Proposition V.1.3), we have

$$\begin{aligned}\int_{B(o;R)} |F - \mathcal{SD}F|^2\, dV &\le \text{const.} \int_0^{5\rho} ds \int_{\mathsf{S}B(o;R)} |\text{grad}\, F|^2(\pi \circ \Phi_s(v))\, d\mu(v) \\ &= \text{const.} \int_0^{5\rho} ds \int_{\Phi_s(\mathsf{S}B(o;R))} |\text{grad}\, F|^2(\pi(v))\, d\mu(v) \\ &\le \text{const.} \int_{\mathsf{S}B(o;R+5\rho)} |\text{grad}\, F|^2(\pi(v))\, d\mu(v) \\ &= \text{const.} \int_{B(o;R+5\rho)} |\text{grad}\, F|^2\, dV.\end{aligned}$$

To summarize,

$$\begin{aligned}\int_{B(o;R)} |F - \mathcal{SD}F|^2\, dV &\le \text{const.} \int_{B(o;R+5\rho)} |\text{grad}\, F|^2\, dV, \\ \int_M |F - \mathcal{SD}F|^2\, dV &\le \text{const.} \int_M |\text{grad}\, F|^2\, dV.\end{aligned}$$

VI.6 Bibliographic Notes

Surveys of applications of isoperimetric inequalities in analysis can be found in Bandle (1980), Kawohl (1985), Mossino (1984), Payne (1967), and the classic and Pólya and Szegö (1951). Also, see the recent survey (Hebey, 1999).

§VI.1 The Nirenberg–Sobolev inequality can be found in Nirenberg (1959, p. 14). For Lemma VI.1.2, see Moser (1964, p. 116), Cheng and Li (1981), and Nash (1958). For Theorem VI.1.1 see Bakry, Coulhon, Ledoux, and Saloff-Coste (1995).

One might wonder whether the Nirenberg– and Nash–Sobolev inequalities imply the Federer–Fleming inequality, that is, whether the L^2 Sobolev inequalities imply the the L^1 Sobolev inequality. Counterexamples are presented in Coulhon and Ledoux (1994).

Theorem VI.1.2 is from Cheeger (1970).

§VI.2 Details for the analogue of the Federer–Fleming theorem (Federer and Fleming, 1959) in the compact case can be found in Chavel (1984, p. 111).

A more delicate isoperimetric function for the compact case is inspired by M. Gromov's proof of the isoperimetric inequality for spheres (Gromov, 1986). One lets $I(M, \beta)$, $\beta \in (0, 1)$, denote the infimum of $A(\partial\Omega)$ among all Ω satisfying $V(\Omega) = \beta V(M)$. See the disucussion in Bérard (1986, Chapter IV).

§**VI.3** As mentioned in V.4, one has a Faber–Krahn theorem in the model spaces of constant sectional curvature. Similarly, one has generalizations of the Faber–Krahn argument in Bérard (1986, Chapter IV).

Theorem VI.3.1 is from Carron (1996). Proposition VI.3.1 is from Aubin (1982, p. 116 ff.). The general Faber–Krahn inequalities were first treated in Grigor'yan (1994c).

§**VI.4** To my knowledge, the earliest proofs of the Federer–Fleming theorem for graphs are in Varopoulos (1985), where he considers the result there for Cayley graphs, and in Dodziuk (1984), where he considers the discrete Cheeger inequality.

One can consider both Riemannian manifolds and graphs with weight functions associated to their measures. See the discussion in Chavel and Feldman (1991) and Coulhon and Saloff-Coste (1995).

§**VI.5** The discussion of discretization and smoothing of functions follows the treatments of Kanai (1986a, 1986b) and Coulhon (1992).

VII
Laplace and Heat Operators

Here, we introduce the Laplace operator on Riemannian manifolds and its associated heat diffusion, and prepare for the study of how geometric isoperimetric inequalities on Riemannian manifolds are reflected in the properties of large time diffusion. We present the necessary definitions and background results, most of which can be found in Chavel (1984). The ones we discuss in some detail are either to correct an argument, or to fill in matters not discussed, there. The manner in which the heat diffusion expresses the geometry of the manifold will be presented in the next chapter.

VII.1 Self-adjoint Operators and Their Semigroups

Definition Let $\mathcal{H}$ be a Hilbert space with inner product $(,)$. Recall that a linear operator $T : \mathcal{D}(T) \to \mathcal{H}$ on the subspace $\mathcal{D}(T)$ in $\mathcal{H}$ is an *extension of the linear operator of* $S : \mathcal{D}(S) \to \mathcal{H}$ if $\mathcal{D}(S) \subseteq \mathcal{D}(T)$ and $T|\mathcal{D}(S) = S$. We write $S \subset T$.

The linear operator $T : \mathcal{D}(T) \to \mathcal{H}$ is *closed* if every sequence (x_k) in $\mathcal{D}(T)$ satisfying

$$x_k \to x, \qquad Tx_k \to y, \qquad x, y \in \mathcal{H},$$

must also satisfy

$$x \in \mathcal{D}(T), \qquad y = Tx.$$

A linear operator $T : \mathcal{D}(T) \to \mathcal{H}$ is *closable* if it has a closed extension. When T is closable, then it has a minimal closed extension $\overline{T}$, called its *minimal extension* or, for short, its *closure*.

Definition Assume $T : \mathcal{D}(T) \to \mathcal{H}$ is a linear operator with dense domain, that is, $\overline{\mathcal{D}(T)} = \mathcal{H}$. The domain of the adjoint T^* of T will consist of those $x \in \mathcal{H}$

for which there exists an element $x^* \in \mathcal{H}$ such that $(x, Ty) = (x^*, y)$ for all $y \in \mathcal{D}(T)$. For such x we define $T^*x = x^*$.

The operator T is called *symmetric* if $(Tx, y) = (x, Ty)$ for all $x, y \in \mathcal{D}(T)$.

In general, T^* is a closed operator. When T is symmetric, then T^* is an extension of T, so T is closable. Then $\overline{T}$ is the smallest closed symmetric extension of T.

Definition We say that T is *self-adjoint* if $T = T^*$. We say that T is *essentially self-adjoint* if T has a unique self-adjoint extension, in which case $\overline{T} = T^*$.

One knows that if T_1, T_2 are self-adjoint, and $\mathcal{D}(T_1) \subseteq \mathcal{D}(T_2)$, then $T_1 = T_2$.

Definition Let T be a symmetric operator on $\mathcal{H}$ with dense domain $\mathcal{D}(T)$, for which there exists a real number ϵ such that

$$(x, Tx) \geq \epsilon(x, x) \qquad \forall\, x \in \mathcal{D}(T). \tag{VII.1.1}$$

We then say that T is *semibounded from below.* We say T is *nonnegative* if $\epsilon \geq 0$.

Let T be semibounded from below, with ϵ as given in (VII.1.1); then the bilinear form on $\mathcal{D}(T)$ defined by

$$(x, y)_T = (x, Ty) + (1 - \epsilon)(x, y)$$

is positive definite, and hence defines an inner product. Complete $\mathcal{D}(T)$ to D_T relative to the inner product $(,)_T$. Then it is known that D_T may be realized as a subspace of $\mathcal{H}$, so $\mathcal{D}(T) \subseteq D_T \subseteq \mathcal{H}$.

Definition Let T be semibounded from below and symmetric, with dense domain. Define the *Friedrichs extension* $\mathfrak{F}_T$ of T to be $T^*|D_T$. That is, the domain of $\mathfrak{F}_T$, $\mathfrak{D}_T$, is given by

$$\mathfrak{D}_T := \mathcal{D}(\mathfrak{F}_T) = \mathcal{D}(T^*) \cap D_T,$$

and $\mathfrak{F}_T$ is given by

$$(\mathfrak{F}_T x, y) = (x, Ty) \qquad \forall\, x \in \mathfrak{D}_T,\ y \in \mathcal{D}(T).$$

Therefore, $\mathcal{D}(T) \subseteq \mathfrak{D}_T \subseteq D_T \subseteq \mathcal{H}$. Also, $\mathfrak{F}_T$ is self-adjoint, and

$$(\mathfrak{F}_T x, y) = (x, Ty) \qquad \forall\, x \in \mathfrak{D}_T,\ y \in D_T.$$

The full result is:

Proposition VII.1.1 *Every semibounded from below symmetric operator with dense domain has at least one self-adjoint extension, its Friedrichs extension. Moreover, $\mathfrak{F}_T$ is the unique self-adjoint extension of T whose domain is contained in D_T.*

VII.1.1 The Spectrum of Self-adjoint Operators

Definition Let $\mathcal{H}$ be a Hilbert space with inner product $(,)$, $T : \mathcal{D}(T) \to \mathcal{H}$ a linear operator on $\mathcal{H}$.

A complex number λ is *in the resolvent set of T* if $\lambda I - T$ maps $\mathcal{D}(T)$ one to one onto all of $\mathcal{H}$, and if $(\lambda I - T)^{-1}$ is bounded. In this case we refer to $(\lambda I - T)^{-1}$ as the *resolvent of T at λ*.

The *spectrum of T*, spec T, is the complement of the resolvent set of T.

The spectrum is always a closed subset of the complex numbers. When T is self-adjoint (and this is the case we always study), spec T is contained in the real axis. For T self-adjoint we have the *spectral theorem,* formulated as follows:

Recall that an *orthogonal projection* is characterized as a self-adjoint transformation P of $\mathcal{H}$ for which $P^2 = P$. In particular,

$$(P\phi, \phi) = (P^2\phi, \phi) = (P\phi, P\phi) \geq 0$$

for all ϕ. A *spectral family* is a family $\{E_\lambda : \lambda \in \mathbb{R}\}$ of orthogonal projections of $\mathcal{H}$ satisfying

1. $E_\lambda \leq E_\mu$ [that is, $(E_\lambda\phi, \phi) \leq (E_\mu\phi, \phi) \quad \forall\, \phi$] when $\lambda < \mu$;
2. $E_{\lambda+0} = E_\lambda$;
3. $E_\lambda \to 0$ as $\lambda \to -\infty$, and $E_\lambda \to I$ as $\lambda \to +\infty$ (all convergence here is in the operator norm).

Then the spectral theorem states that any self-adjoint transformation T of $\mathcal{H}$ possesses a uniquely determined spectral family $\{E_\lambda : \lambda \in \mathbb{R}\}$ for which the representation

$$T = \int_{-\infty}^{\infty} \lambda\, dE_\lambda$$

is valid. That is, for any $\phi \in \mathcal{D}(T)$ we have

$$(T\phi, \phi) = \int_{-\infty}^{\infty} \lambda\, (dE_\lambda\phi, \phi),$$

where $(dE_\lambda\phi, \phi)$ is now a Lebesgue–Stieljes measure (with respect to λ) on $\mathbb{R}$, supported on spec T. The domain of T, $\mathcal{D}(T)$, consists of those $\phi \in \mathcal{H}$ for which

$$\int_{-\infty}^{\infty} \lambda^2 (dE_\lambda\phi, \phi) < +\infty.$$

Furthermore, for any Borel function $f(\lambda)$, defined a.e.-$[dE_\lambda]$ and finite, we have

$$f(T) = \int_{-\infty}^{\infty} f(\lambda)\, dE_\lambda.$$

This last equation is a theorem if one takes the functional calculus of self-adjoint operators as already well defined, and is a definition if one wishes to first develop the functional calculus of self-adjoint operators from the spectral family formulation of the spectral theorem.

In particular,

$$\|T\phi\|^2 = (T^2\phi, \phi) = \int_{-\infty}^{\infty} \lambda^2 (dE_\lambda\phi, \phi).$$

Moreover, if f and g are two such functions satisfying

$$\mathcal{D}(f(T)g(T)) = \mathcal{D}(g(T)) \cap \mathcal{D}(fg(T)),$$

then

$$f(T)g(T) = \int_{-\infty}^{\infty} f(\lambda)g(\lambda)\, dE_\lambda.$$

Definition Within spec T we distinguish a variety of subsets.

First, a number λ is in the *point spectrum* if λ is an eigenvalue of T, that is, there exists a nontrivial element f of $\mathcal{H}$ for which $Tf = \lambda f$.

Next, assume T is self-adjoint. Then spec $T \subseteq \mathbb{R}$. The *discrete spectrum of* T consists of those $\lambda \in$ spec T for which there exists $\epsilon > 0$ such that

$$\dim (E_{\lambda+\epsilon} - E_{\lambda-\epsilon})(\mathcal{H}) < \infty.$$

So λ is an eigenvalue of T of finite multiplicity, *and* is an isolated element of spec T.

We refer to the complement of the discrete spectrum as the *essential spectrum.*

Therefore, the essential spectrum consists of those $\lambda \in$ spec T such that

$$\dim (E_{\lambda+\epsilon} - E_{\lambda-\epsilon})(\mathcal{H}) = \infty$$

for all $\epsilon > 0$.

Thus $\lambda \in \operatorname{spec} T$ precisely when there exists a sequence $\{\phi_n\} \subseteq T$, $\|\phi_n\| = 1 \; \forall \, n$, such that $(\lambda - T)\phi_n \to 0$ as $n \to \infty$. And λ is in the essential spectrum precisely when the sequence can be chosen to be orthonormal. For any $\lambda \in \operatorname{spec} T$, the sequence $\{\phi_n\}$ is referred to as a *sequence of normalized approximate eigenfunctions of* λ.

The spectral theorem implies that if $\lambda_0 = \inf \operatorname{spec} T > -\infty$, then

$$\begin{aligned}(T\phi, \phi) &= \int_{-\infty}^{\infty} \lambda (dE_\lambda \phi, \phi) \\ &= \int_{\lambda_0}^{\infty} \lambda (dE_\lambda \phi, \phi) \\ &\geq \int_{\lambda_0}^{\infty} \lambda_0 (dE_\lambda \phi, \phi) \\ &= \lambda_0 \|\phi\|^2.\end{aligned}$$

Therefore, T is semibounded from below, with best constant $\epsilon \geq \lambda_0$. Conversely, if T is semibounded from below with constant ϵ, λ is an element of $\operatorname{spec} T$, and $\{\phi_n\}$ is a sequence of normalized approximate eigenfunctions of λ, then

$$0 = \lim_{n \to \infty} ((\lambda I - T)\phi_n, \phi_n) = \lambda - \lim_{n \to \infty} (T\phi_n, \phi_n) \leq \lambda - \epsilon;$$

so the spectrum is bounded from below, with $\epsilon \leq \lambda_0$. We conclude that T is semibounded from below if and only if the spectrum is bounded from below, in which case we have

$$\text{(VII.1.2)} \qquad \inf \operatorname{spec} T = \inf_{\phi \neq 0} \frac{(T\phi, \phi)}{\|\phi\|^2}.$$

Note that the argument only involves those $\phi \in \mathcal{D}(T)$, but the quadratic form $\phi \mapsto (T\phi, \phi)$ is defined on $\mathcal{D}((T - \epsilon)^{1/2})$, and the infimum, above, does not change if we allow ϕ to vary over $\mathcal{D}((T - \epsilon)^{1/2}) \supseteq \mathcal{D}(T - \epsilon) = \mathcal{D}(T)$.

VII.1.2 Quadratic Forms

In defining the Friedrichs extension, above, we started with a semibounded symmetric operator T on a dense domain $\mathcal{D}(T)$, with which we associated a quadratic form $(\,,\,)_T$. We then used the quadratic form $(\,,\,)_T$ to define a self-adjoint extension $\mathfrak{F}_T$ of the original T. Moreover, the domain of the quadratic form is precisely $\mathcal{D}((T - \epsilon)^{1/2})$ [where ϵ is given by (VII.1.1)]. Here we note that one can start with the quadratic form as the fundamental object.

Definition Let $\mathcal{D}$ denote a dense domain in a Hilbert space $\mathcal{H}$. Then a *sesquilinear form* Q' on $\mathcal{D}$ is a map $Q' : \mathcal{D} \times \mathcal{D} \to \mathbb{C}$ such that

1. $Q'(x, y)$ is linear in x,
2. $Q'(x, y) = \overline{Q'(y, x)}$.

We say Q' is *semibounded from below* if there exists $\epsilon \in \mathbb{R}$ such that

$$Q'(x, x) \geq \epsilon \|x\|^2 \tag{VII.1.3}$$

for all $x \in \mathcal{D}$. We say Q' is *nonnegative* if $\epsilon \geq 0$.

Given Q' semibounded from below, with ϵ given by (VII.1.3). Complete $\mathcal{D}$ with respect to the inner product

$$(x, y)_{Q'} = Q'(x, y) + (1 - \epsilon)(x, y).$$

Assume the resulting space may be identified with a closed domain $D_Q \supseteq \mathcal{D}$ in $\mathcal{H}$. Denote the new inner product on D_Q by $(\,,\,)_Q$, and define on D_Q the quadratic form

$$Q(x, y) = (x, y)_Q - (1 - \epsilon)(x, y).$$

Determine the Friedrichs operator $\mathfrak{F}_Q$ by

$$Q(x, y) = (\mathfrak{F}_Q x, y);$$

so the domain $\mathfrak{D}_Q$ of $\mathfrak{F}_Q$ consists of those $x \in D_Q$ for which there exists $x^* \in \mathcal{H}$ such that $(x^*, y) = Q(x, y)$ for all $y \in D_Q$. Then we define $\mathfrak{F}_Q x = x^*$.

One knows that $\mathfrak{F}_Q$ is self-adjoint, and that $D_Q = \mathcal{D}({\mathfrak{F}_Q}^{1/2})$.

VII.1.3 1-Parameter Semigroups

Definition Given a Banach space $\mathcal{B}$, a family $\{T_t : t \geq 0\}$ of bounded linear operators on $\mathcal{B}$ is said to be a C^0 *1-parameter semigroup* (*semigroup*, for short) if:

1. $T_0 = I$;
2. if $0 \leq s, t \leq \infty$, then $T_t T_s = T_{t+s}$;
3. the map $(t, f) \mapsto T_t f$ from $[0, \infty) \times \mathcal{B}$ to $\mathcal{B}$ is jointly continuous.

We say that the semigroup T_t is *contractive* if $\|T_t\| \leq 1$ for all $t \geq 0$.

Given (1) and (2) in the definition of a semigroup, to verify (3) of the definition it suffices to verify that

$$\lim_{t \downarrow 0} T_t f = f$$

for all $f \in \mathcal{B}$.

Definition With every semigroup T_t we associate its (*infinitesmal*) *generator*, Z, defined by

$$Zf = \lim_{t \downarrow 0} \frac{T_t f - f}{t},$$

the *domain* of Z, $\mathcal{D}(Z)$, being those $f \in \mathcal{B}$ for which the limit exists.

Proposition VII.1.2

(a) *$\mathcal{D}(Z)$ is a dense linear subspace of $\mathcal{B}$, and $T_t(\mathcal{D}(Z)) \subseteq \mathcal{D}(Z)$ for all $t \geq 0$. Moreover,*

$$T_t Z = Z T_t$$

on all of $\mathcal{D}(Z)$, for all $t \geq 0$.

(b) *If $f \in \mathcal{D}(Z)$, then $F(t) = T_t f$ is C^1 on $[0, +\infty)$, and*

$$F'(t) = Z F(t).$$

(c) *Furthermore, Z is a closed linear operator, with $\mathcal{D}(Z)$ complete with respect to the norm*

$$\|\!| f \|\!| = ||f|| + ||Zf||.$$

Moreover, T_t acts as a semigroup on $\mathcal{D}(Z)$ for this norm.

(d) *Conversely to (b), if a path $\Phi(t)$ in $\mathcal{D}(Z)$ satisfies*

$$\Phi' = Z\Phi$$

on some interval $[0, \alpha)$, then

$$\Phi(t) = T_t(\Phi(0)) \qquad \forall\, t \in [0, \alpha).$$

Thus, the infinitesmal generator uniquely determines the semigroup.

(e) *A densely defined operator Z on the Banach space $\mathcal{B}$ is the infinitesmal generator of a contractive semigroup if and only if all $\lambda > 0$ lie in the resolvent set of Z and*

$$\|(\lambda - Z)^{-1}\| \leq \lambda^{-1}$$

for all $\lambda > 0$.

Proposition VII.1.3

(a) *If $\mathcal{B}$ is a Hilbert space, and T_t is a self-adjoint semigroup acting on $\mathcal{B}$, that is, T_t is self-adjoint for each $t > 0$, then Z is also self-adjoint. If Z is*

nonnegative self-adjoint, then the spectral theorem implies that $-Z$ *is the infinitesmal generator of a self-adjoint contraction semigroup.*

(b) *If* Z *is a symmetric operator with dense domain* $\mathcal{D}$ *in the Hilbert space* $\mathcal{B}$, *and for every* $f \in \mathcal{D}$ *there exists* $\epsilon = \epsilon(f) > 0$ *such that the heat equation*

$$F'(t) = -ZF(t) \tag{VII.1.4}$$

has a solution satisfying $F(0) = f$ *and* $F(t) \in \mathcal{D}$ *for all* $t \in [0, \epsilon(f)]$, *then* Z *is essentially self-adjoint on* $\mathcal{D}$, *and the solution to* (VII.1.4) *subject to the given conditions is unique.*

VII.2 The Laplacian

VII.2.1 The Laplacian Acting on C^2 Functions

Let M be an n-dimensional Riemannian manifold. If $\mathbf{x}: U \to \mathbb{R}^n$ is a chart on M, then we have

$$g_{ij} = \langle \partial_i, \partial_j \rangle, \qquad G = (g_{ij}), \qquad G^{-1} = (g^{ij}), \qquad g = \det G > 0.$$

Recall that, for any C^1 function f on M, we have

$$\langle \operatorname{grad} f, \xi \rangle = df(\xi) = \xi f \qquad \forall\, \xi \in TM.$$

For C^1 functions $f,\ h$ on M we have

$$\begin{aligned} \operatorname{grad}(f + h) &= \operatorname{grad} f + \operatorname{grad} h, \\ \operatorname{grad} fh &= f \operatorname{grad} h + h \operatorname{grad} f. \end{aligned}$$

If $\mathbf{x}: U \to \mathbb{R}^n$ is a chart on M, then

$$\operatorname{grad} f = \sum_{j,k} \frac{\partial (f \circ \mathbf{x}^{-1})}{\partial x^j} g^{jk} \frac{\partial}{\partial x^k}.$$

Definition For any C^1 vector field X on M we define the *divergence of* X *with respect to the Riemannian metric*, div X, by

$$\operatorname{div} X = \operatorname{tr}(\xi \mapsto \nabla_\xi X).$$

(Recall, ∇ denotes the Levi-Civita connection of the Riemannian metric.)

For the C^1 function f and vector fields $X,\ Y$ on M we have

$$\begin{aligned} \operatorname{div}(X + Y) &= \operatorname{div} X + \operatorname{div} Y, \\ \operatorname{div} fX &= \langle \operatorname{grad} f, X \rangle + f \operatorname{div} X. \end{aligned}$$

If $\mathbf{x}: U \to \mathbb{R}^n$ is a chart on M, and

$$X|U = \sum_j \xi^j \frac{\partial}{\partial x^j},$$

then

$$\operatorname{div} X = \frac{1}{\sqrt{g}} \sum_{j=1}^{n} \frac{\partial(\sqrt{g}\xi^j)}{\partial x^j}.$$

Given $x \in M$, and $r \in [0, c(\xi))$, $\xi \in \mathsf{S}_x$ geodesic spherical coordinates on M about x, with

$$dV = \sqrt{\mathbf{g}}(r;\xi)\, dr\, d\mu_x(\xi)$$

the Riemannian measure in spherical coordinates (where $d\mu_x$ denotes the standard measure on S_x), then for the radial vector field $\partial/\partial r$ we have

$$\operatorname{div} \frac{\partial}{\partial r} = \frac{\partial_r \sqrt{\mathbf{g}}(r;\xi)}{\sqrt{\mathbf{g}}(r;\xi)}.$$

One verifies that if X has compact support on M then we have the Riemannian divergence theorem:

$$\int_M \operatorname{div} X\, dV = 0.$$

In particular, if f is a function and X is a C^1 vector field on M, at least one of which has compact support, then

(VII.2.1) $$\int f \operatorname{div} X\, dV = -\int \langle \operatorname{grad} f, X\rangle\, dV.$$

Definition Let f be a C^2 function on M. Then we define the Laplacian of f, Δf, by

$$\Delta f = \operatorname{div} \operatorname{grad} f.$$

Thus, in a chart $\mathbf{x}: U \to \mathbb{R}^n$,

$$\Delta f = \frac{1}{\sqrt{g}} \sum_{j,k=1}^{n} \frac{\partial}{\partial x^j} \left\{ \sqrt{g} g^{jk} \frac{\partial (f \circ \mathbf{x}^{-1})}{\partial x^k} \right\}.$$

Furthermore, for C^2 functions f and h on M we have

$$\Delta(f + h) = \Delta f + \Delta h,$$
$$\operatorname{div} f \operatorname{grad} h = f \Delta h + \langle \operatorname{grad} f, \operatorname{grad} h\rangle$$

(this last formula only requires that $f \in C^1, h \in C^2$), which implies

$$\Delta fh = f\Delta h + 2\langle \operatorname{grad} f, \operatorname{grad} h\rangle + h\Delta f.$$

One has *Green's formulae*: Let $f : M \to \mathbb{R} \in C^2(M), h : M \to \mathbb{R} \in C^1(M)$, with at least one of them compactly supported. Then

$$\int_M \{h\Delta f + \langle \operatorname{grad} h, \operatorname{grad} f\rangle\}\, dV = 0.$$

If both f and h are C^2, then

$$\int_M \{h\Delta f - f\Delta h\}\, dV = 0.$$

Let M be oriented, Ω a domain in M with C^∞ boundary $\partial\Omega$, and ν the outward unit vector field along $\partial\Omega$ that is pointwise orthogonal to $\partial\Omega$ (there is only one such vector field). Then for any compactly supported C^1 vector field X on M we have

$$\iint_\Omega \operatorname{div} X\, dV = \int_{\partial\Omega} \langle X, \nu\rangle\, dA.$$

The corresponding Green's formulae are: Given M, Ω, and ν as just described, and given $f \in C^2(M)$, $h \in C^1(M)$, at least one of them compactly supported. Then

$$\text{(VII.2.2)} \qquad \iint_\Omega \{h\Delta f + \langle \operatorname{grad} h, \operatorname{grad} f\rangle\}\, dV = \int_{\partial\Omega} h\langle \nu, \operatorname{grad} f\rangle\, dA.$$

If both f and h are C^2, then

$$\text{(VII.2.3)} \qquad \iint_\Omega \{h\Delta f - f\Delta h\}\, dV = \int_{\partial\Omega} \{h\langle \nu, \operatorname{grad} f\rangle - f\langle \nu, \operatorname{grad} h\rangle\}\, dA.$$

Notation We generally write $\partial f/\partial\nu$ for $\langle \nu, \operatorname{grad} f\rangle$.

VII.2.2 The Laplacian as an Operator on L^2

Until now, we have only considered the pointwise action of the Laplacian on functions which are C^2. We now wish to view the Laplacian as an operator on the Hilbert space $L^2 = L^2(M, dV)$. For any two functions f, g in L^2 we have the inner product and norm

$$(f, g) = \int_M fg\, dV, \qquad \|f\|^2 = \int_M f^2\, dV.$$

We may also speak of L^2 vector fields, and the associated inner product and norm

$$(X, Y) = \int_M \langle X, Y \rangle \, dV, \qquad \|X\|^2 = \int_M |X|^2 \, dV.$$

Recall that, for any subset A of M, $[A]_r$ denotes the set of all points in M with distance from A less than or equal to r.

We now consider the action of the Laplacian Δ on L^2. Recall that Green's formula states that if $f \in C^2$, $h \in C^1$, one of which is compactly supported, then

$$\text{(VII.2.4)} \qquad (-\Delta f, h) = (\operatorname{grad} f, \operatorname{grad} h).$$

Let $\Delta_c = \Delta | C_c^\infty$. Then $-\Delta_c$ is a nonnegative symmetric operator on the dense subspace C_c^∞ of L^2, with associated quadratic form

$$Q_c' = D[\phi, \psi] := (\operatorname{grad} \phi, \operatorname{grad} \psi), \qquad \phi, \psi \in C_c^\infty.$$

We refer to $D[\,,\,]$ as the *Dirichlet energy integral*. Complete C_c^∞ to $\mathfrak{H}_c$ relative to the metric

$$(\phi, \psi)_c = D[\phi, \psi] + (\phi, \psi)$$

on C_c^∞, and define the Friedrichs extension $\mathfrak{F}_c = \mathfrak{F}_{\Delta_c}$ of Δ_c on the domain $\mathfrak{D}_c = \mathfrak{D}_{\Delta_c}$. So $C_c^\infty \subset \mathfrak{D}_c \subset \mathfrak{H}_c$. Henceforth, unless otherwise noted, the self-adjoint Laplacian on L^2 will be the Friedrichs extension $\mathfrak{F}_c$. We will refer to $\mathfrak{F}_c$ as the *Friedrichs extension of the Laplacian on M*.

The identification of $\mathfrak{H}_c$ goes as follows: Let H denote the collection of C^∞ functions f on M for which both $f, \operatorname{grad} f \in L^2$. Endow H with the inner product

$$(f, g)_H = D[f, g] + (f, g), \qquad f, g \in H,$$

and complete H, with respect to the inner product $(\,,\,)_H$, to the Hilbert space $\mathfrak{H}$. Then $\mathfrak{H}_c$ is the closure of C_c^∞ in $\mathfrak{H}$.

For the elements of $\mathfrak{H}$ themselves, we consider weak derivatives. Namely,

Definition We say that $f \in L^2$ *has a weak derivative* if there exists an L^2 vector field X such that

$$(f, \operatorname{div} Y) = -(X, Y)$$

for all compactly supported C^∞ vector fields Y on M. The vector field X, should it exist, must be unique, and we denote it by $X = \operatorname{Grad} f$ [see (VII.2.1) above].

By the Meyers–Serrin theorem, $\mathfrak{H}$ is the Hilbert space consisting of functions $f \in L^2$ possessing weak derivatives, with the inner product $(\,,\,)_{\mathfrak{H}}$ given by

$$(f, g)_{\mathfrak{H}} = (\operatorname{Grad} f, \operatorname{Grad} g) + (f, g), \qquad f, g \in \mathfrak{H}.$$

Until now, M was an arbitrary Riemannian manifold. Assume, now, that M has C^∞ boundary Γ and compact closure. Then we may consider the Laplacian acting on functions with prescribed boundary conditions, chosen so that the action is symmetric, from which we then determine the new Friedrichs extension for the domain of functions in question. The simplest is:

Definition (Vanishing Dirichlet Boundary Conditions) Let

$$\mathcal{D}_{\text{dir}} = \{f \in C^\infty(\overline{M}) : f|\Gamma = 0\}, \qquad \Delta_{\text{dir}} = \Delta|\mathcal{D}_{\text{dir}}.$$

Then $-\Delta_{\text{dir}}$ is nonnegative symmetric, and has associated quadratic form and inner product given by

$$\mathcal{Q}'_{\text{dir}}(\phi, \psi) = D[\phi, \psi], \qquad (\phi, \psi)_{\text{dir}} = D[\phi, \psi] + (\phi, \psi)$$

for all $\phi, \psi \in \mathcal{D}_{\text{dir}}$. Complete the inner product $(,)_{\text{dir}}$ to the subspace D_{dir}, and thereby determine the Friedrichs extension $\mathfrak{F}_{\text{dir}}$ of $-\Delta_{\text{dir}}$, on $\mathfrak{D}_{\text{dir}}$, the *Dirichlet Laplacian on M*.

Lemma VII.2.1 *Let M have C^∞ boundary Γ and compact closure. Then given any $\alpha, \delta > 0$, there exists a function $\rho \in C_c^\infty(M)$ such that the support of ρ has distance from Γ greater than δ, that is,* $\operatorname{supp} \rho \subset M \setminus [\Gamma]_\delta$, *and*

$$\rho|M \setminus [\Gamma]_{3\delta} = 1, \qquad \rho \le 1 + \alpha, \qquad |\operatorname{grad} \rho| \le (1 + \alpha)/\delta.$$

Proof Since $\overline{M}$ is compact, it has a constant c and a finite covering $\{U_\iota : \iota = 1, \ldots, \ell\}$ by charts, for which the Riemannian metric in any of the charts satisfies

$$c^{-1} \sum_j (\xi^j)^2 \le \sum_{i,j} g_{ij} \xi^i \xi^j \le c \sum_j (\xi^j)^2$$

for all $\xi = (\xi^1, \ldots, \xi^n)$. Then there exists a constant K such that for every $\iota = 1, \ldots, \ell$, and every C^∞ function F compactly supported on U_ι we have

$$\text{(VII.2.5)} \qquad \|F\|_M{}^2 + \|\operatorname{grad} F\|_M{}^2 \le K\{\|F\|_{\mathbb{R}^n}{}^2 + \|\operatorname{grad} F\|_{\mathbb{R}^n}{}^2\},$$

where the subscripts $\mathbb{R}^n$ and M refer to the respective L^2 spaces. Fix the cover $\mathcal{U} = \{U_\iota : \iota = 1, \ldots, \ell\}$, with subordinate partition of unity $\{\varphi_\iota\}$.

Now consider the function $\varrho_\delta : M \to \mathbb{R}$ defined by

$$\varrho_\delta(q) = \begin{cases} 1, & q \in M \setminus [\Gamma]_{3\delta}, \\ d(q, \Gamma)/\delta - 2, & q \in [\Gamma]_{3\delta} \setminus [\Gamma]_{2\delta}, \\ 0, & q \in [\Gamma]_{2\delta}. \end{cases}$$

Then

$$\varrho_\delta = \sum_{\iota=1}^{\ell} \varphi_\iota \varrho_\delta.$$

Subject each $\varphi_\iota \varrho_\delta$ to mollification by j_{ϵ_ι}, as in §I.3; namely, for the index $\epsilon = (\epsilon_1, \ldots, \epsilon_\ell)$, define

$$\begin{aligned} \rho(p) = \varrho_{\delta,\epsilon}(p) &= \sum_{\iota=1}^{\ell} \left(j_{\epsilon_\iota} * (\varphi_\iota \varrho_\delta) \circ \mathbf{x}_\iota{}^{-1}\right)(\mathbf{x}_\iota(p)) \\ &= \sum_{\iota=1}^{\ell} \int j_{\epsilon_\iota}(\mathbf{y}_\iota - \mathbf{x}_\iota(p))\left((\varphi_\iota \varrho_\delta) \circ \mathbf{x}_\iota{}^{-1}\right)(\mathbf{y}_\iota)\, d\mathbf{v}_n(\mathbf{y}_\iota). \end{aligned}$$

Then one can easily use Theorem I.3.3 and (VII.2.5) to prove the lemma. ■

Lemma VII.2.2 *Let M have C^∞ boundary Γ and compact closure, and let $f : \overline{M} \to \mathbb{R}$ be Lipschitz on $\overline{M}$, with $f|\Gamma = 0$. Then, given any $\epsilon > 0$, there exist $\delta > 0$ and a function $\phi \in C_c^\infty(M)$ such that the support of ϕ has distance from Γ greater than δ, that is,* $\operatorname{supp} \phi \subset M \setminus [\Gamma]_\delta$, *and*

$$\|f - \phi\|_{\mathfrak{H}} < \epsilon.$$

Corollary VII.2.1 *The Dirichlet Laplacian $\mathfrak{F}_{\mathrm{dir}}$ on M coincides with the Friedrichs extension $\mathfrak{F}_c$ of Δ_c on M. In particular, their associated quadratic forms coincide, possessing the common domain $\mathfrak{H}_c$.*

Proof of Lemma VII.2.2. Let $\rho : M \to [0, \infty) \in C_c^\infty$ satisfy the properties of Lemma VII.2.1, and let $\phi = f\rho$. Then

$$\|f - \phi\|^2 = \int_{\Gamma_{4\delta}} f^2(1 - \rho)^2\, dV \le \text{const.}_f \delta,$$

by a standard calculation in Fermi coordinates based on Γ. Similarly,

$$\operatorname{grad}(f - \phi) = \operatorname{grad} f(1 - \rho) = (1 - \rho)\operatorname{grad} f + f \operatorname{grad}(1 - \rho),$$

which implies

$$|\text{grad}\,(f-\phi)|^2(p) \le 2\{|1-\rho|^2|\text{grad}\,f|^2 + |f|^2|\text{grad}\,\rho|^2\}(p)$$
$$\le \text{const.}_f \left(1 + \frac{d(p,\Gamma)^2}{\delta^2}\right)$$

[the $d(p, \Gamma)^2$ in the numerator of the second term of the parentheses follows from f vanishing on Γ]. But then calculation in Fermi coordinates, based on Γ, implies

$$\|\text{grad}\,f - \text{grad}\,\phi\|^2 = \int_{\Gamma_{4\delta}} |\text{grad}\,f(1-\rho)|^2\,dV \le \text{const.}_f\,\delta,$$

which implies the proposition. ■

Remark VII.2.1 We shall show in Theorem VII.4.2 below that the Laplacian with vanishing Dirichlet boundary condition is essentially self-adjoint.

Definition A function $f \in L^2$ on M has Δf *acting as an* L^2 *distribution on* M if there exists $g \in L^2$ such that

$$(g, \phi) = (f, \Delta\phi) \qquad \forall\, \phi \in C_c^\infty.$$

We write Δf for g.

Theorem VII.2.1 *If M is a complete Riemannian manifold, then $\mathfrak{H}_c = \mathfrak{H}$. Given any function $f \in C^\infty$ for which $f, \Delta f \in L^2$, we also have* grad $f \in L^2$ *and*

$$(\text{VII.2.6}) \qquad (-\Delta f, f) = (\text{grad}\,f, \text{grad}\,f).$$

Finally, $\Delta_c = \Delta|C_c^\infty$ is essentially self-adjoint, so the Friedrichs extension $\mathfrak{F}_c$ of Δ_c is the unique self-adjoint extension of Δ_c.

Corollary VII.2.2 *Let M be a complete Riemannian manifold. Then M has no L^2 nonconstant harmonic function, that is, any function f in $L^2(M)$ satisfying $\Delta f = 0$ on all of M must be constant. Moreover, given any $\lambda < 0$, there is no solution $u \in L^2$ to the equation*

$$\Delta u + \lambda u = 0$$

except for $u = 0$ identically on M.

Proof of Theorem VII.2.1. Assume $f \in \mathfrak{H}(M) \cap C^\infty(M)$. Fix $o \in M$, and for each ϵ, $R > 0$, consider the function $f_R = f\phi_R$, where $\phi_R \in C_c^\infty(B(o; R+3))$

satisfies

$$\phi_R | B(o; R) = 1, \qquad \phi_R, |\mathrm{grad}\, \phi_R| \leq 1 + \epsilon;$$

such a function ϕ_R is constructed in Lemma VII.2.1. Then

$$\|f - f_R\|^2 = \int f^2(1 - \phi_R)^2\, dV \leq \text{const.} \int_{M \setminus B(o;R)} f^2\, dV \to 0$$

as $R \to \infty$, since $f \in L^2$. Similarly,

$$\begin{aligned}
&\int |\mathrm{grad}\, f - \mathrm{grad}\, f_R|^2\, dV \\
&\quad = \int |\mathrm{grad}\, f(1 - \phi_R)|^2\, dV \\
&\quad \leq 2 \int \{f^2 |\mathrm{grad}\, \phi_R|^2 + |1 - \phi_R|^2 |\mathrm{grad}\, f|^2\}\, dV \\
&\quad = 2 \int_{M \setminus B(o;R)} \{f^2 |\mathrm{grad}\, \phi_R|^2 + |1 - \phi_R|^2 |\mathrm{grad}\, f|^2\}\, dV \\
&\quad \leq \text{const.} \int_{M \setminus B(o;R)} \{f^2 + |\mathrm{grad}\, f|^2\}\, dV \\
&\quad \to 0
\end{aligned}$$

as $R \to \infty$. Therefore f is approximated in $\mathfrak{H}$ by functions in $\mathfrak{H}_c$. Since $\mathfrak{H} \cap C^\infty$ is dense in $\mathfrak{H}$ (by definition), we obtain $\mathfrak{H}_c = \mathfrak{H}$.

Assume we are given the function $f \in C^\infty$ such that $f, \Delta f \in L^2$. Then

$$\int \mathrm{div}\, ({\phi_R}^2 f \,\mathrm{grad}\, f) = 0,$$

which implies

$$\begin{aligned}
\int {\phi_R}^2 \{|\mathrm{grad}\, f|^2 + f \Delta f\} &= -2 \int \phi_R f \langle \mathrm{grad}\, \phi_R, \mathrm{grad}\, f \rangle \\
&\leq \frac{1}{2} \int {\phi_R}^2 |\mathrm{grad}\, f|^2 + 2 \int f^2 |\mathrm{grad}\, \phi_R|^2,
\end{aligned}$$

which implies

$$\begin{aligned}
\frac{1}{2} \int_{B(o;R)} |\mathrm{grad}\, f|^2 &\leq \frac{1}{2} \int {\phi_R}^2 |\mathrm{grad}\, f|^2 \\
&\leq \int 2 f^2 |\mathrm{grad}\, \phi_R|^2 - {\phi_R}^2 f \Delta f \\
&\leq \text{const.} \int f^2 + |f \Delta f| \\
&< +\infty
\end{aligned}$$

for all R. This implies $\operatorname{grad} f \in L^2$. But

$$\int {\phi_R}^2\{|\operatorname{grad} f|^2 + f\Delta f\} = -2\int \phi_R f\langle \operatorname{grad}\phi_R, \operatorname{grad} f\rangle$$
$$= -2\int_{B(o;R+3)\setminus B(o;R)} \phi_R f\langle \operatorname{grad}\phi_R, \operatorname{grad} f\rangle$$
$$\leq \text{const.}\int_{B(o;R+3)\setminus B(o;R)} |f||\operatorname{grad} f|,$$

As $R \to \infty$ we have

$$\int {\phi_R}^2\{|\operatorname{grad} f|^2 + f\Delta f\} \to \int |\operatorname{grad} f|^2 + f\Delta f,$$
$$\int_{B(o;R+3)\setminus B(o;R)} |f||\operatorname{grad} f| \to 0,$$

which implies (VII.2.6).

For the essential self-adjointness of Δ_c we consider L^2 to be a complex Hilbert space with inner product

$$(f, g) = \int_M f\overline{g}\, dV,$$

and with the Laplacian defined by operating on real and imaginary parts of a complex-valued function. It is standard that Δ_c is essentially self-adjoint if and only if

$$\ker\{\Delta_c \pm i\}^* = 0,$$

that is, the subspace of functions $f \in L^2$ that satisfy

$$(f, (\Delta \pm i)\phi) = 0 \qquad \forall\, \phi \in C_c^\infty$$

is trivial. So assume $\{\Delta_c \pm i\}^* f = 0$. Then $(\Delta \pm i)f = 0$ as an L^2 distribution. Elliptic regularity then implies that $f \in C^\infty$, and Δf in the classical sense is the L^2-distributional Laplacian of f. Therefore

$$\Delta f \pm if = 0$$

on all of M. Multiply both sides of the equation by $\overline{f}$ and integrate over M. Then

$$-\int_M \overline{f}\Delta f\, dV = \int_M |\operatorname{grad} f|^2\, dV$$

for all $f \in L^2 \cap C^\infty$ for which $\Delta f \in L^2$ [the purpose of this comment is to prove that $(\Delta f, f)$ is real]; therefore $\Delta f + if = 0$ implies

$$i \int_M |f|^2 = \int_M |\text{grad } f|^2,$$

which implies $f = 0$. A similar argument applies to a solution of $\Delta f - if = 0$, and the theorem is proven. ∎

VII.3 The Heat Equation and Its Kernels

We are given a fixed Riemannian manifold M of dimension $n \geq 1$, with associated Laplacian Δ acting on C^2 functions on M. The *heat operator* L associated with the Laplacian acts on functions $f = f(x, t)$ in $C^0(M \times (0, \infty))$, and is given by

$$L = \Delta - \frac{\partial}{\partial t},$$

where f is C^2 in the *space* variable $x \in M$, and C^1 in the *time* variable $t \in (0, \infty)$. The *homogeneous heat equation*, or *heat equation*, is given by $Lu = 0$, that is,

$$\Delta u = \frac{\partial u}{\partial t},$$

and the *inhomogeneous heat equation* is given by $Lu = -F$, where $F : M \times (0, \infty) \to \mathbb{R} \in C^0$.

The physical interpretation of the equation is given by considering the Riemannian manifold as a homogeneous isotropic medium (with conveniently normalized physical constants so that, for example, heat and temperature can be considered one and the same), and $u(x, t)$ is the temperature of $x \in M$ at time t. Then the inhomogeneous heat equation describes the evolution of the temperature distribution, with $F(x, t)$ the instantaneous rate with respect to time at which (by some external source) heat is supplied to, or withdrawn from, x at time t. The argument is predicated on Newtonian heat conduction, namely, given any domain Ω in M with compact closure and C^∞ boundary, the instantaneous change of the total temperature in Ω with respect to time is equal to the total change due to the supply of heat to, or withdrawal of heat from, Ω, added to the spatial rate of change of the heat distribution across the boundary of Ω, $\partial\Omega$. That is,

$$\frac{\partial}{\partial t} \iint_\Omega u(x, t)\, dV(x) = \iint_\Omega F(x, t)\, dV(x) + \int_{\partial\Omega} \frac{\partial u}{\partial \nu}(w, t)\, dA(w).$$

By Green's formula we have

$$\iint_{\Omega} \frac{\partial u}{\partial t}\, dV = \iint_{\Omega} \{F + \Delta u\}\, dV,$$

which implies

$$\iint_{\Omega} \{Lu + F\}\, dV = 0$$

for all such domains Ω in M. This then implies the inhomogeneous heat equation on all of M.

Notation Henceforth, if there is no comment otherwise, given any function $f : M \times (0, +\infty) \to \mathbb{R}$, we denote the spatial component of the gradient of f by $\operatorname{grad} f$.

Definition A *heat kernel* is a continuous function $p : M \times M \times (0, \infty)$ such that for every $y \in M$ the function $u(x, t) = p(x, y, t)$ is a solution to the heat equation with initial data

$$\lim_{t \downarrow 0} u(x, t) = \delta_y(x),$$

the delta function concentrated at y, namely,

$$\lim_{t \downarrow 0} \int_M \phi(x) p(x, y, t)\, dV(x) = \phi(y)$$

for every bounded continuous $\phi : M \to \mathbb{R}$.

Intuitively, $p(\ , y,\ )$ is the solution of the heat equation resulting from an initial temperature distribution having total temperature equal to 1 concentrated at y. We use the linearity of the heat equation (otherwise known as superposition of solutions) as follows: If at time $t = 0$ the initial temperature distribution was concentrated at y with total temperature α, then the solution to the heat equation will be $u = \alpha p(\ , y,\ )$. If at time $t = 0$ the initial temperature distribution was concentrated at the points y and z with total temperatures α and β, respectively, then the solution to the heat equation will be $u = \alpha p(\ , y,\ ) + \beta p(\ , z,\ )$. Therefore if at time $t = 0$ we are given the temperature distribution $f(y)$, then by summing "spatially" the contribution of each point to the initial data, we obtain

$$u(x, t) = \int_M p(x, y, t) f(y)\, dV(y),$$

with

$$\lim_{t \downarrow 0} u(x, t) = \lim_{t \downarrow 0} \int_M p(x, y, t) f(y)\, dV(y) = f(x).$$

If, in addition, heat or refrigeration is supplied to M, as described by $F(x, t)$ above, then the contribution of $F(\ , \tau)$, $\tau \in (0, t)$, to the temperature distribution at time t is given by

$$\int_M p(x, y, t - \tau) F(y, \tau)\, dV(y),$$

since we think of the function $F(\ , \tau)$ as initial data for heat diffusion starting at time τ and lasting for $t - \tau$ time units. So the total contribution to the temperature distribution, at time t, by F is given by summing "temporally," namely,

$$\int_0^t d\tau \int_M p(x, y, t - \tau) F(y, \tau)\, dV(y).$$

Therefore, we expect the solution to the inhomogeneous heat equation with initial temperature distribution f and external source F to be given by

$$u(x, t) = \int_M p(x, y, t) f(y)\, dV(y) + \int_0^t d\tau \int_M p(x, y, t - \tau) F(y, \tau)\, dV(y).$$

Remark VII.3.1 Note that we have argued that $p(x, y, t)$ is symmetric with respect to x and y as $t \downarrow 0$, since $p(x, \ , t) \to \delta_x$, intuitively, and $p(, y, t) \to \delta_y$, by hypothesis, as $t \downarrow 0$. We shall see shortly that in all cases that we study, we have the symmetry $p(x, y, t) = p(y, x, t)$ on all of $M \times M \times (0, +\infty)$.

VII.3.1 The Heat Kernel of Euclidean Space

In order to calculate the heat kernel on $\mathbb{R}^n$, we first derive a candidate, using the Fourier transform. Then we check that our candidate is indeed legitimate.

Given a function $f : \mathbb{R}^n \to \mathbb{C}$ in $L^1(\mathbb{R}^n)$, define its n-dimensional Fourier transform, $\widehat{f}$, by

$$\widehat{f}(\xi) = (2\pi)^{-n/2} \int_{\mathbb{R}^n} f(x) e^{-ix \cdot \xi}\, d\mathbf{v}_n(x),$$

where

$$x := (x^1, \ldots, x^n), \qquad \xi := (\xi^1, \ldots, \xi^n).$$

The basic properties of the Fourier transform are:

1. $|\widehat{f}(\xi)| \leq ||f||_1/(2\pi)^{n/2}$.
2. $g(x) = f(x - y) \Rightarrow \hat{g}(\xi) = e^{-iy \cdot \xi} \widehat{f}(\xi); \qquad \{e^{iy \cdot x} \widehat{f}\}(\xi) = \widehat{f}(\xi - y)$.

3. $\{f(\lambda x)\}\widehat{\ }(\xi) = \lambda^{-n}\widehat{f}(\xi/\lambda)$; $\{f * g\}\widehat{\ }(\xi) = \widehat{f}(\xi)\widehat{g}(\xi)$, where $f * g$ denotes the convolution (normalized here to be)

$$(f * g)(x) = (2\pi)^{-n/2} \int f(y - x)g(y)\, d\mathbf{v}_n(y).$$

4. One also has the specific example

$$\phi(x) = e^{-|x|^2/2} \quad \Rightarrow \quad \widehat{\phi}(\xi) = e^{-|\xi|^2/2}.$$

5. Consider the collection $\mathcal{S}$ of functions on $\mathbb{R}^n$ that, with all their partial derivatives of all orders, have rapid decrease on $\mathbb{R}^n$. For functions f in $\mathcal{S}$ we have

$$\{\partial f/\partial x^j\}\widehat{\ }(\xi) = i\xi^j \widehat{f}(\xi), \qquad (\partial \widehat{f}/\partial \xi^j)(\xi) = \{-ix^j f\}\widehat{\ }(\xi).$$

The Riemann–Lebesgue lemma implies, with these last two formulae, that the Fourier transforms maps $\mathcal{S}$ into $\mathcal{S}$.
6. Finally, one has the *Fourier inversion formula:*

$$f(x) = (2\pi)^{-n/2} \int_{\mathbb{R}^n} \widehat{f}(\xi)e^{ix\cdot\xi}\, d\mathbf{v}_n(\xi).$$

Now, consider a solution $u(x, t)$ to the heat equation on $\mathbb{R}^n$, with initial data $\phi(x)$. Since this is a formal calculation in search of a candidate solution, we treat all functions as though they were in $\mathcal{S}$, with respect to the space variable for every t. Let $v(\xi, t)$ be the Fourier transform of $u(x, t)$ in the space variables, that is,

$$v(\xi, t) = (2\pi)^{-n/2} \int_{\mathbb{R}^n} u(x, t)e^{-ix\cdot\xi}\, d\mathbf{v}_n(x).$$

Then

$$\begin{aligned}
\frac{\partial v}{\partial t}(\xi, t) &= (2\pi)^{-n/2} \int_{\mathbb{R}^n} \frac{\partial u}{\partial t}(x, t)e^{-ix\cdot\xi}\, d\mathbf{v}_n(x) \\
&= (2\pi)^{-n/2} \int_{\mathbb{R}^n} \Delta u(x, t)e^{-ix\cdot\xi}\, d\mathbf{v}_n(x) \\
&= (2\pi)^{-n/2} \sum_{j=1}^{n} \int_{\mathbb{R}^n} \frac{\partial^2 u}{(\partial x^j)^2}(x, t)e^{-ix\cdot\xi}\, d\mathbf{v}_n(x) \\
&= \sum_{j=1}^{n} (i\xi^j)^2 v(\xi, t) \\
&= -|\xi|^2 v(\xi, t),
\end{aligned}$$

that is,

$$\frac{\partial v}{\partial t}(\xi, t) = -|\xi|^2 v(\xi, t).$$

For any fixed ξ, view the equation as an ordinary differential equation in t, from which one has

$$v(\xi, t) = c(\xi)e^{-|\xi|^2 t}.$$

To evaluate $c(\xi)$, let $t = 0$. Then one obtains $c(\xi) = \widehat{\phi}(\xi)$, the Fourier transform of the initial data of ϕ. We conclude

$$v(\xi, t) = \widehat{\phi}(\xi)e^{-|\xi|^2 t}.$$

If one now inverts the Fourier transform, to recapture $u(x, t)$, one obtains

$$u(x, t) = \frac{1}{(4\pi t)^{n/2}} \int_{\mathbb{R}^n} e^{-|y-x|^2/4t} \phi(y)\, d\mathbf{v}_n(y).$$

Our candidate, therefore, for the heat kernel on $\mathbb{R}^n$ is

$$\mathbf{E}(x, y, t) = \frac{1}{(4\pi t)^{n/2}} e^{-|y-x|^2/4t}.$$

One can directly verify that $\mathbf{E}$ is C^∞ on $\mathbb{R}^n \times \mathbb{R}^n \times (0, \infty)$ and is a solution to the heat equation on $\mathbb{R}^n$ in x and t. We now check that

$$\lim_{t \downarrow 0} \mathbf{E}(x, y, t) = \delta_y(x).$$

First start with any nonnegative integrable function $\rho : \mathbb{R}^n \to \mathbb{R}$ with $\int \rho(x)\, d\mathbf{v}_n(x) = 1$, and set $\rho_\epsilon(x) = \epsilon^{-n}\rho(x/\epsilon)$. Then $\int \rho_\epsilon(x)\, d\mathbf{v}_n(x) = 1$, and it is standard that for any bounded continuous function $f(x)$ on $\mathbb{R}^n$ we have

$$\lim_{\epsilon \downarrow 0} \int \rho_\epsilon(x - y) f(x)\, d\mathbf{v}_n(x) = f(y)$$

[see Proposition 1(b), where we have j_ϵ compactly supported on $\mathbb{R}^n$]. So, for any fixed y,

$$\lim_{\epsilon \downarrow 0} \rho_\epsilon(x - y) = \delta_y(x).$$

In our situation, we have

$$\rho(x) = \pi^{-n/2} e^{-|x|^2}, \qquad \mathbf{E}(x, y, t) = \rho_{2\sqrt{t}}(x - y),$$

which implies the claim; so $\mathbf{E}$ is a heat kernel on $\mathbb{R}^n$. Two important properties, which we see repeatedly, are the symmetry and positivity of the heat kernel, namely,

$$\mathbf{E}(x, y, t) = \mathbf{E}(y, x, t) > 0, \tag{VII.3.1}$$

for all $x, y \in M$ and $t > 0$.

For a function $f(x)$ on $\mathbb{R}^n$, we have the following results for solutions to the *initial value problem for the homogeneous heat equation*:

$$\Delta u = \frac{\partial u}{\partial t}, \qquad \lim_{t \downarrow 0} u(x, t) = f(x). \tag{VII.3.2}$$

Proposition VII.3.1

(a) If $f(x)$ is bounded continuous, then the function

$$u(x, t) = \int_{\mathbb{R}^n} \mathbf{E}(x, y, t) f(y)\, d\mathbf{v}_n(y) \tag{VII.3.3}$$

is a solution of (VII.3.2).

(b) We always have

$$\int_{\mathbb{R}^n} \mathbf{E}(x, y, t)\, d\mathbf{v}_n(y) = 1 \tag{VII.3.4}$$

for all $(x, t) \in \mathbb{R}^n \times (0, \infty)$. If f is a function in L^1, and $u(x, t)$ is given by (VII.3.3), *then*

$$\int_{\mathbb{R}^n} u(x, t)\, d\mathbf{v}_n(x) = \int_{\mathbb{R}^n} f(x)\, d\mathbf{v}_n(x) \tag{VII.3.5}$$

for all $t > 0$.

(c) If $f \in C^0$ is supported on the compact set K, and $u(x, t)$ is given by (VII.3.3), *then*

$$|u(x, t)| \leq (4\pi t)^{-n/2} e^{-d^2(x, K)/4t} \left\{ \int |f(y)|\, d\mathbf{v}_n(y) \right\}.$$

(d) If $f(x)$ is continuous on $\mathbb{R}^n$, vanishing at infinity, that is,

$$\lim_{|x| \to \infty} f(x) = 0,$$

and $u(x, t)$ is given by (VII.3.3), *then*

$$\lim_{|x| \to \infty} u(x, t) = 0, \qquad \lim_{|x| \to \infty} (\operatorname{grad} u)(x, t) = 0.$$

(e) If $f(x)$ is bounded continuous on $\mathbb{R}^n$, then from among all functions $v(x, t)$ on $\mathbb{R}^n \times (0, \infty)$ satisfying

$$|v(x, t)| \leq \text{const.},$$

where the constant is independent of (x, t), *the function* $u(x, t)$ *given by* (VII.3.3) *is the unique solution to the initial value problem* (VII.3.2).

For the initial-value problem for the inhomogeneous heat equation we have

Proposition VII.3.2 *Given* $F : \mathbb{R}^n \times [0, \infty) \to \mathbb{R} \in C^1$ *such that* $\operatorname{supp} F \subseteq K \times [0, \infty)$ *for some compact* K *in* $\mathbb{R}^n$, *then the function*

$$u(x, t) = \int_{\mathbb{R}^n} \mathbf{E}(x, y, t) f(y)\, d\mathbf{v}_n(y) + \int_0^t d\tau \int_{\mathbb{R}^n} \mathbf{E}(x, y, t - \tau) F(y, \tau)\, d\mathbf{v}_n(y)$$

satisfies

$$\text{(VII.3.6)} \qquad \Delta u - \frac{\partial u}{\partial t} = -F(x, t), \qquad \lim_{t \downarrow 0} u(x, t) = f(x).$$

VII.3.2 Preliminary Principles

Proposition VII.3.3 (Duhamel's Principle) *Let* M *be a Riemannian manifold with compact closure and* C^∞ *boundary (possibly empty), and let* $u, v : \overline{M} \times (0, t) \to \mathbb{R} \in C^1$ *be* C^2 *with respect to the space variable in* M. *Then for any* $[\alpha, \beta] \subset (0, t)$ *we have*

$$\begin{aligned} &\iint_M \{u(z, t - \beta) v(z, \beta) - u(z, t - \alpha) v(z, \alpha)\}\, dV(z) \\ &\quad = \int_\alpha^\beta d\tau \iint_M \{Lu(z, t - \tau) v(z, \tau) - u(z, t - \tau) Lv(z, \tau)\}\, dV(z) \\ &\qquad + \int_\alpha^\beta d\tau \int_{\partial M} \left\{ -\frac{\partial u}{\partial \nu_w}(w, t - \tau) v(w, \tau) \right. \\ &\qquad \left. + u(w, t - \tau) \frac{\partial v}{\partial \nu_w}(w, \tau) \right\} dA(w). \end{aligned}$$

Proposition VII.3.4 (Strong Maximum Principle) *Let* u *be a bounded continuous function on* $M \times [0, T]$ *that is* C^2 *on* $M \times (0, T)$ *and that satisfies*

$$\Delta u - \frac{\partial u}{\partial t} \geq 0$$

on $M \times (0, T)$. *If there exists* (x_0, t_0) *in* $M \times (0, T]$ *– note the half-open time interval – such that*

$$u(x_0, t_0) = \sup_{M \times [0, T]} u,$$

then

$$u|M \times [0, t_0] = u(x_0, t_0).$$

Furthermore, if M *has* C^∞ *boundary,* $u \in C^0(\overline{M} \times [0, T])$*, and* $(w, t_1) \in \partial M \times (0, T]$ *satisfies*

$$u(w, t_1) = \sup_{M \times [0,T]} u,$$

then

$$\frac{\partial u}{\partial \nu_w}(w, t_1) > 0.$$

Remark VII.3.2 If one is given $\Delta u - \partial u/\partial t \leq 0$, then one has a corresponding minimum principle. For solutions of the heat equation, both principles are valid.

VII.3.3 Properties and Types of Heat Kernels

Definition Given a Riemannian manifold with heat kernel $p(x, y, t)$, we say that $p(x, y, t)$ is *symmetric* if

$$p(x, y, t) = p(y, x, t)$$

on all of $M \times M \times (0, +\infty)$. We say that $p(x, y, t)$ is *positive* if

$$p(x, y, t) > 0$$

on all of $M \times M \times (0, +\infty)$.

We say that $p(x, y, t)$ satisfies the *conservation of heat property*, or that $p(x, y, t)$ is *stochastically complete*, if

$$\int_M p(x, y, t)\, dV(x) = 1$$

for all $(y, t) \in M \times (0, +\infty)$.

For M noncompact, we say that $p(x, y, t)$ satisfies the *Feller property* if

$$\lim_{x \to \infty} p(x, y, t) = 0$$

for all $(y, t) \in M \times (0, +\infty)$.

Remark VII.3.3 If p is a heat kernel satisfying the symmetry property, then

$$(\Delta_x p)(x, y, t) = (\Delta_y p)(x, y, t).$$

Definition Let M be a compact Riemannian manifold with no boundary, a *closed* manifold. Then any heat kernel on such a manifold will be referred to as a *closed heat kernel*.

Definition Let M be a Riemannian manifold with compact closure and C^∞ (nonempty) boundary. Then a *Dirichlet heat kernel* $q : M \times M \times (0, \infty) \in C^0$ is a heat kernel on M that can be extended to a continuous function on $\overline{M} \times \overline{M} \times (0, \infty)$ such that

$$q(\,, y, t)|\partial M = 0$$

for all $(y, t) \in M \times (0, \infty)$. Thus the vanishing Dirichlet boundary data will correspond to absolute refrigeration of the boundary.

Theorem VII.3.1 *Let p be either a closed or a Dirichlet heat kernel on a Riemannian manifold M. Then p is symmetric in the two space variables in M. One has uniqueness of heat kernels in both cases, that is, a closed Riemannian manifold has at most one heat kernel, and a Riemannian manifold with compact closure and C^∞ boundary has at most one Dirichlet heat kernel.*

Proof Fix one of the two cases. Let p_1, p_2 be two kernels on M, set

$$u(z, \tau) = p_1(z, x, \tau), \qquad v(z, \tau) = p_2(z, y, \tau),$$

and apply Duhamel's principle. Then the boundary integrals in Duhamel's principle (should the boundary be nonempty) vanish. Let $\alpha \downarrow 0$, $\beta \uparrow t$. Then one obtains

$$p_2(x, y, t) = p_1(y, x, t).$$

If we apply the argument to any heat kernel $p_1 = p_2 = p$, then we have

$$p(x, y, t) = p(y, x, t)$$

for all $x, y \in M$. So any heat kernel is symmetric in the space variables. But given arbitrary heat kernels p_1, p_2, we may apply the symmetry to p_1 and thereby obtain $p_1 = p_2$. ■

Theorem VII.3.2 *Let p be a closed heat kernel on a Riemannian manifold M. Then p satisfies the conservation of heat property.*

Proof Simply note that

$$\begin{aligned}\frac{\partial}{\partial t}\int_M p(x, y, t)\, dV(x) &= \int_M \frac{\partial}{\partial t} p(x, y, t)\, dV(x)\\ &= \int_M \Delta_x p(x, y, t)\, dV(x) = 0,\end{aligned}$$

which implies the claim. ■

Given any Riemannian manifold M, set

$$\mathcal{E}(x, y, t) = (4\pi t)^{-n/2} e^{-d^2(x,y)/4t}.$$

We do not presume that $\mathcal{E}$ will actually be a heat kernel on M, but it should be some sort of approximate heat kernel. The question is in what sense, and with what validity. The answer, for our purposes, goes as follows:

Proposition VII.3.5 *Assume M is a closed Riemannian manifold. Then M has a closed heat kernel, which we know is unique.*

Similarly, assume M is a Riemannian manifold with C^∞ boundary and compact closure. Then M has a Dirichlet heat kernel, which we know is unique.

In either case, the heat kernel is locally Euclidean in the sense that, for any compact K in M,

$$\lim_{t \downarrow 0} \frac{p(x, y, t)}{\mathcal{E}(x, y, t)} = 1 + O_K(d(x, y)) \tag{VII.3.7}$$

uniformly for all $x, y \in K$ satisfying

$$d(x, y) \leq \min\left\{\frac{\operatorname{inj} M}{4}, \frac{d(K, \partial M)}{2}\right\}.$$

Theorem VII.3.3 *Assume p is a closed or Dirichlet heat kernel on a Riemannian manifold M. Then p is always positive.*

Proof From (VII.3.7) we know that the heat kernel $p(x, y, t)$ assumes positive values, and moreover is unbounded from above on $M \times M \times (0, T]$ for any $T > 0$.

If we are given the closed heat kernel p, and there exists (x_0, y_0, t_0) in $M \times M \times (0, +\infty)$ such that $p(x_0, y_0, t_0) \leq 0$, then, for fixed y_0, $u(x, t) = p(x, y_0, t)$ has a nonpositive minimum value on $M \times (0, t_0]$, which implies by the strong minimum principle that there exist $t_1 \in (0, t_0)$ and a nonpositive constant δ such that $u(x, t) = \delta$ on all of $M \times (0, t_1]$, which implies a contradiction.

If we are given the Dirichlet heat kernel p, then for any fixed y_0 in M, the function $u(x, t) = p(x, y_0, t)$ vanishes on the boundary ∂M. Therefore, if u is not positive everywhere on M, then there there exist (x_0, t_0) in $M \times (0, +\infty)$ such that $p(x_0, y_0, t_0) \leq 0$. The argument is then the same as above. ∎

Remark VII.3.4 Let p be a Dirichlet heat kernel on M. Since $p(x, y, t) > 0$, then

$$\frac{\partial p}{\partial \nu_w}(w, y, t) < 0$$

on all of $\partial M \times M \times (0, +\infty)$. In particular,

$$\frac{\partial}{\partial t} \int_M p(x, y, t)\, dV(x) < 0,$$

which implies

$$\int_M p(x, y, t)\, dV(x) < 1, \tag{VII.3.8}$$

on all of $M \times (0, +\infty)$.

Definition Let M be an arbitrary noncompact Riemannian manifold. We say that the positive heat kernel p is the *minimal positive heat kernel* if, for any other positive heat kernel $P(x, y, t)$ on M, we have $p \leq P$ on all of $M \times M \times (0, +\infty)$.

Certainly, the minimal positive heat kernel is unique, by definition. For the existence of the minimal positive heat kernel one starts with domains Ω in M with compact closure and C^∞ boundary, and associates with each Ω its Dirichlet heat kernel q_Ω. An application of the maximum principle implies that for domains $\Omega_1 \subset\subset \Omega_2$ we have

$$q_{\Omega_1} < q_{\Omega_2} \qquad \text{on } \Omega_1 \times \Omega_1 \times (0, +\infty).$$

The minimal positive heat kernel of M, $p(x, y, t)$, is given by

$$p = \sup_\Omega q_\Omega,$$

where we set q_Ω to be identically equal to 0 on the complement of $\Omega \times \Omega \times (0, +\infty)$ in $M \times M \times (0, +\infty)$. For any exhaustion $\Omega_j \uparrow M$ by relatively compact subsets of M, we have the monotone convergence of $q_{\Omega_j} \uparrow p$ uniformly on compact subsets of M as $j \to \infty$. All derivatives of q_{Ω_j} converge to those of p uniformly on compact subsets of M as $j \to \infty$.

Then p is positive on all of $M \times M \times (0, +\infty)$, it is the minimal positive heat kernel of M, it is symmetric, and it satisfies

$$\int_M p(x, y, t)\, dV(x) \leq 1 \tag{VII.3.9}$$

for all $(y, t) \in M \times (0, +\infty)$. So, for ϕ bounded continuous, the function $u(x, t)$ given by

$$u(x, t) = \int_M p(x, y, t)\phi(y)\, dV(y)$$

satisfies the heat equation, with

$$\lim_{t\downarrow 0} u(x,t) = \phi(x)$$

for all $x \in M$.

Definition Given a Riemannian manifold M, we say that M *satisfies the uniqueness property for the heat equation* if for any bounded continuous function f on M there is at most one bounded solution to the heat equation on M satisfying

$$\lim_{t\downarrow 0} u(x,t) = f(x).$$

Remark VII.3.5 Of course, if M satisfies the uniqueness property, then M has a unique heat kernel.

Remark VII.3.6 Assume M satisfies the uniqueness property. Since for any bounded continuous $f(x)$ the function

$$u(x,t) = \int_M p(x,y,t) f(y)\, dV(y) \tag{VII.3.10}$$

is a solution to the initial value problem for initial data f, we conclude that every solution to the initial value problem for the heat equation on M with bounded continuous initial data f admits the representation (VII.3.10). In particular, the solution $u(x,t) \equiv 1$ admits the representation

$$1 = \int_M p(x,y,t)\, dV(y),$$

which implies M is stochastically complete.

Theorem VII.3.4 *Let M be a compact Riemannian manifold. Then M satisfies the uniqueness property.*

Also, if M has compact closure and C^∞ boundary, then given the boundary data

$$u(w,t) = U(w,t) \qquad \forall\,(w,t) \in \partial M \times (0,\infty)$$

for some function $U(w,t)$, $t > 0$, M satisfies the uniqueness property.

Proof Let $u(x,t)$ be a solution to the heat equation on M, in either of the two cases.

Then

$$\begin{aligned}
\frac{1}{2}\frac{\partial}{\partial t}\iint_M u^2(x,t)\,dV(x)
&= \frac{1}{2}\iint_M \frac{\partial u^2}{\partial t}(x,t)\,dV(x) \\
&= \iint_M \left(u\frac{\partial u}{\partial t}\right)(x,t)\,dV(x) \\
&= \iint_M (u\,\Delta\,u)(x,t)\,dV(x) \\
&= -\iint_M |\mathrm{grad}\,u|^2(x,t)\,dV(x) + \int_{\partial M}\left(u\frac{\partial u}{\partial \nu_w}\right)(w,t)\,dA(w) \\
&= -\iint_M |\mathrm{grad}\,u|^2(x,t)\,dV(x) \\
&\le 0.
\end{aligned}$$

Therefore, if we have two solutions v_1, v_2 to the same initial–boundary value problem, namely,

$$v_1(x,0) = v_2(x,0) = \phi(x) \qquad \forall\, x \in M$$

and (if there are nonempty boundary data)

$$v_1(w,t) = v_2(w,t) \qquad \forall\,(w,t) \in \partial M \times (0,\infty),$$

then the solution $u = v_1 - v_2$ has both vanishing initial and boundary data. But its L^2 integral does not increase with respect to time. So it stays constantly equal to 0. So $v_1 = v_2$. Therefore, solutions to the initial–boundary value problems are unique. ∎

Remark VII.3.7 If M has C^∞ boundary and compact closure, then the Dirichlet heat kernel of M is equal to the minimal positive heat kernel of M, that is, $p = q_M$.

Proposition VII.3.6 *Assume M is Riemannian complete, with Ricci curvature bounded from below. Then M has the uniqueness property and the Feller property.*

Corollary VII.3.3 *Assume M is Riemannian complete, with Ricci curvature bounded from below. Then M is stochastically complete.*

Remark VII.3.8 If M does not satisfy the uniqueness property, then the counterexample to uniqueness cannot come from the minimal positive heat kernel and solutions of the form (VII.3.10), because (VII.3.9) implies there is only one solution of the initial value problem with initial data f of the form (VII.3.10). Indeed, (VII.3.10) and (VII.3.9) imply

$$|u(x,t)| \leq \|f\|_\infty \int_M p(x,y,t)\,dV(y) \leq \|f\|_\infty.$$

So $f = 0$ implies $u = 0$, which is uniqueness.

Note the remark applies to any positive heat kernel P on M satisfying the inequality (VII.3.9) for all $(x,t) \in M \times (0,\infty)$.

On the other hand, if the minimal positive heat kernel p of M is stochastically complete, then p is the unique heat kernel among all positive heat kernels on M satisfying (VII.3.9). Indeed, if P is a positive heat kernel on M satisfying (VII.3.9), then $P \geq p$ on all of $M \times M \times (0,\infty)$, which implies

$$1 = \int_M p(x,y,t)\,dV(y) \leq \int_M P(x,y,t)\,dV(y) \leq 1,$$

which implies the claim.

Definition Let M be a Riemannian manifold that satisfies the uniqueness property for the heat equation. Then to every bounded continuous function ϕ on M one assigns the *heat flow* $t \mapsto H_t\phi$ of ϕ, defined by

$$(H_t\phi)(x) = u(x,t),$$

where $u(x,t)$ is the solution of the initial value problem with initial data ϕ.

Theorem VII.3.5 *Let M be a Riemannian manifold that satisfies the uniqueness property for the heat equation. Then the heat flow H_t is given by*

$$(H_t\phi)(x) = \int_M p(x,y,t)\phi(y)\,dV(y),$$

where p is the unique heat kernel of M. Furthermore, the heat flow H_t satisfies the semigroup property, *namely,*

$$H_{t+s} = H_t \circ H_s$$

for all $t, s > 0$.

Proof Given a bounded continuous function f, both

$$u(x,\tau) = H_{\tau+s}f(x), \qquad v(x,\tau) = H_\tau \circ H_s f(x)$$

are solutions to the heat equation. Both have the same initial values $u(x,0) = v(x,0) = H_s f(x)$, which is bounded continuous. The uniqueness then implies that $u(x,\tau) = v(x,\tau)$ for all $(x,\tau) \in M \times (0,+\infty)$, which implies the claim. ∎

Theorem VII.3.6 *Let M be a Riemannian manifold with compact closure and C^∞ boundary, and let Q_t denote the heat flow on M subject to the constraint that the solution vanishes on $\partial M \times (0,\infty)$. Then Q_t is given by*

$$(Q_t\phi)(x) = \int_M q(x,y,t)\phi(y)\,dV(y),$$

where q is the Dirichlet heat kernel of M. Furthermore, the heat flow Q_t satisfies the semigroup property.

Proof The same as the previous theorem. ∎

Definition Given a Riemannian manifold M with heat kernel p. For any $t > 0$, the heat kernel defines an integral operator P_t by

$$(P_t f)(x) = \int_M p(x,y,t)f(y)\,dV(y)$$

for any bounded continuous function f on M. Then

$$\lim_{t\downarrow 0}(P_t f)(x) = f(x)$$

for all $x \in M$.

Theorem VII.3.7 *Let p be a closed, Dirichlet, or minimal positive heat kernel on the Riemannian manifold M. Then p satisfies*

$$p(x,y,t+s) = \int_M p(x,z,t)p(z,y,s)\,dV(z) \qquad \text{(VII.3.11)}$$

for all x, y and t, s.

Proof In the first two cases, M satisfies the uniqueness property, which implies the semigroup property.

For the minimal positive heat kernel, we first note that the Dirichlet heat kernel q of a domain Ω in M satisfies (VII.3.11). Next, for an arbitrary noncompact M, and an exhaustion Ω_j of M by domains with compact closure and C^∞ boundary possessing Dirichlet heat kernels q_j, one obtains (VII.3.11) for p by passing to the limit of (VII.3.11) for q_j. ∎

VII.4 The Action of the Heat Semigroup

Theorem VII.4.1 *Let M be a Riemannian manifold with minimal positive heat kernel p (this includes closed and Dirichlet heat kernels). We denote by $\mathcal{C}$ the space of functions $C_c^\infty(M)$. Then for each positive k, the action of P_t on $\mathcal{C}$ extends to a contractive, continuous semigroup on $L^k(M)$. Furthermore, P_t is positive self-adjoint on $L^2(M)$. Thus, any eigenvalue of P_t must be positive, with C^∞ eigenfunction.*

Proof First, (VII.3.9) and the symmetry of p imply

$$\int_M p(x, y, t)\, dV(y) \leq 1 \tag{VII.4.1}$$

for all $(x, t) \in M \times (0, +\infty)$. For $f \in \mathcal{C}$ we have

$$\begin{aligned}
\|P_t f\|_1 &= \int_M dV(x) \left| \int_M p(x, y, t) f(y)\, dV(y) \right| \\
&\leq \int_M dV(x) \int_M p(x, y, t) |f(y)|\, dV(y) \\
&= \int_M |f(y)|\, dV(y) \int_M p(x, y, t)\, dV(x) \\
&\leq \int_M |f(y)|\, dV(y) \\
&= \|f\|_1,
\end{aligned}$$

that is, $\|P_t f\|_1 \leq \|f\|_1$; in particular, $\|P_t\|_{1\to1} \leq 1$. Similarly, Hölder's inequality and (VII.4.1) imply

$$\begin{aligned}
\|P_t f\|_k{}^k &= \int_M \left| \int_M p(x, y, t) f(y)\, dV(y) \right|^k dV(x) \\
&\leq \int_M dV(x) \int_M p(x, y, t) |f(y)|^k\, dV(y) \\
&\leq \int_M |f(y)|^k\, dV(y),
\end{aligned}$$

that is, $\|P_t f\|_k \leq \|f\|_k$ for all $k \in [1, \infty)$, $f \in \mathcal{C}$. ■

Because the subspace $\mathcal{C}$ is dense in L^k, for each $k \in [1, \infty)$, we may extend the bounded operator $P_t : \mathcal{C} \to C^\infty$ to a bounded operator on L^k satisfying

$$\|P_t\|_{k\to k} \leq 1$$

for all $k \in [1, \infty]$ (the case $k = \infty$ is in Remark VII.3.8).

Next, given $k \in [1, \infty)$, we approximate $f \in L^k$ by $\phi \in \mathcal{C}$. Then

$$P_t f - f = P_t\{f - \phi\} + P_t\phi - \phi + \phi - f,$$

which implies

$$\begin{aligned}\limsup_{t\downarrow 0} \|P_t f - f\|_k &\le \limsup_{t\downarrow 0} \{\|P_t\{f - \phi\}\|_k + \|P_t\phi - \phi\|_k + \|\phi - f\|_k\} \\ &= O(\|f - \phi\|_k)\end{aligned}$$

as $\phi \to f$ in L^k. Thus $P_t f \to f$ in L^k for all $k \in [1, \infty)$. Therefore P_t is continuous at $t = 0$.

The semigroup property

$$P_t \circ P_s = P_{t+s}$$

remains valid on L^k for all k.

Because P_t is bounded, with symmetric kernel, P_t is self-adjoint on L^2.

Next, consider P_t acting on L^2, t fixed. Then

$$(P_t f, f) = (P_{t/2}P_{t/2}f, f) = (P_{t/2}f, P_{t/2}f) = \|P_{t/2}f\|^2 \ge 0$$

(we mean L^2 norm unless otherwise indicated). If there exists f such that $P_t f = 0$, then the above shows that $P_{t/2}f = 0$. One uses the continuity of the semigroup to show that this implies that $f = 0$. Therefore the quadratic form associated with P_t (as a self-adjoint operator) is positive definite, and all eigenvalues of P_t must be positive. Moreover, any eigenfunction of P_t must be C^∞.

As discussed in §VII.1.3, the semigroup $P_t : L^\ell \to L^\ell$ has generator $\overline{\Delta}_\ell$ defined by

$$\overline{\Delta}_\ell f = \lim_{t\downarrow 0} \frac{P_t f - f}{t},$$

with domain $\mathcal{D}(\overline{\Delta}_\ell)$ consisting of functions for which the limit on the right hand side exists in L^ℓ. Then

$$P_t : \mathcal{D}(\overline{\Delta}_\ell) \to \mathcal{D}(\overline{\Delta}_\ell),$$

$\mathcal{D}(\overline{\Delta}_\ell)$ is dense in L^ℓ, and $\mathcal{D}(\overline{\Delta}_\ell)$ is complete with respect to the norm

$$|||f|||_\ell = \|f\|_\ell + \|\overline{\Delta}_\ell f\|_\ell.$$

Notation We shall write $\overline{\Delta}$ for $\overline{\Delta}_2$.

Henceforth we shall fix $P_t : L^2 \to L^2$. Since P_t is self-adjoint for all $t > 0$, we have $\overline{\Delta}$ is self-adjoint. It remains to identify $\overline{\Delta}$. We know that it is a self-adjoint extension of Δ_c, the Laplacian acting on C_c^∞.

Theorem VII.4.2 *Assume M has compact closure with C^∞ boundary, Q_t the heat semigroup associated with the Dirichlet heat kernel of M. Then $\overline{\Delta}$ is equal to the Friedrichs extension $\mathfrak{F}_c$ of Δ_c, the Laplacian acting on $C_c^\infty(M)$.*

Proof Let $\mathcal{D}_{\text{dir}}$ consist of those functions $f \in C^\infty(\overline{M})$, which satisfy $f|\partial M = 0$, so $\mathcal{D}_{\text{dir}}$ determines the Laplacian Δ_{dir} associated with vanishing Dirichlet boundary data, with Friedrichs extension $\mathfrak{F}_{\text{dir}} = \mathfrak{F}_c$ (see Corollary VII.2.1). Since $Q_t(\mathcal{D}_{\text{dir}}) \subseteq \mathcal{D}_{\text{dir}}$ for all $t > 0$, Proposition VII.1.3 implies that Δ_{dir} is essentially self-adjoint. So, when M has compact closure and C^∞ boundary, $\mathfrak{F}_c$ is the unique self-adjoint extension of Δ_{dir}. In particular, the Friedrichs extension $\mathfrak{F}_c = \overline{\Delta}$. ∎

Remark VII.4.1 It is worth noting that once we have the self-adjoint Dirichlet Laplacian realized as the Friedrichs extension of the function space $C_c^\infty(M)$, we have removed any mention of the boundary, its regularity, and the associated boundary data from the characterization of the self-adjoint extension. Therefore, for an arbitrary Riemannian manifold M one may refer to the Friedrichs extension of $C_c^\infty(M)$ as the Dirichlet Laplacian. So one is always speaking of a specific self-adjoint extension of the Laplacian. The issue of genuinely identifying the domain of the Dirichlet Laplacian is then a separate question, should one be interested in it, or should one require the specific information. Then the details of the geometry of M and its boundary would come to the fore.

Assume M is a complete Riemannian manifold. Then, by Theorem VII.2.1, $\overline{\Delta}$ is the unique self-adjoint extension of Δ_c. In particular, $\overline{\Delta}$ agrees with the Friedrichs extension $\mathfrak{F}_c$.

Theorem VII.4.3 *Let p denote the minimal postive heat kernel of an arbitrary Riemannian manifold M, with associated semigroup P_t. Then*

$$\|\text{grad}\, P_t\phi\|^2 = -(P_t\phi, \Delta P_t\phi) \tag{VII.4.2}$$

for all $\phi \in C_c^\infty$, $t > 0$.

Proof (Here, Δ still denotes the pointwise Laplace operator.)

Start with an exhaustion of M, $D_j \uparrow M$, consisting of relatively compact domains with C^∞ boundary, and their attendant Dirichlet heat kernels $q_j = q_{D_j} \uparrow p$. Associate with each q_j its semigroup Q_t^j. Then for any function $\phi \in C_c^\infty(M)$, we have

$$Q_t^j\phi \to P_t\phi, \qquad \text{grad}\, Q_t^j\phi \to \text{grad}\, P_t\phi, \qquad \Delta Q_t^j\phi \to \Delta P_t\phi,$$

as $j \to \infty$, uniformly on compact subsets of M. This implies

$$\begin{aligned}\|\operatorname{grad} Q_t^j \phi\|^2 &= -(Q_t^j \phi, \Delta Q_t^j \phi) \\ &= -(Q_t^j \phi, Q_t^j \Delta \phi) \\ &= -(Q_{2t}^j \phi, \Delta \phi) \\ &\to -(P_{2t}\phi, \Delta\phi) \\ &= -(P_t\phi, \Delta P_t \phi)\end{aligned}$$

[the fourth line follows from $q_j \uparrow p$, and the fifth line follows from Proposition VII.1.2(a)]. For any compact K in M and $\epsilon > 0$ there exists j_0 such that for all $j \geq j_0$ we have

$$\begin{aligned}\int_K |\operatorname{grad} P_t\phi|^2 \, dV &\leq \int_K |\operatorname{grad} Q_t^j \phi|^2 \, dV + \epsilon \\ &\leq \int_M |\operatorname{grad} Q_t^j \phi|^2 \, dV + \epsilon,\end{aligned}$$

which implies

$$\begin{aligned}\int_K |\operatorname{grad} P_t\phi|^2 \, dV &\leq \liminf_{j\to\infty} \int_M |\operatorname{grad} Q_t^j \phi|^2 \, dV + \epsilon \\ &= -(P_t\phi, \Delta P_t\phi) + \epsilon.\end{aligned}$$

In particular, letting $\epsilon \downarrow 0$ and $K \uparrow M$, we obtain

$$\|\operatorname{grad} P_t\phi\|^2 \leq -(P_t\phi, \Delta P_t\phi).$$

So $\operatorname{grad} P_t\phi \in L^2$.

Therefore, given $\phi \in C_c^\infty$, we have $P_t\phi \in \mathfrak{H} = \mathcal{D}(\Delta^{1/2})$, the domain of the quadratic form associated with Δ. Next, we note that the sequence $Q_t^j\phi$ is a Cauchy sequence in $\mathfrak{H}$. Indeed, we certainly have $Q_t^j\phi$ a Cauchy sequence in L^2. As above, we also have

$$\begin{aligned}\|\operatorname{grad} Q_t^j\phi - \operatorname{grad} Q_t^k\phi\|^2 &= -(Q_t^j\phi - Q_t^k\phi, \Delta Q_t^j\phi - \Delta Q_t^k\phi) \\ &= -(Q_t^j\phi - Q_t^k\phi, Q_t^j\Delta\phi - Q_t^k\Delta\phi) \\ &\to 0\end{aligned}$$

as $j, k \to \infty$. So $Q_t^j\phi$ is Cauchy in $\mathfrak{H}$. Therefore, there exists $\Phi \in \mathfrak{H}$ such that $Q_t^j\phi \to \Phi$ in $\mathfrak{H}$ as $j \to \infty$. Then Φ has an L^2 weak derivative, that is, there exists an L^2 vector field Ψ on M such that

$$(\Psi, X) = -(\Phi, \operatorname{div} X)$$

for all C^∞ vector fields X on M with compact support. But then for such a vector field X we have

$$\begin{aligned}(\operatorname{grad} P_t\phi, X) &= -(P_t\phi, \operatorname{div} X)\\ &= \lim_{j\to\infty} -\left(Q_t^j\phi, \operatorname{div} X\right)\\ &= \lim_{j\to\infty} \left(\operatorname{grad} Q_t^j\phi, X\right)\\ &= \lim_{j\to\infty} (\Psi, X),\end{aligned}$$

which implies $\Psi = \operatorname{grad} P_t\phi$, which implies $Q_t^j\phi \to P_t\phi$ in $\mathfrak{H}$. Therefore

$$\|\operatorname{grad} P_t\phi\|^2 = \lim_{j\to\infty} \left\|\operatorname{grad} Q_t^j\phi\right\|^2 = -(P_t\phi, \Delta P_t\phi),$$

which conludes the proof of the theorem. ■

Theorem VII.4.4 *Let M be an arbitrary Riemannian manifold, $\mathfrak{F}_c$ the Friedrichs extension of Δ_c, and p the minimal positive heat kernel, with associated semigroup P_t and infinitesmal generator $\overline{\Delta}$. Then*

$$\overline{\Delta} = \mathfrak{F}_c.$$

Proof Let T_t be the 1-parameter semigroup generated by $\mathfrak{F}_c$, $\phi \in C_c^\infty(M)$, and set

$$v(x,t) = (T_t\phi)(x) - (P_t\phi)(x).$$

Then v is a solution of the heat equation with vanishing initial data. We want to show that v vanishes on all of $M \times (0,\infty)$.

The spectral theorem, Lebesgue's dominated convergence theorem, and the estimate

$$\frac{e^x - 1}{x} \le 2, \qquad 0 < x \ll 1,$$

combine to imply that we may differentiate $\int v^2(x,t)\,dV(x)$, under the integral sign, with respect to t. Then (see the proof of Theorem VII.3.4)

$$\begin{aligned}\frac{1}{2}\frac{\partial}{\partial t}\int v^2(x,t)\,dV(x) &= \int \frac{\partial v}{\partial t}(x,t)v(x,t)\,dV(x)\\ &= \int \Delta v(x,t)v(x,t)\,dV(x).\end{aligned}$$

By Theorem VII.4.3, for each $t > 0$, $P_t\phi$ is in the domain of $\mathfrak{F}_c$, and therefore

v is in the domain of $\mathfrak{F}_c$, which implies

$$\frac{1}{2}\frac{\partial}{\partial t}\int v^2(x,t)\,dV(x) = \int \Delta v(x,t)v(x,t)\,dV(x)$$
$$= -\int |\mathrm{grad}\, v|^2(x,t)\,dV(x) \leq 0,$$

which implies the claim. ∎

Notation We henceforth write, when there is no confusion, Δ for $\overline{\Delta} = \mathfrak{F}_c$. Thus, if $\{E_\lambda : \lambda \geq 0\}$ is a spectral family for Δ, then

$$\Delta = -\int_0^\infty \lambda\, dE_\lambda,$$

with associated quadratic form

$$D[f, f] = (-\Delta f, f) = \int_0^\infty \lambda\,(dE_\lambda f, f)$$

possessing domain $\mathcal{D}((-\Delta)^{1/2}) \supseteq \mathcal{D}(\Delta)$; and

$$P_t = \int_0^\infty e^{-\lambda t}\, dE_\lambda.$$

Theorem VII.4.5 *Let p be a closed or Dirichlet heat kernel on a Riemannian manifold M. Then $-\Delta$ has discrete spectrum, and $p(x, y, t)$ has the absolutely, uniformly convergent Sturm–Liouville expansion*

(VII.4.3) $$p(x, y, t) = \sum_{j=1}^{\infty} e^{-\lambda_j t}\phi_j(x)\phi_j(y),$$

where $\{\phi_1, \phi_2, \ldots\}$ is a complete orthonormal basis of $L^2(M)$ consisting of eigenfunctions of $-\Delta$, with the eigenvalue of ϕ_j equal to λ_j. Also, $\phi_j \in C^\infty$ for all j.

Proof Indeed, under these hypotheses, for every fixed t, $p(x, y, t)$ is a Hilbert–Schmidt kernel (that is, symmetric and square-integrable on $M \times M$), which implies the discreteness of the spectrum and the expansion (VII.4.3) in $L^2(M \times M)$. The stronger statement of convergence of (VII.4.3) follows from Mercer's theorem (Riesz and Nagy, 1955, p. 245). ∎

Remark VII.4.2 Let p be the heat kernel of a compact Riemannian manifold M. The constant functions clearly are annihilated by Δ, so $\lambda_1 = 0$. On the other hand, if $\Delta\phi = 0$, then $0 = (-\Delta\,\phi, \phi) = \|\mathrm{grad}\,\phi\|^2$, which implies $\phi = \mathrm{const.}$

This implies $\lambda_2 > 0$. In particular,

$$\phi_1 = \frac{1}{\sqrt{V(M)}},$$

which implies (using L^2–L^∞ estimates of eigenfunctions, as in Li (1980) for example)

$$\lim_{t\uparrow+\infty} p(x, y, t) = \frac{1}{V(M)}.$$

So the heat is evenly distributed throughout a compact manifold as $t \uparrow +\infty$.

Notation For any Riemannian manifold M we let

$$\lambda(M) = \inf \operatorname{spec} \; - \Delta.$$

Proposition VII.4.1 *Assume $\lambda(M)$ has an L^2 eigenfunction. Then the eigenfunction never vanishes, and the eigenspace of $\lambda(M)$ is 1-dimensional.*

Proposition VII.4.2 *Given two domains $\Omega_1 \subseteq \Omega_2$ in a Riemannian manifold M, let $-\Delta_1$ and $-\Delta_2$ denote their respective Laplacians; then $\lambda(\Omega_1) \geq \lambda(\Omega_2)$. When $\Omega_1 \subset\subset \Omega_2$ then the inequality is strict. In particular, $\lambda(\Omega) > \lambda(M)$ for any relatively compact Ω in M.*

VII.5 Simplest Examples

Example VII.5.1 If $M = \mathbb{R}^n$, then

$$ds^2 = \sum_{j=1}^{n} (dx^j)^2 = |dx|^2, \qquad dV = dx^1 \cdots dx^n,$$

$$p(x, y, t) = \frac{1}{(4\pi t)^{n/2}} e^{-|x-y|^2/4t}$$

as discussed in §VII.3.1.

Example VII.5.2 Let $M = \mathbb{H}^n = \mathbb{M}_{-1}$, the n-dimensional hyperbolic space of constant sectional curvature -1, and $x \in M$. Then any point $y \in M$ is determined by its distance r from x, together with the unit tangent vector ξ at x of the unique geodesic from x to y. So $(r; \xi)$ determine geodesic spherical coordinates on all of M, relative to which

$$ds^2 = dr^2 + \sinh^2 r \, |d\xi|^2,$$

where $|d\xi|^2$ denotes the Riemannian metric on S_x (the unit tangent sphere to

M at x) with its standard metric. The volume element is given by

$$dV = \sinh^{n-1} r \, dr d\mu_x(\xi),$$

where $d\mu_x(\xi)$ denotes the $(n-1)$-volume element of S_x. The Laplacian is given by

$$(\Delta f)(r;\xi) = \frac{\partial^2 f}{\partial r^2} + (n-1)\coth r \frac{\partial f}{\partial r} + \sinh^{-2} r \, \Delta_\xi(f_r),$$

where Δ_ξ denotes the Laplacian on S_x, and f_r denotes the restriction of f to the metric sphere in M centered at x with radius r, viewed as a function on S_x. The simplest explicit formula for the heat kernel is when $n = 3$, namely, on $\mathbb{H}^3$ the heat kernel is given by

$$p(x, y, t) = \frac{e^{-t}}{(4\pi t)^{3/2}} \frac{d(x, y)}{\sinh d(x, y)} e^{-d(x,y)^2/4t}.$$

Example VII.5.3 If $M = M_1 \times M_2$ is the Riemannian product of M_1 and M_2, then for any chart $\mathbf{x} = (\mathbf{x}_1, \mathbf{x}_2)$ on M, where x_j is a chart on M_j, $j = 1, 2$, we have for the matrix G^* of the product Riemannian metric

$$G^* = \begin{pmatrix} G_1 & \\ & G_2 \end{pmatrix},$$

where G_j denotes the matrix of the Riemannian metric on M_j. For $F : M \to \mathbb{R}$ given by

$$F(q_1, q_2) = f_1(q_1) f_2(q_2), \qquad (q_1, q_2) \in M_1 \times M_2,$$

we have

$$\Delta F(q_1, q_2) = \Delta_1 f_1(q_1) f_2(q_2) + f_1(q_1) \Delta_2 f_2(q_2)$$

(with obvious notation). For the heat kernel we have

$$p((x_1, x_2), (y_1, y_2), t) = p_1(x_1, y_1, t) p_2(x_2, y_2, t).$$

Example VII.5.4 If p is a closed or Dirichlet heat kernel on M, then by Theorem VII.4.5, $p(x, y, t)$ has the Sturm–Liouville eigenvalue–eigenfunction expansion

(VII.5.1) $$p(x, y, t) = \sum_{j=1}^{\infty} e^{-\lambda_j t} \phi_j(x) \phi_j(y),$$

where $\{\lambda_1 < \lambda_2 \leq \cdots \uparrow +\infty\}$ denotes the spectrum of $-\Delta$ on M (it is discrete), with eigenvalues repeated according to their multiplicity, and $\{\phi_j : j = 1, \ldots\}$

is a complete orthonormal basis of L^2 such that ϕ_j is an eigenfunction of λ_j for each j.

Example VII.5.5 The 2-dimensional jungle gym JG^2 as described in Example V.2.4. We shall see that for sufficiently large time one has

$$p(x, y, t) \leq \text{const.} t^{-3/2} e^{-d^2(x,y)/\text{const.}t}.$$

VII.6 Bibliographic Notes

General references for the material in this chapter are Davies (1980, 1989, 1995) and Chavel (1984, 1994). [Formula (38) of Chavel (1984, p. 150) is *only* valid when $\kappa < 0$.] See also the articles of Chavel, Davies, and Grigor'yan in the 1998 Edinburgh Lectures (Davies and Safaro. pp. 30–94, 14–225).

§**VII.1** Additional background for this section may be found in Yosida (1978) and Reed and Simon (1975).

§**VII.2** The Meyers–Serrin theorem is proved in Adams (1975), Aubin (1982), and Gilbarg and Trudinger (1977). Theorem VII.2.1 was first proved in Gaffney (1954); our proof follows Karp (1984). Corollary VII.2.2 is from Yau (1976). Elliptic regularity is treated in Gilbarg and Trudinger (1977, p. 176).

§**VII.3** One can find a proof of the strong maximum principle in Protter and Weinberger (1984, §III.3.4) for the case where M is diffeomeorphic to a domain in Euclidean space. A standard continuation argument then extends the theorem to arbitrary Riemannian manifolds.

The proof of Theorem VII.3.3 corrects an incorrect proof given in Chavel (1984, p. 139).

The uniqueness property and stochastic completeness for Ricci curvature bounded from below (Proposition VII.3.6) was first proved in Dodziuk (1983), and the Feller property in Yau (1978). The last two results are valid in much greater generality; for stochastic completeness, see Azencott (1974), Davies (1992), Gaffney (1959), Grigor'yan (1987), Hasminkii (1960), Ichihara (1982, 1984, 1986), Karp and Li (unpublished), Oshima (1989), Pang (1988, 1996), Takeda (1989, 1991), Varopoulos (1983). An extensive survey of these and other results can be found in Grigor'yan (1999). For the Feller property, see Azencott (1974), Davies (1992), Karp and Li (unpublished), Pang (1988, 1996), Gaffney (1959), Hasminkii (1960).

§**VII.4** Theorem VII.4.4 is proved in Dodziuk (1983). Proposition VII.4.1 is from Sullivan (1987).

§**VII.5** Explicit calculations for the heat kernel on a hyperbolic space of any dimension can be found in Davies and Mandouvalos (1987) and Grigor'yan and Noguchi (1998).

For the heat kernel on Lie groups, see the articles of Arede (1985) and Fegan (1976, 1983). For the heat kernel on homogeneous spaces, see Benabdallah (1973). For the special case of spheres, see Fischer, Jungster, and Williams (1984) and Ndumu (1986).

VIII

Large Time Heat Diffusion

In this chapter we present the main result of this half of the book – Theorem VIII.5.4 – that, in a complete Riemannian manifold of bounded geometry, a positive modified isoperimetric constant $\mathfrak{I}_{\nu,\rho}$ implies an upper bound for the minimal positive heat kernel p of the type

$$p(x, y, t) \leq \text{const.} t^{-\nu/2} \qquad \forall\, t \geq T_0 > 0,$$

or, equivalently,

$$\limsup_{t \uparrow +\infty} t^{\nu/2} p(x, y, t) < +\infty \tag{VIII.0.1}$$

for all $x, y \in M$. So the result states that if M is "ν-dimensional" in the sense of isoperimetric inequalities, then it is at least "ν-dimensional" in the sense of heat diffusion. The method of proof is to pass from the geometric inequality $\mathfrak{I}_{\nu,\rho} > 0$ to a modified Nirenberg–Sobolev inequality, which is then shown to be equivalent to the large time upper bound (VIII.0.1).

The transition from $\mathfrak{I}_{\nu,\rho} > 0$ to the modified Nirenberg–Sobolev inequality is achieved by shifting the problem to any discretization of the manifold (Theorem VIII.5.2). A by-product of this argument is the invariance of (VIII.0.1) under compact perturbations of M, of the type discussed in Theorem V.3.1.

The final section of the chapter closes the book with an alternate proof of Theorem VIII.5.4 (see Example VI.3.1 and Example VIII.6.1 below) using Grigor'yan's replacement of the Nirenberg–Sobolev inequalities with Faber–Krahn isoperimetric inequalities as the primary tool for establishing upper bounds on the heat kernel. The full possibilities of this method seem to go well beyond the result presented here; so the best may be yet to come.

We assume, throughout this chapter, that M is a Riemannian manifold and p its minimal positive heat kernel, with attendant semigroup P_t acting on $L^2(M)$, which is contractive, continuous, and self-adjoint. The infinitesimal generator of the semigroup is the Friedrichs extension of Δ_c, the Laplacian acting on $C_c^\infty(M)$ considered as a dense subspace of $L^2(M)$.

The semigroup P_t also has an action on L^k, $k \geq 1$, which is contractive and continuous. Since $L^k \cap L^2$ is dense in L^k, we shall be able to study the norms of

$$P_t : L^k \cap L^2 \to L^\ell \cap L^2$$

as though they were mappings from L^k to L^ℓ, with their associated norms $\|P_t\|_{k\to\ell}$.

VIII.1 The Main Problem

Proposition VIII.1.1 *Assume $x \in M$ such that the exponential map* exp *is defined on the closed disk $\overline{\mathrm{B}(x;R)}$ in M_x. Then for*

$$d(x,w) < R/2, \quad 0 < r < \frac{d(x,w)}{2},$$

we have

$$\text{(VIII.1.1)} \quad p(x,w,s) \leq \text{const.} \frac{s^{-n/2} + sr^{-(n+2)}}{\sqrt{\mathfrak{I}_n(B(x;r))\mathfrak{I}_n(B(w;r))}} e^{-\{d(x,w)-2r\}^2/4s},$$

where $\mathfrak{I}_n(B(z;\rho))$ denotes the n-dimensional isoperimetric constant of $B(z;\rho)$.

Recall that, for any Riemannian manifold M, we set

$$\mathcal{E}(x,y,t) = (4\pi t)^{-n/2} e^{-d^2(x,y)/4t}.$$

Theorem VIII.1.1 *Assume M is a complete Riemannian manifold. Then for every compact $K \subset\subset M$ there exist positive constants (depending on K) such that*

$$\text{(VIII.1.2)} \qquad \limsup_{t\downarrow 0} \left| \frac{p(x,y,t)}{\mathcal{E}(x,y,t)} - 1 \right| \leq \text{const.} d(x,y)$$

is valid for all $x, y \in K$.

Proof We already know the result for M compact, so we assume that M is complete and noncompact.

Given x and y in M, let Ω be any domain in M, with C^∞ boundary and compact closure, containing x, y, and let q denote the Dirichlet heat kernel of Ω. Then $p(x,y,t) > q(x,y,t)$ by the maximum principle, which implies by

Proposition VII.3.5 that

$$\liminf_{t\downarrow 0} \frac{p(x, y, t)}{\mathcal{E}(x, y, t)} \geq 1 - \text{const.}d(x, y).$$

To prove

$$\limsup_{t\downarrow 0} \frac{p(x, y, t)}{\mathcal{E}(x, y, t)} \leq 1 + \text{const.}d(x, y)$$

we argue as follows: Let Ω and q be as above. Then Duhamel's principle implies

$$\begin{aligned}
&p(x, y, t) - q(x, y, t) \\
&\quad = -\int_0^t ds \int_{\partial\Omega} p(x, w, s)\frac{\partial q}{\partial \nu_w}(w, y, t-s)\, dA(w) \\
&\quad \leq \left\{ \sup_{w\in\partial\Omega,\ s\in[0,t]} p(x, w, s)\right\} \int_0^t ds \int_{\partial\Omega} -\frac{\partial q}{\partial \nu_w}(w, y, t-s)\, dA(w) \\
&\quad = \left\{ \sup_{w\in\partial\Omega,\ s\in[0,t]} p(x, w, s)\right\} \left\{1 - \int_0^t ds \iint_{\Omega} q(z, y, t)\, dV(z)\right\} \\
&\quad \leq \sup_{w\in\partial\Omega,\ s\in[0,t]} p(x, w, s).
\end{aligned}$$

But one has, for any $r \in (0, d(x, w)/2)$, the upper bound (VIII.1.1). Therefore pick Ω so that $d(x, \partial\Omega)$ is very large compared to $d(x, y)$, fix r very small, and let $t \downarrow 0$. Then one has, for some $\alpha \in (0, 1)$,

$$\begin{aligned}
\limsup_{t\downarrow 0} \frac{p(x, y, t)}{\mathcal{E}(x, y, t)} &\leq \limsup_{t\downarrow 0} \frac{q(x, y, t)}{\mathcal{E}(x, y, t)} \\
&\quad + \limsup_{t\downarrow 0} \frac{\text{const.}_{x,\partial\Omega} t^{-n/2} e^{-(1-\alpha)d^2(x,\partial\Omega)/4t}}{\mathcal{E}(x, y, t)} \\
&\leq 1 + \text{const.}d(x, y),
\end{aligned}$$

which is the claim. ■

Remark VIII.1.1 One can see (VIII.1.2) explicitly in both the example of $\mathbb{H}^3$ (Example VII.5.2) and the product metrics (Example VII.5.3), in the sense that if one has this asymptotic behavior on each of the factors, then one also has it on the product.

Remark VIII.1.2 The theorem remains valid even if M is not complete, but one must formulate it differently. Namely, given any $\Omega \subset\subset M$, $x \in \Omega$, the inequality (VIII.1.2) will be valid for y sufficiently close to x.

For large time considerations the behavior of the heat diffusion takes into account the large scale structure of the manifold, and then anything can happen. Hyperbolic space displays an example of when the heat kernel decays exponentially with respect to time, and Euclidean space when the rate of decay with respect to time is given by the dimension of the manifold. For M compact, one has $p(x, y, t) \to 1/V(M)$ as $t \uparrow +\infty$. Therefore, for $M = \mathbb{S}^{n-k} \times \mathbb{R}^k$, the heat kernel decays at the rate $t^{-k/2}$, slower than that indicated by the dimension n of M. Finally, we have examples of polynomial rate of decay that are faster than that indicated by n, the dimension of M, namely, the jungle gym of Example VII.5.5.

Main Problem Study the large time decay of the heat kernel, as a consequence of geometric hypotheses on M.

Except for Theorem VIII.1.4 (which is included for motivation), we shall restrict ourselves to the study of upper bounds.

In general, upper bounds on the heat kernel are studied in two stages: The first is referred to as *on-diagonal* upper bounds, namely, upper bounds for $p(x, x, t)$. By the semigroup property, the Cauchy–Schwarz inequality, and the symmetry property we have

$$\begin{aligned} p(x, y, t) &= \int_M p(x, z, t/2)p(z, y, t/2)\, dV(z) \\ &\leq \left\{\int_M p^2(x, z, t/2)\, dV(z)\right\}^{1/2} \left\{\int_M p^2(z, y, t/2)\, dV(z)\right\}^{1/2} \\ &= \sqrt{p(x, x, t)}\sqrt{p(y, y, t)}, \end{aligned}$$

that is,

$$p(x, y, t) \leq \sqrt{p(x, x, t)}\sqrt{p(y, y, t)}. \tag{VIII.1.3}$$

Therefore, as soon as we have any upper bound valid for $p(x, x, t)$ for all x, we automatically have an upper bound on $p(x, y, t)$. But this upper bound could not contain a Gaussian term of the form

$$e^{-d^2(x,y)/\text{const.}t}.$$

Then the second stage is to produce new arguments to sharpen the upper bounds based solely on the on-diagonal upper bounds. The introduction of the Gaussian correction to the upper bounds is also referred to as the *off-diagonal correction.* We shall not pursue this second stage here. It has recently been definitively established that the correction in upper bounds does not depend on the specific geometry of the manifold; rather, it is characteristic of the heat equation that upper on-diagonal bounds on the heat kernel automatically imply off-diagonal corrections.

VIII.1.1 General Considerations

Theorem VIII.1.2 *For on-diagonal upper bounds have*

$$\sup_x p(x, x, t) = \|P_t\|_{1\to\infty}. \tag{VIII.1.4}$$

Proof First, given $f \in L^1$, we have

$$|P_t f|(x) = \left|\int_M p(x, y, t) f(y)\, dV(y)\right| \le \sup_y p(x, y, t)\|f\|_1,$$

which implies

$$\begin{aligned}\sup_x |P_t f|(x) &\le \sup_{x,y} p(x, y, t)\|f\|_1 \\ &\le \|f\|_1 \sup_{x,y} \sqrt{p(x, x, t)}\sqrt{p(y, y, t)} \\ &= \|f\|_1 \sup_x p(x, x, t).\end{aligned}$$

Therefore, $\|P_t\|_{1\to\infty} \le \sup_x \, p(x, x, t)$. For the opposite inequality, recall that an *approximate identity at* $x \in M$ is a family of nonnegative functions ϕ_h, $h > 0$, satisfying

$$\int_M \phi_h\, dV = 1 \;\; \forall\, h > 0, \qquad \lim_{h\downarrow 0} \int_{M\setminus B(x;\epsilon)} \phi_h\, dV = 0$$

for all $\epsilon > 0$. Then $\phi_h \to \delta_x$, the delta function concentrated at x, as $h \downarrow 0$. For such an approximate identity ϕ_h at x we have

$$\begin{aligned}\int_M p(x, y, t)\phi_h(y)\, dV(y) &= P_t\phi_h(x) \\ &\le \|P_t\phi_h\|_\infty \\ &\le \|P_t\|_{1\to\infty}\|\phi_h\|_1 \\ &= \|P_t\|_{1\to\infty}.\end{aligned}$$

Now let $h \downarrow 0$. Then we obtain $p(x, x, t) \le \|P_t\|_{1\to\infty}$ for all $x \in M$. This implies (VIII.1.4). ■

Recall that $\lambda(M)$ denotes the infimum of the spectrum of the Laplacian $-\Delta$ on M.

Theorem VIII.1.3 *For all $x \in M$ we have that $t \mapsto e^{\lambda(M)t} p(x, x, t)$ is a decreasing function of t.*

Proof Let D be a relatively compact domain in M with C^∞ boundary and Dirichlet heat kernel q. Then the Sturm–Liouville eigenvalue–eigenfunction expansion of q, and $\lambda(D) > \lambda(M)$ combine to imply $e^{\lambda(M)t} q(x, x, t)$ is a decreasing function of t.

Now pick an exhaustion of M, $D_j \uparrow M$ as $j \uparrow +\infty$, by domains that are relatively compact in M and that possess C^∞ boundary. Let q_j denote the Dirichlet heat kernel of D_j. Since $q_j \uparrow p$, the lemma follows immediately. ■

Corollary VIII.1.1 *The norm $\| P_t \|_{1\to\infty}$ decreases with respect to time.*

Proposition VIII.1.2 *Let $\lambda = \lambda(M)$. For all x, y in M we have the existence of the limit*

$$\lim_{t\uparrow+\infty} e^{\lambda t} p(x, y, t) := \mathcal{F}(x, y), \tag{VIII.1.5}$$

for which we have the following alternative: Either $\mathcal{F}$ vanishes identically on all of $M \times M$, in which case λ possesses no L^2 eigenfunctions; or $\mathcal{F}$ is strictly positive on all of $M \times M$, in which case λ possesses a positive normalized L^2 eigenfunction ϕ (normalized in the sense that its L^2 norm is equal to 1) for which

$$\lim_{t\uparrow+\infty} e^{\lambda t} p(x, y, t) = \phi(x)\phi(y) \tag{VIII.1.6}$$

locally uniformly on all of $M \times M$.

The simplest example of the case $\mathcal{F} = 0$ is $\mathbb{R}^n$, $n \geq 1$, discussed above — just note that $\lambda = 0$, so $e^{\lambda t} p(x, y, t) = p(x, y, t) \to 0$ as $t \uparrow +\infty$. When M is noncompact with compact closure and smooth boundary, then one always has $\mathcal{F}$ strictly positive, and (VIII.1.6) follows from the Sturm–Liouville expansion (VII.5.1) of p.

We note some easy consequences of the proposition.

Corollary VIII.1.2 *We always have*

$$\lim_{t\uparrow+\infty} \frac{\ln p(x, y, t)}{t} = -\lambda \tag{VIII.1.7}$$

locally uniformly on $M \times M$; and when M has finite volume V, we have

(VIII.1.8)
$$\lim_{t \uparrow +\infty} p(x, y, t) = 1/V$$

locally uniformly on $M \times M$.

Proof We wish to show

$$\lim_{t \uparrow +\infty} \frac{\ln e^{\lambda t} p(x, y, t)}{t} = 0.$$

If the limit function $\mathcal{F}$ is positive, then the result is obvious. So we are only concerned with the situation where $\mathcal{F}$ is identically equal to 0. For any domain D in M, let q_D denote the Dirichlet heat kernel of D, and λ_D the lowest Dirichlet eigenvalue of D. Then, of course, we have

$$\frac{\ln e^{\lambda t} q_D(x, y, t)}{t} \leq \frac{\ln e^{\lambda t} p(x, y, t)}{t}.$$

We let $t \uparrow +\infty$. Then

$$\lambda - \lambda_D \leq \liminf_{t \uparrow +\infty} \frac{\ln e^{\lambda t} p(x, y, t)}{t}.$$

Now let $D \uparrow M$. We conclude that

$$0 \leq \liminf_{t \uparrow +\infty} \frac{\ln e^{\lambda t} p(x, y, t)}{t}.$$

But since $\mathcal{F} = 0$, we have $\ln e^{\lambda t} p(x, y, t) < 0$ for large t, which implies

$$\limsup_{t \uparrow +\infty} \frac{\ln e^{\lambda t} p(x, y, t)}{t} \leq 0,$$

which implies (VIII.1.7).

When M has finite volume then $\lambda = 0$ with normalized L^2–eigenfunction $\phi(x) = 1/\sqrt{V(M)}$ (for all x), which implies (VIII.1.8). ■

Corollary VIII.1.3 *For any M we have*

(VIII.1.9)
$$\lim_{t \uparrow +\infty} p(x, y, t) = 0$$

if and only if M has infinite volume.

Proof If M has finite volume, then (VIII.1.8) implies that $\lim p$, as $t \uparrow +\infty$, is nonzero. If, on the other hand, M has infinite volume, then (a) for $\lambda > 0$ simply use (VIII.1.5); and (b) for $\lambda = 0$, if $\mathcal{F}$ were positive, we would have the existence of an L^2 harmonic function on M, which is impossible by Corollary VII.2.2. ■

Corollary VIII.1.4 *Suppose M noncompact is a covering of a compact Riemannian manifold. Then $\mathcal{F}$ is identically equal to zero. Consequently, if the covering is nonamenable – by Brooks (1981), $\lambda > 0$ – then p tends to 0 faster than $e^{-\lambda t}$.*

Proof If $\lambda = 0$, one uses the above corollary, because M has infinite volume. If $\lambda > 0$ and $\mathcal{F} > 0$, then, as mentioned above, the L^2 eigenspace of λ, which is nontrivial, is 1-dimensional (by Proposition VII.4.1). But this is impossible, by the invariance of the eigenspace under the action of the deck transformation group. ■

Remark VIII.1.3 A fundamental distinction emerges between the cases $\lambda = 0$ and the cases $\lambda > 0$. If one subjects the complete Riemannian manifold M to a compact perturbation, then in the case of bounded geometry (that is, Ricci curvature bounded from below and positive injectivity radius) $\lambda = 0$ remains invariant under the perturbation. Moreover (Theorem VIII.5.2 below) the rate of polynomial decay of the heat kernel also remains invariant under the perturbation. However, in the case $\lambda > 0$ the value of λ does not necessarily remain invariant under the perturbation. Simply consider $\mathbb{H}^n$, n-dimensional hyperbolic space of constant sectional curvature -1, with $\lambda = (n-1)^2/4$, and subject it to a compact perturbation in which it has a flat disk with lowest Dirichlet eigenvalue strictly less than $(n-1)^2/4$. Then the new Riemannian metric has $\lambda < (n-1)^2/4$ with appropriately slower heat kernel decay for large time.

Our interest in what follows will usually involve polynomial decay with respect to time. So we will be interested in the case $\lambda = 0$.

VIII.1.2 Volume Growth Considerations

The most elementary geometric intuition is that, because the total amount of heat in the space cannot increase with respect to time, the larger the space, the quicker its heat kernel decays. Indeed, if M is compact, then $1/V(M)$ is the limit if the heat kernel as $t \uparrow +\infty$. In general, the intuition is valid, if not sufficiently precise. Most, if not all, of our geometric results will aim toward a deeper understanding of the effect of volume growth of a Riemmanian manifold on heat kernel decay.

Theorem VIII.1.4 *Given $o \in M$ and $\beta > 0$, for which one has the heat kernel lower bound*

(VIII.1.10) $$p(o, x, t) \geq \text{const.} t^{-\nu/2} e^{-d^2(o,x)/\beta t}$$

for all $x \in M$ and $t \geq T > 0$, then one has the volume growth upper bound

(VIII.1.11) $$V(o; r) \leq \text{const.} r^{\nu}$$

for sufficiently large $r > 0$.

Proof The heat kernel lower bound implies

$$1 \geq \int_{B(o;\sqrt{t})} p(o, x, t)\, dV(x) \geq \text{const.} t^{-\nu/2} e^{-1/\beta} V(o; \sqrt{t}),$$

which implies the volume upper bound. ■

Remark VIII.1.4 There are results for a partial converse (Coulhon and Grigor'yan, 1997): If (VIII.1.11) is valid for a given $o \in M$ and sufficiently large r, then

(VIII.1.12) $$p(o, o, t) \geq \text{const.} (t \ln t)^{-\nu/2}$$

for large t. Examples exist to show the estimate is sharp. Also, one can improve (VIII.1.12) to

$$p(x, x, t) \geq \text{const.} t^{-\nu/2}$$

assuming one is also given the upper bound

$$p(x, x, t) \leq \text{const.} t^{-\nu/2}$$

for large t. [See earlier versions in Benjamini, Chavel, and Feldman (1996)].

Example VIII.1.1 Does a volume growth lower bound imply a heat kernel upper bound? No. Endow $M = \mathbb{R}^2$ with the Riemannian metric

$$ds^2 = \phi(y)\{dx^2 + dy^2\},$$

where

$$0 < \phi(-y) = \phi(y) \in C^{\infty}, \qquad \phi(y) = y^{-2} \text{ for } |y| > 1.$$

Then one easily sees that M has exponential volume growth. But general considerations show, since the metric is conformal to the Euclidean plane (this is just for the 2-dimensional case), that

$$\int_1^\infty p(x,x,t)\,dt = +\infty,$$

which precludes a rate of decay of the form $t^{-\nu/2}$, $\nu > 2$.

We shall need the following heat kernel upper bound in the sequel.

Proposition VIII.1.3 *Let M be complete with Ricci curvature bounded from below by the constant $-K$, $K \geq 0$. Then for any $\alpha > 1$, one has the upper bound*

$$p(x,y,t) \leq \frac{C(\epsilon)^\alpha}{V^{1/2}(x;\sqrt{t})V^{1/2}(y;\sqrt{t})} \exp\left\{\frac{-d^2(x,y)}{(4+\epsilon)t} + c(n)\epsilon Kt\right\},$$

(VIII.1.13)

where the constant $C(\epsilon) \to +\infty$ as $\epsilon \to 0$.

Remark VIII.1.5 An application of Hölder's inequality implies that for any k, ℓ and respective conjugates k', ℓ' we have

$$\|P_t\|_{k\to\ell} = \|P_t{}^*\|_{\ell'\to k'}.$$

Because P_t is self-adjoint for all t, we have

$$\|P_t\|_{k\to\ell} = \|P_t\|_{\ell'\to k'}. \tag{VIII.1.14}$$

We shall use this repeatedly throughout our arguments.

VIII.2 The Nash Approach

Theorem VIII.2.1 *Assume there exists $\nu > 0$ such that the Nash–Sobolev inequality*

$$\|\text{grad}\, f\|_2 \geq \text{const.}\|f\|_2{}^{1+2/\nu}\|f\|_1{}^{-2/\nu} \tag{VIII.2.1}$$

is valid for all $f \in C_c^\infty$. Then

$$\|P_t\|_{1\to\infty} \leq \text{const.}t^{-\nu/2}, \tag{VIII.2.2}$$

for all $t > 0$.

Proof Set $u(t) = \|P_t f\|_2{}^2$. We then obtain

$$\frac{1}{2}u'(t) = \left(P_t f, \frac{\partial}{\partial t}P_t f\right) = (P_t f, \Delta P_t f) = -\|\text{grad}\, P_t f\|^2,$$

by Theorem VII.4.3. Then

$$\begin{aligned}\frac{1}{2}u'(t) &= -\|\text{grad } P_t f\|_2{}^2 \\ &\le -\text{const.}\|P_t f\|_2{}^{2+4/\nu}\|P_t f\|_1{}^{-4/\nu} \\ &\le -\text{const.}u(t)^{1+2/\nu}\|f\|_1{}^{-4/\nu},\end{aligned}$$

that is,

$$-\frac{u'(t)}{u(t)^{1+2/\nu}} \ge \text{const.}\|f\|_1{}^{-4/\nu},$$

which one integrates to obtain $u(t) \le \text{const.}\|f\|_1{}^2 t^{-\nu/2}$.

So $\|P_t\|_{1\to 2} \le \text{const.}t^{-\nu/4}$, which implies, by duality (VIII.1.14), $\|P_t\|_{2\to\infty} \le \text{const.}t^{-\nu/4}$, which implies (VIII.2.2). ■

Remark VIII.2.1 One can pass from $u(t) \le \text{const.}\|f\|_1{}^2 t^{-\nu/2}$ to (VIII.2.2) without using duality (III.2.5). Namely, for any $f \in C_c^\infty$ we have

$$\int \left\{\int p(w,y,t)f(y)\,dV(y)\right\}^2 dV(w) \le \text{const.}\|f\|_1{}^2 t^{-\nu/2},$$

that is,

$$\begin{aligned}\text{const.}\|f\|_1{}^2 t^{-\nu/2} &\ge \int \left\{\int p(w,y,t)f(y)\,dV(y)\right\}^2 dV(w) \\ &= \int \left\{\int p(w,y,t)f(y)\,dV(y)\right\} \\ &\quad \times \left\{\int p(w,z,t)f(z)\,dV(z)\right\} dV(w).\end{aligned}$$

Given any $x \in M$, let ϕ_h be an approximation of the identity concentrated at x, and let $f = \phi_h$. Then, letting $h \downarrow 0$, we obtain

$$\text{const.}t^{-\nu/2} \ge \int p(w,x,t)p(w,x,t)\,dV(w) = p(x,x,2t),$$

which implies the theorem. ■

Theorem VIII.2.2 *Conversely, assume* (VIII.2.2) *for all* $t > 0$. *Then* (VIII.2.1) *is valid on* C_c^∞.

Proof Let $\phi \in C_c^\infty$. Then the spectral theorem implies

$$\begin{aligned}(P_t\phi,\phi) &= \int_0^\infty e^{-\lambda t}\,(dE_\lambda\phi,\phi) \\ &\ge \int_0^\infty (1-\lambda t)\,(dE_\lambda\phi,\phi) = \|\phi\|_2{}^2 - t\|\text{grad}\,\phi\|_2{}^2,\end{aligned}$$

which implies

$$\begin{aligned}\|\phi\|_2{}^2 &\le t\|\operatorname{grad}\phi\|_2{}^2 + (P_t\phi, \phi)\\ &\le t\|\operatorname{grad}\phi\|_2{}^2 + \|P_t\phi\|_\infty\|\phi\|_1\\ &\le t\|\operatorname{grad}\phi\|_2{}^2 + \text{const.}t^{-\nu/2}\|\phi\|_1{}^2\end{aligned}$$

for all $t > 0$. Now minimize the right hand side with respect to t. The minimum is achieved at $t_0 = \text{const.}\|\phi\|_1{}^2/\|\operatorname{grad}\phi\|_2{}^2$, which implies the Nash–Sobolev inequality (VIII.2.1). ■

Remark VIII.2.2 Here is a proof without using the spectral theorem. One simply has

$$\frac{d}{dt}\|P_t\phi\|_2{}^2 = -\|\operatorname{grad} P_t\phi\|_2{}^2$$

and

$$\frac{d}{dt}\|\operatorname{grad} P_t\phi\|_2{}^2 = -\frac{\partial}{\partial t}(P_t\phi, \Delta P_t\phi) = -2\|\Delta P_t\phi\|^2 \le 0,$$

which implies $t \mapsto \|\operatorname{grad} P_t\phi\|_2{}^2$ is nonincreasing, and

$$\|\phi\|_2{}^2 = \|P_t\phi\|_2{}^2 + \int_0^t \|\operatorname{grad} P_t\phi\|^2\,dt.$$

The Riesz–Thorin interpolation theorem (see Proposition VIII.3.1 immediately below) implies, from (VIII.2.2), $\|P_t\|_{1\to 2} \le \text{const.}t^{-\nu/4}$ for all $t > 0$. Therefore

$$\|\phi\|_2{}^2 \le \text{const.}t^{-\nu/2}\|P_t\phi\|_1{}^2 + t\|\operatorname{grad}\phi\|_2{}^2 \le \text{const.}t^{-\nu/2}\|\phi\|_1{}^2 + t\|\operatorname{grad}\phi\|_2{}^2,$$

and we minimize as above.

VIII.3 The Varopoulos Approach

We first state some background results.

Proposition VIII.3.1 (Riesz–Thorin Interpolation Theorem) *Let $T : L^2 \to L^2$ satisfy*

$$\begin{aligned} T &: L^{p_0} \to L^{q_0}, & M_0 &= \|T\|_{p_0\to q_0},\\ T &: L^{p_1} \to L^{q_1}, & M_1 &= \|T\|_{p_1\to q_1},\end{aligned}$$

where $p_0, q_0, p_1, q_1 \in [0, \infty]$. For $\lambda \in (0, 1)$, set

$$\frac{1}{p_\lambda} = \frac{1-\lambda}{p_0} + \frac{\lambda}{p_1}, \qquad \frac{1}{q_\lambda} = \frac{1-\lambda}{q_0} + \frac{\lambda}{q_1}.$$

Then

$$T : L^{p_\lambda} \to L^{q_\lambda}, \qquad \|T\|_{p_\lambda \to q_\lambda} = {M_0}^{1-\lambda} {M_1}^{\lambda}.$$

Proposition VIII.3.2 (Maximal Theorem) *Let $T_t : L^2 \to L^2$ be a self-adjoint contraction semigroup. To every function $f \in L^2$ associate the function*

$$f^*(x) = \sup_{t>0} |T_t f|(x).$$

Then

$$\|f^*\|_p \le \text{const.}_p \|f\|_p$$

for all $f \in L^p \cap L^2$, $p \in [1, +\infty]$.

Definition Let T_t be a self-adjoint contraction semigroup with generator $-A$ (so A is nonnegative). We define

$$\mathcal{A}_{-\mu} := \frac{1}{\Gamma(\mu)} \int_0^\infty t^{\mu-1} T_t \, dt = \frac{1}{\Gamma(\mu)} \int_0^\infty \lambda^{-\mu} \, dE_\lambda,$$

(where $\{E_\lambda\}$ is a spectral family for A) with domain consisting of those $\phi \in L^2$ for which

$$\int_0^\infty \lambda^{-2\mu} (dE_\lambda \phi, \phi) < +\infty.$$

The definition is motivated by the fact that

$$\int_0^\infty e^{-\lambda t} t^{\mu-1} \, dt = \Gamma(\mu) \lambda^{-\mu},$$

for any given $\mu \in (0, 1)$.

Theorem VIII.3.1 *Let M be a Riemannian manifold, $A = -\Delta$. Then*

$$\mathcal{A}_{-\mu} = A^{-\mu}$$

if and only if M has infinite volume, if and only if $p(x, y, t) \to 0$ as $t \uparrow +\infty$.

Proof We have

$$\begin{aligned}
\mathcal{A}_{-\mu} A^\mu &= \frac{1}{\Gamma(\mu)} \int_0^\infty t^{\mu-1} \, dt \int_0^\infty \lambda^\mu e^{-\lambda t} \, dE_\lambda \\
&= \frac{1}{\Gamma(\mu)} \int_0^\infty \lambda^\mu \, dE_\lambda \int_0^\infty t^{\mu-1} e^{-\lambda t} \, dt \\
&= \int_{0^+}^\infty dE_\lambda \\
&= I - E_0,
\end{aligned}$$

where E_0 is the eigenspace of $\lambda = 0$, either the line of constant functions (Proposition VII.4.1) or the origin in L^2. If M has infinite volume, then L^2 has no nontrivial constant function. If M has finite volume, then the constant functions are all in L^2. ∎

Theorem VIII.3.2 *Let $\nu > 2$. Then* (VIII.2.2) *is valid for all $t > 0$ if and only if*

(VIII.3.1) $$\|f\|_{2\nu/(\nu-2)} \leq \text{const.}\|\text{grad } f\|_2$$

for all $f \in C_c^\infty$, equivalently, if and only if

(VIII.3.2) $$\|(-\Delta)^{-1/2} f\|_{2\nu/(\nu-2)} \leq \text{const.}\|f\|_2$$

for all $f \in \mathcal{D}((-\Delta)^{-1/2})$.

Proof Certainly, given (VIII.3.1) on C_c^∞ we have (VIII.2.1) by the proof Lemma VI.1.2, which implies (VIII.2.2) for all $t > 0$. For the converse, set $A = -\Delta$, and assume (VIII.2.2) on L^2. By Theorem VIII.3.1 we have the representation

$$A^{-1/2} f = \frac{1}{\sqrt{\pi}} \int_0^\infty t^{1/2-1} P_t f \, dt$$

for all $f \in \mathcal{D}((-\Delta)^{-1/2})$, which implies, for any $T > 0$,

$$\begin{aligned} |A^{-1/2} f|(x) &\leq \left\{ \sup_{t>0} |P_t f|(x) \right\} \frac{1}{\sqrt{\pi}} \int_0^T t^{-1/2} \, dt \\ &\quad + \frac{1}{\sqrt{\pi}} \int_T^\infty t^{-1/2} \|P_t f\|_\infty \, dt \\ &\leq \frac{2T^{1/2}}{\sqrt{\pi}} f^*(x) + \frac{\|f\|_2}{\sqrt{\pi}} \int_T^\infty t^{-1/2} \|P_t\|_{2\to\infty} \, dt. \end{aligned}$$

Now use the Riesz–Thorin interpolation theorem, with

$$p_0 = 1, \qquad q_0 = \infty, \qquad p_1 = q_1 = \infty, \qquad \lambda = 1/2.$$

Then

$$p_\lambda = 2, \qquad q_\lambda = \infty,$$

which implies

$$\|P_t\|_{2\to\infty} \leq {M_0}^{1-\lambda} {M_1}^{\lambda} \leq \text{const.} t^{-\nu/4},$$

which implies

$$|A^{-1/2}f|(x) \le \frac{2T^{1/2}}{\sqrt{\pi}} f^*(x) + c_\nu T^{1/2-\nu/4}\|f\|_2.$$

Minimize the right hand side with respect to T. Then the minimum is realized at

$$T_0 = \text{const.}\{f^*(x)\}^{-4/\nu}\|f\|_2{}^{4/\nu},$$

which implies

$$|A^{-1/2}f|(x) \le c_\nu |f^*|^{1-2/\nu}(x)\|f\|_2{}^{2/\nu},$$

which implies

$$\begin{aligned}\int_M |A^{-1/2}f|^{2\nu/(\nu-2)}(x)\,dV(x) &\le c_\nu \|f^*\|_2{}^2\|f\|_2{}^{4/(\nu-2)}\\ &\le c_\nu\|f\|_2{}^{2+4/(\nu-2)}\\ &= c_\nu\|f\|_2{}^{2\nu/(\nu-2)}.\end{aligned}$$

We therefore have (VIII.3.2). ■

Remark VIII.3.1 We could have easily used Theorems VI.1.1 and VIII.2.2 to prove the above theorem. But the method is useful for other considerations, as well. For example, the proof of the theorem also implies a short time result:

Theorem VIII.3.3 *Let $\nu > 2$. Then* (VIII.2.2) *is valid for all $t \in (0, 1]$ if and only if*

(VIII.3.3) $$\|f\|_{2\nu/(\nu-2)} \le \text{const.}\{\|\text{grad}\, f\|_2 + \|f\|_2\}$$

for all $f \in C_c^\infty$, equivalently, if and only if

(VIII.3.4) $$\|(-\Delta)^{-1/2}f\|_{2\nu/(\nu-2)} \le \text{const.}\{\|f\|_2 + \|(-\Delta)^{-1/2}f\|_2\}$$

for all $f \in \mathcal{D}((-\Delta)^{-1/2})$.

Proof Assume $\|P_t\|_{1\to\infty} \le ct^{-\nu/2}$ for all $t \in (0, 1]$. Consider the semigroup

$$T_t = e^{-t}P_t.$$

Then, by Theorem VIII.1.1, $\|P_t\|_{1\to\infty}$ decreases with respect to time, which implies T_t satisfies $\|T_t\|_{1\to\infty} \le ct^{-\nu/2}$ for *all* $t > 0$. Now the infinitesimal generator of T_t is $\Delta - I$, which implies (VIII.3.3) by the previous theorem. The converse is similar. ■

A direct consequence of the Li–Yau upper bounds (Proposition VIII.1.3) and Croke's inequality (V.2.15) is

Proposition VIII.3.3 *If M is n-dimensional Riemannian complete with bounded geometry, then*

$$\|P_t\|_{1\to\infty} \le \text{const.}t^{-n/2}$$

for all $t \in (0, 1]$.

In particular, when $n > 2$, we have

$$\|f\|_{2n/(n-2)} \le \text{const.}\{\|\text{grad } f\|_2 + \|f\|_2\} \qquad \text{(VIII.3.5)}$$

for all $f \in C_c^\infty$.

VIII.4 Coulhon's Modified Sobolev Inequality

What if we are only given (VIII.2.2) for all $t \ge 1$?

Theorem VIII.4.1 *If* (VIII.2.2) *is valid for all $t \ge 1$, then for each $f \in \mathcal{D}((-\Delta)^{-1/2})$ we have, instead of* (VIII.3.2),

$$(-\Delta)^{-1/2} f = g + h,$$

where $g \in L^2$ and $h \in L^{2\nu/(\nu-2)}$, with

$$\|g\|_2 \le \text{const.}\|f\|_2, \qquad \|h\|_{2\nu/(\nu-2)} \le \text{const.}\|f\|_2.$$

Proof First, set $A = -\Delta$, and write

$$A^{-1/2} = P_1 A^{-1/2} + (I - P_1)A^{-1/2}.$$

Certainly,

$$P_1 A^{-1/2} = \int_0^\infty t^{-1/2} P_{t+1}\, dt = A^{-1/2} P_1,$$

which implies (using the proof of Theorem VIII.3.2),

$$\|P_1 A^{-1/2} f\|_{2\nu/(\nu-2)} \le \text{const.}\|f\|_2;$$

so we pick $h = P_1 A^{-1/2}$ and $g = (I - P_1)A^{-1/2}$. Now

$$(I - P_1)A^{-1/2} = -\left\{\int_0^1 \left(\frac{d}{ds} P_s\right) ds\right\} A^{-1/2} = \int_0^1 P_s A^{1/2}\, ds$$

and

$$P_s A^{1/2} = \int_0^\infty \lambda^{1/2} e^{-\lambda s}\, dE_\lambda = s^{-1/2} \int_0^\infty (\lambda s)^{1/2} e^{-\lambda s}\, dE_\lambda.$$

This implies, for $f \in \mathcal{D}(P_s A^{1/2})$,

$$\| P_s A^{1/2} f \|^2 \le s^{-1} \int_0^\infty (\lambda s) e^{-2\lambda s} (dE_\lambda f, f) \le \text{const.} s^{-1} \| f \|_2{}^2,$$

that is,

$$\| P_s A^{1/2} \|_{2\to 2} \le \text{const.} s^{-1/2}, \tag{VIII.4.1}$$

which implies

$$\| (I - P_1) A^{-1/2} \|_{2\to 2} \le \text{const.}$$

■

What about a converse? We first have

Theorem VIII.4.2 (Extrapolation Theorem) *Given $1 \le \alpha < \beta \le +\infty$ for which*

$$\| P_t \|_{\alpha\to\beta} \le \text{const.} t^{-\epsilon}$$

for all $t \ge 1$. Then

$$\| P_t \|_{1\to\infty} \le \text{const.} t^{-\delta},$$

where $\delta = \epsilon\{\alpha^{-1} - \beta^{-1}\}^{-1}$, for all $t \ge 1$.

Proof Fix θ such that

$$\frac{1}{\alpha} = \theta + \frac{1-\theta}{\beta}, \qquad \theta = \frac{\alpha^{-1} - \beta^{-1}}{1 - \beta^{-1}},$$

so $\theta,\ 1-\theta \in (0, 1)$. First note that, for $t \ge 2$,

$$\begin{aligned}
\| P_t f \|_\beta &= \| P_{t/2} P_{t/2} f \|_\beta \\
&\le \text{const.} t^{-\epsilon} \| P_{t/2} f \|_\alpha \\
&\le \text{const.} t^{-\epsilon} \| P_{t/2} f \|_1{}^\theta \| P_{t/2} f \|_\beta{}^{1-\theta} \\
&\le \text{const.} t^{-\epsilon} \| f \|_1{}^\theta \| P_{t/2} f \|_\beta{}^{1-\theta} \\
&\le \text{const.} t^{-\epsilon} \| P_{t/2} \|_{1\to\beta}{}^{1-\theta} \| f \|_1
\end{aligned}$$

– the third line is Hölder's inequality, applied to

$$\begin{aligned}
\phi &= (P_{t/2} f)^{\alpha\theta}, \qquad & p &= 1/\alpha\theta, \\
\psi &= (P_{t/2} f)^{\alpha(1-\theta)}, \qquad & q &= \beta/\alpha(1-\theta)
\end{aligned}$$

– which implies

$$\|P_t\|_{1\to\beta} \le \text{const.}t^{-\epsilon}\|P_{t/2}\|_{1\to\beta}{}^{1-\theta},$$

which, in turn, yields

$$t^{\epsilon/\theta}\|P_t\|_{1\to\beta} \le \text{const.}\left\{(t/2)^{\epsilon/\theta}\|P_{t/2}\|_{1\to\beta}\right\}^{1-\theta}.$$

So we wish to solve the inequality

$$\sigma(t) \le \text{const.}\{\sigma(t/2)\}^{1-\theta}, \qquad t \ge 2.$$

Well, we have

$$\sigma(2^k t) \le (\text{const.})^{1+\cdots+(1-\theta)^{k-1}}\sigma(t)^{(1-\theta)^k}$$

for all $t \ge 1$. Therefore, if we vary $t \in [1, 2]$ and $k = 0, 1, 2, \ldots,$ we obtain $\sigma(t) \le \text{const.}$, $t \ge 1$, which implies

(VIII.4.2) $$\|P_t\|_{1\to\beta} \le \text{const.}t^{-\epsilon/\theta}$$

for all $t \ge 1$.

Next, duality then implies

$$\|P_t\|_{\beta'\to\infty} \le \text{const.}t^{-\epsilon/\theta}$$

for all $t \ge 1$, where $\beta' = \beta/(\beta - 1)$ denotes the conjugate of β. Now duality also implies

$$\|P_t\|_{\beta'\to\alpha'} \le \text{const.}t^{-\epsilon}$$

for all $t \ge 1$. Then for Θ given by

$$\frac{1}{\beta'} = \Theta + \frac{1-\Theta}{\alpha'},$$

we have (using Hölder's inequality applied to $|P_t f| = |P_t f|^{\Theta}|P_t f|^{1-\Theta}$)

$$\|P_t f\|_{\beta'} \le \|P_t f\|_1{}^{\Theta}\|P_t f\|_{\alpha'}{}^{1-\Theta} \le \|f\|_1{}^{\Theta}\text{const.}t^{-\epsilon(1-\Theta)}\|P_{t/2} f\|_{\beta'}{}^{1-\Theta}$$

(VIII.4.3)

for all $t \ge 2$. So

$$\|P_t f\|_{1\to\beta'} \le \text{const.}t^{-\epsilon(1-\Theta)}\|P_{t/2} f\|_{1\to\beta'}{}^{1-\Theta}$$

for all $t \ge 2$. Set $\epsilon' = \epsilon(1 - \Theta)$. Then the argument for (VIII.4.2) also implies

$$\|P_t\|_{1\to\beta'} \le \text{const.}t^{-\epsilon'/\Theta},$$

which implies

$$\|P_t\|_{1\to\infty} \le \|P_{t/2}\|_{1\to\beta}\|P_{t/2}\|_{\beta\to\infty} \le \text{const.}t^{-(\epsilon/\theta+\epsilon'/\Theta)},$$

which implies the theorem. ■

Theorem VIII.4.3 *Let M have infinite volume. Suppose we are given*

(VIII.4.4) $$(-\Delta)^{-1/2} : L^2 \to L^2 + L^{2\nu/(\nu-2)},$$

is bounded in the sense of Theorem VIII.4.1, *and $P_1 : L^1 \to L^\infty$ is also bounded. Then* (VIII.2.2) *is valid for all $t \geq 1$.*

Proof We first have by the Riesz–Thorin interpolation theorem [with $\alpha = 2$, and $\beta = 2\nu/(\nu - 2)$] that $P_1 : L^2 \to L^{2\nu/(\nu-2)}$ is bounded, which implies $P_1(-\Delta)^{-1/2} : L^2 \to L^{2\nu/(\nu-2)}$ is bounded, which implies

$$\begin{aligned}\|P_{t+1}\|_{2\to 2\nu/(\nu-2)} &= \|P_1(-\Delta)^{-1/2}(-\Delta)^{1/2}P_t\|_{2\to 2\nu/(\nu-2)} \\ &\leq \|P_1(-\Delta)^{-1/2}\|_{2\to 2\nu/(\nu-2)}\|(-\Delta)^{1/2}P_t\|_{2\to 2} \\ &\leq \text{const.}t^{-1/2}\end{aligned}$$

by (VIII.4.1), which implies, by the extrapolation theorem [with $\alpha = 2$, $\beta = 2\nu/(\nu - 2)$, $\epsilon = 1/2)$],

$$\|P_{t+1}\|_{1\to\infty} \leq \text{const.}t^{-\nu/2}$$

for all $t > 0$. ∎

VIII.5 The Denouement: Geometric Applications

Let **G** be a graph with vertices $\mathcal{G}$. Recall that for every $\xi \in \mathcal{G}$, we denote its collection of neighbors by $\mathrm{N}(\xi)$, and its valence by $m(\xi) = \text{card } \mathrm{N}(\xi)$. The collection of edges of **G** is denoted by $\mathcal{G}_e$, with oriented edges denoted by $[\xi, \eta]$ where $\eta \in \mathrm{N}(\xi)$. The measure of functions on $\mathcal{G}$ is given by $d\mathrm{V}(\xi) = m(\xi)d\iota(\xi)$, and of functions on $\mathcal{G}_e$ by $d\mathrm{A}([\xi, \eta]) = d\iota([\xi, \eta])$, where $d\iota$ always denotes counting measure.

Lemma VIII.5.1 *For any function f on $\mathcal{G}$ we have*

$$\sum_{\xi\in\mathcal{G}} f(\xi)m(\xi) = \frac{1}{2}\sum_{[\xi,\eta]\in\mathcal{G}_e} f(\xi) + f(\eta).$$

Proof We have

$$\sum_{\xi\in\mathcal{G}} f(\xi)m(\xi) = \sum_{\xi\in\mathcal{G}}\sum_{\eta\in\mathrm{N}(\xi)} f(\xi) = \sum_{[\xi,\eta]\in\mathcal{G}_e} f(\xi) = \frac{1}{2}\sum_{[\xi,\eta]\in\mathcal{G}_e} f(\xi) + f(\eta),$$

which is the claim. ∎

Definition We define the *Laplacian of the function* $f : \mathcal{G} \to \mathbb{R}$, denoted by $\Delta_{\mathbf{G}} f$, by

$$\Delta_{\mathbf{G}} f(\xi) = \frac{1}{m(\xi)} \sum_{\eta \in \mathrm{N}(\xi)} \{f(\eta) - f(\xi)\}.$$

So $\Delta_{\mathbf{G}} f$ is a function on $\mathcal{G}$.

Then for functions f and h we have

$$\begin{aligned} -\int_{\mathcal{G}} f \Delta_{\mathbf{G}} h \, d\mathrm{V} &= -\sum_{\xi} \sum_{\eta \in \mathrm{N}(\xi)} f(\xi)\{h(\eta) - h(\xi)\} \\ &= -\frac{1}{2} \sum_{[\xi,\eta]} f(\xi)\{h(\eta) - h(\xi)\} + f(\eta)\{h(\xi) - h(\eta)\} \\ &= \frac{1}{2} \sum_{[\xi,\eta]} \{f(\eta) - f(\xi)\}\{h(\eta) - h(\xi)\} \\ &= \frac{1}{2} \int_{\mathcal{G}_e} \langle \mathfrak{D} f, \mathfrak{D} h \rangle \, d\mathrm{A}, \end{aligned}$$

that is,

$$-\int_{\mathcal{G}} f \Delta_{\mathbf{G}} h \, d\mathrm{V} = \frac{1}{2} \int_{\mathcal{G}_e} \langle \mathfrak{D} f, \mathfrak{D} h \rangle \, d\mathrm{A}; \tag{VIII.5.1}$$

and

$$-\int_{\mathcal{G}} f \Delta_{\mathbf{G}} f \, d\mathrm{V} = \frac{1}{2} \sum_{[\xi,\eta]} |f(\eta) - f(\xi)|^2 \leq \sum_{[\xi,\eta]} f^2(\eta) + f^2(\xi) = 2 \int_{\mathcal{G}} f^2 \, d\mathrm{V}.$$

We conclude

Theorem VIII.5.1 $\Delta_{\mathbf{G}}$ *is self-adjoint nonpositive bounded operator on* $L^2(\mathcal{G}, d\mathrm{V})$, *with heat semigroup*

$$\mathrm{P}_t = \exp t \Delta_{\mathbf{G}},$$

with associated heat kernel $\mathrm{p}(\xi, \eta, t)$.

Remark VIII.5.1 Note that we are using two definitions of $|\mathfrak{D} f|^2$. In §VI.5 we defined $|\mathfrak{D} f|^2$ as a function on $\mathcal{G}$, the vertices of $\mathbf{G}$, namely,

$${}_{\mathrm{V}}|\mathfrak{D} f|_s{}^s(\xi) := \sum_{\eta \in \mathrm{N}(\xi)} |\mathfrak{D} f([\xi, \eta])|^s$$

for the vertex $\xi \in \mathcal{G}$. Here $|\mathfrak{D} f|^2$ is viewed as a function on the oriented edges $\mathcal{G}_e$ of $\mathbf{G}$. But as noted at the very beginning of §VI.5, the definitions are equivalent.

So for all arguments involving Sobolev inequalities, we may work with either definition.

Example VIII.5.1 Consider the integer lattice $\mathbb{Z}^k$ in $\mathbb{R}^k$. The graph structure on $\mathbb{Z}^k$ is given by: $\eta \in \mathrm{N}(\xi)$ if and only if $|\eta - \xi| = 1$ in $\mathbb{R}^k$. Then the heat kernel p on the associated graph is given by

$$\mathrm{p}(\xi, \eta, t) = e^{-t} \prod_{j=1}^{k} \mathbf{I}_{\xi^j - \eta^j}(t/k),$$

where $\mathbf{I}_\nu(z) = e^{-i\nu\pi/2}\mathbf{J}_\nu(iz)$ denotes the modified Bessel function.

Proof We only consider the case $k = 1$. The higher dimensional case is similar. Let $f = f(n)$, $n \in \mathbb{Z}$. The Laplacian here is given by

$$(\Delta_{\mathbf{G}} f)(n) = \frac{f(n+1) + f(n-1)}{2} - f(n).$$

For $u = u(n, t)$, where n ranges over the integers and t ranges over positive time, consider the heat equation

$$\text{(VIII.5.2)} \qquad \Delta_{\mathbf{G}} u = \frac{\partial u}{\partial t}, \qquad t > 0,$$

with initial values given by

$$\lim_{t \downarrow 0} u(n, t) = \phi(n),$$

where $\phi(n)$ is a given function on the integers. Consider the function on the circle, having time t as parameter,

$$U(\theta, t) = \sum_{n=-\infty}^{\infty} u(n, t) e^{in\theta};$$

then (VIII.5.2) becomes

$$\frac{\partial U}{\partial t}(\theta, t) = \{\cos\theta - 1\} U(\theta, t),$$

which implies

$$U(\theta, t) = e^{-(1-\cos\theta)t} \sum_{n=-\infty}^{\infty} \phi(n) e^{in\theta}.$$

One now uses the integral representation

$$\mathbf{J}_n(z) = \frac{1}{2\pi} \int_{-\pi}^{\pi} e^{i(z\sin\theta - n\theta)}\, d\theta$$

to show

$$u(n,t) = e^{-t} \sum_{m=-\infty}^{\infty} \phi(m) i^{-(n-m)} \mathbf{J}_{n-m}(it)$$

for the solution to the initial value problem, which implies the claim. Note that

$$\mathbf{I}_\nu(t) = (2\pi t)^{-1/2} e^t \{1 - O_\nu(t^{-1})\}, \qquad t \uparrow +\infty,$$

which implies, on $\mathbb{Z}^k$,

$$\|\mathrm{P}_t\|_{1\to\infty} \sim \text{const.} t^{-k/2}, \qquad t \uparrow +\infty.$$

Remark VIII.5.2 The Varopoulos and Coulhon criteria apply as well to graphs with heat diffusion with continuous time, and with "heat diffusion" with discrete time, namely, random walks on graphs.

Theorem VIII.5.2 *Let M be Riemannian complete with bounded geometry, and let $\nu \geq 1$. Then* (VIII.4.4) *is satisfied on M if and only if the discrete Nirenberg–Sobolev inequality* (VIII.3.1) *is valid on every discretization of M.*

In particular, the heat semigroup P_t of M satisfies

(VIII.5.3) $$\|P_t\|_{1\to\infty} = O(t^{-\nu/2}), \qquad t \uparrow +\infty$$

if and only if

(VIII.5.4) $$\|\mathrm{P}_t\|_{1\to\infty} = O(t^{-\nu/2}), \qquad t \uparrow +\infty$$

for the heat semigroup P_t of any discretization $\mathbf{G}$ of M.

Also, the estimate (VIII.5.3) *remains invariant under compact perturbations of M.*

Proof First, because M has bounded geometry, one knows, by Proposition VIII.3.3, that $P_1 : L^1 \to L^\infty$ is bounded.

Next, assume that $\nu > 2$. Let $\mathbf{G}$ be a discretization of M, and assume the Nirenberg–Sobolev inequality (VIII.3.1) is valid on $\mathbf{G}$. Consider the associated discretization operator $\mathcal{D}$ and smoothing operator $\mathcal{S}$ on functions defined on M and $\mathcal{G}$, respectively.

Given any smooth function $F : M \to \mathbb{R}$, consider $\Phi = (-\Delta_M)^{-1/2} F$, and write

$$\Phi = \{\Phi - \mathcal{S}\mathcal{D}\Phi\} + \mathcal{S}\mathcal{D}\Phi.$$

Then (see §VI.5.5)

$$\|\Phi - \mathcal{S}\mathcal{D}\Phi\|_2 \leq \text{const.} \|\text{grad}\, \Phi\|_2 = \text{const.} \|F\|_2,$$

and (by $\mathcal{D}$:iii of §VI.5.1)

$$\begin{aligned}\|\mathcal{SD}\Phi\|_{2\nu/(\nu-2)} &\leq \text{const.}\|\mathcal{D}\Phi\|_{2\nu/(\nu-2)}\\ &\leq \text{const.}\|\mathfrak{D}\mathcal{D}\Phi\|_2\\ &\leq \text{const.}\|\text{grad}\,\Phi\|_2\\ &= \text{const.}\|F\|_2.\end{aligned}$$

Therefore, if one starts with the discrete Nirenberg–Sobolev inequality (VIII.3.1) on **G**, one also has the inequality (VIII.4.4) on any discretization of M.

Conversely, assume (VIII.4.4) on M; and note that $\mathcal{SD}: L^1(M) \to L^\infty(M)$ is bounded, which implies by Proposition VIII.3.1 (the Riesz–Thorin interpolation theorem), that

$$\mathcal{SD}: L^2(M) \to L^{2\nu/(\nu-2)}(M)$$

is bounded, which implies [with (VIII.4.4)]

$$\mathcal{SD}(-\Delta_M)^{-1/2}: L^2(M) \to L^{2\nu/(\nu-2)}(M),$$

is bounded. Therefore, given $\phi: \mathcal{G} \to [0, +\infty)$ then [by ($\mathcal{S}$:ii) of §VI.5.2 and ($\mathcal{DS}$:ii) of §VI.5.3]

$$\begin{aligned}\|\phi\|_{2\nu/(\nu-2)} &\leq \text{const.}\|\mathcal{SDS}\phi\|_{2\nu/(\nu-2)}\\ &\leq \text{const.}\|\mathcal{SD}(-\Delta_M)^{-1/2}(-\Delta_M)^{1/2}\mathcal{S}\phi\|_{2\nu/(\nu-2)}\\ &\leq \text{const.}\|(-\Delta_M)^{1/2}\mathcal{S}\phi\|_2\\ &= \text{const.}\|\text{grad}\,\mathcal{S}\phi\|_2\\ &\leq \text{const.}\|\mathfrak{D}\phi\|_2,\end{aligned}$$

which is the discrete Nirenberg–Sobolev inequality on **G** for nonnegative functions. For arbitrary functions, one has

$$\|\,\mathfrak{D}|\phi|\,\|_2 \leq \|\mathfrak{D}\phi\|_2,$$

and now can easily obtain the result for general ϕ.

By Theorem VIII.3.2 (applied to graphs) and Theorems VIII.4.1 and VIII.4.3, we obtain the equivalence of the large time upper bounds on the heat semigroups of M and **G** when $\nu > 2$.

Now assume $1 \leq \nu \leq 2$, and let M_1 denote the Cartesian product of M with $\mathbb{R}^2$. Then the upper bounds for the heat kernel and the Coulhon criteria are to be considered for $\nu_1 = \nu + 2$, for which the theorem is valid. For $(x, y) \in M \times \mathbb{R}^2 := M_1$ (with obvious notations) we have

$$p(x, x, t) = \text{const.}t^{2/2} p_1((x, y), (x, y), t)$$

and for $(\xi, \eta) \in \mathbf{G} \times \mathbb{Z}^2$ we also have

$$\mathrm{p}(\xi, \xi, t) \sim \text{const.} t^{2/2} \mathrm{p}_1((\xi, \eta), (\xi, \eta), t) \quad \text{as } t \uparrow +\infty.$$

So

$$p(x, x, t) \leq t^{-\nu/2} \iff p_1((x, y), (x, y), t) \leq t^{-\nu_1/2},$$
$$\mathrm{p}(\xi, \xi, t) \leq t^{-\nu/2} \iff \mathrm{p}_1((\xi, \eta), (\xi, \eta), t) \leq t^{-\nu_1/2}.$$

One can now use the above argument for ν_1 to obtain the result for ν.

Finally, the estimate (VIII.5.3) remains valid under compact perturbation of a manifold of bounded geometry since any discretization of M remains a discretization under compact perturbation. ■

Theorem VIII.5.3 *Let M satisfy $\mathfrak{I}_\nu(M) > 0$ for some $\nu > 1$. Then* (VIII.2.2) *is valid for all $t > 0$.*

Proof Let $\nu \geq 2$. By Theorems VI.1.1 and VI.1.2, the hypothesis implies the Nash–Sobolev inequality (VIII.2.1), which implies, by Theorem VIII.2.1, the upper bound (VIII.2.2) for all $t > 0$.

For $\nu \in (1, 2)$, one argues as in the similar case of the above theorem. ■

Corollary VIII.5.1 *For the 2-dimensional jungle gym in $\mathbb{R}^3$ we have*

$$\| P_t \|_{1 \to \infty} \leq \text{const.} t^{-3/2} \qquad \forall t > 0.$$

For the 3-dimensional Heisenberg group with left-invariant Riemannian metric, we have

$$\| P_t \|_{1 \to \infty} \leq \text{const.} t^{-4/2} \qquad \forall t > 0.$$

Proof Indeed, $\mathfrak{I}_3$ is positive for the 2-dimensional jungle gym in $\mathbb{R}^3$, since one has a simultaneous discretization of $\mathbb{R}^3$ and the jungle gym. Also, $\mathfrak{I}_4$ is positive for the 3-dimensional Heisenberg group (see Pansu (1982)). ■

Theorem VIII.5.4 *If M is Riemannian complete with bounded geometry, and $\mathfrak{I}_{\nu,\rho}(M) > 0$ for some $\nu \geq 1$ and $\rho > 0$, then we have the upper bound* (VIII.5.3).

In particular, for every complete Riemannian manifold with bounded geometry we have

$$\| P_t \|_{1 \to \infty} = O(t^{-1/2}), \qquad t \uparrow +\infty. \tag{VIII.5.5}$$

Proof First assume that $\nu > 2$. Then, by Theorem V.3.1, for any discretization **G** of M we have $\mathrm{I}_\nu(\mathbf{G}) > 0$, which implies (VIII.5.4) by the discrete version of Theorem VIII.3.2. But this then implies (VIII.5.3).

If $\nu \le 2$, then consider $M_1 = M \times \mathbb{R}^2$. This implies $\mathfrak{I}_{\nu+2,\rho}(M_1) > 0$ by Proposition V.3.6, which implies $\mathrm{I}_{\nu+2}(\mathbf{G}_1) > 0$ for the discretization $\mathbf{G}_1 = \mathbf{G} \times \mathbb{Z}^2$ of M_1 (where **G** is a discretization of M), which implies (VIII.5.4), which implies (VIII.5.3) for M_1. But this then implies (VIII.5.3) for M. ■

VIII.6 Epilogue: The Faber–Krahn Method

Recall (§VI.3) that, given a positive increasing function $g(v)$, $v \ge 0$, we say that a domain Ω in a Riemannian manifold M satisfies a *geometric g-isoperimetric inequality* if

$$A(\partial D) \ge g(V(D))$$

for all $D \subset\subset \Omega$.

Also, given a positive decreasing function $\Lambda(v)$, $v \ge 0$, we say that a domain Ω in a Riemannian manifold M satisfies an *eigenvalue Λ-isoperimetric inequality* if

$$\lambda(D) \ge \Lambda(V(D))$$

(where λ denotes the lowest Dirichlet eigenvalue) for all $D \subset\subset \Omega$.

Proposition VI.3.2 stated that if Ω satisfies a geometric g-isoperimetric inequality, with $g(v)/v$ a decreasing function of v, then Ω satisfies an eigenvalue Λ-isoperimetric inequality with

$$\Lambda(v) = \frac{1}{4}\left(\frac{g(v)}{v}\right)^2.$$

Definition Given a positive decreasing function $\Lambda(v)$ on $[0, +\infty)$, define the function $\mathfrak{E}(t)$ by

$$t = \int_0^{\mathfrak{E}(t)} \frac{dv}{v\Lambda(v)}. \tag{VIII.6.1}$$

So we must assume that $v \mapsto \{v\Lambda(v)\}^{-1}$ is integrable near $v = 0$. Equivalently, $\mathfrak{E}(t)$ is defined as the solution to the initial value problem

$$\mathfrak{E}' = \mathfrak{E}\Lambda(\mathfrak{E}), \qquad \mathfrak{E}(0) = 0.$$

Lemma VIII.6.1 *Suppose Ω satisfies the eigenvalue Λ-isoperimetric inequality, $u \in C_c^\infty(\Omega)$, $u \ge 0$. Then for any $\delta \in (0, 1)$ we have*

$$\frac{\|\mathrm{grad}\, u\|_2{}^2}{\|u\|_2{}^2} \ge (1-\delta)\Lambda\left(\frac{2\|u\|_1{}^2}{\delta\|u\|_2{}^2}\right).$$

Proof For any $\tau > 0$ we have

$$\begin{aligned}\int_\Omega u^2 - 2\tau u + \tau^2 &= \int_\Omega (u-\tau)^2 \\ &= \int_{u>\tau} + \int_{u<\tau} \\ &\le \int_{u>\tau} (u-\tau)^2 + \int_{u<\tau} \tau^2 \\ &\le \int_{u>\tau} (u-\tau)^2 + \int_\Omega \tau^2,\end{aligned}$$

which implies

$$\|u\|_2{}^2 - 2\tau\|u\|_1 = \int_\Omega u^2 - 2\tau u \le \int_{u>\tau} (u-\tau)^2,$$

which implies

$$\begin{aligned}\int_\Omega |\text{grad}\, u|^2 &\ge \int_{u>\tau} |\text{grad}\, u|^2 \\ &\ge \lambda(\{u>\tau\}) \int_{u>\tau} (u-\tau)^2 \\ &\ge \Lambda(V(\{u>\tau\})) \int_{u>\tau} (u-\tau)^2.\end{aligned}$$

Now

$$V(\{u>\tau\}) \le \frac{1}{\tau}\int_{u>\tau} u \le \frac{1}{\tau}\int_\Omega u,$$

which implies

$$\int_\Omega |\text{grad}\, u|^2 \ge \Lambda(\|u\|_1/\tau)\int_{u>\tau} (u-\tau)^2 \ge \Lambda(\|u\|_1/\tau)\left\{\|u\|_2{}^2 - 2\tau\|u\|_1\right\}.$$

Pick $\tau = \delta\|u\|_2{}^2/2\|u\|_1$, and the claim follows. ■

Theorem VIII.6.1 *If M satisfies the eigenvalue Λ-isoperimetric inequality, then*

(VIII.6.2) $$\|P_t\|_{1\to\infty} \le \frac{\text{const.}}{\mathfrak{C}(\text{const.}t)}$$

for all $t > 0$.

Proof Fix $y \in M$, and define

$$u(x,t) = p(x,y,t);$$

then $\int_M u(x,t)\,dV(x) \le 1$. Define

$$\mathfrak{J}(t) = \int_M u^2(x,t)\,dV(x).$$

Then

$$\mathfrak{J}'(t) = -2\|\operatorname{grad} u\|_2{}^2 \le -2(1-\delta)\mathfrak{J}(t)\Lambda(2/\delta\mathfrak{J}(t))$$

by the previous lemma. Therefore,

$$\int_{\mathfrak{J}(t_0)}^{\mathfrak{J}(t)} \frac{d\sigma}{\sigma\,\Lambda(2/\delta\sigma)} \le -2(1-\delta)(t-t_0),$$

which implies, letting $t_0 \downarrow 0$,

$$\int_{\mathfrak{J}(t)}^{\infty} \frac{d\sigma}{\sigma\,\Lambda(2/\delta\sigma)} \ge 2(1-\delta)t.$$

Now substitute $v = 2/\delta\sigma$; then

$$\int_0^{2/\delta\mathfrak{J}(t)} \frac{dv}{v\,\Lambda(v)} \ge 2(1-\delta)t,$$

which implies

$$\mathfrak{J}(t) \le \frac{2}{\delta\mathfrak{E}(2(1-\delta)t)},$$

which is the theorem. ■

Example VIII.6.1 (Example VI.3.1 continued.) Let M satisfy a geometric g-isoperimetric inequality, with $g(v)$ given by

$$g(v) = \begin{cases} \alpha v^{1-1/n}, & v \le v_0, \\ \beta v^{1-1/\nu}, & v \ge v_0, \end{cases}$$

where $\alpha v_0{}^{1-1/n} = \beta v_0{}^{1-1/\nu}$. Then Theorem VI.3.2 implies that M satisfies a eigenvalue Λ-isoperimetric inequality, where

$$\Lambda(v) = \frac{1}{4}\begin{cases} \alpha v^{-2/n}, & v \le v_0, \\ \beta v^{-2/\nu}, & v \ge v_0, \end{cases}$$

Therefore, one has the existence of $t_0 > 0$ such that

$$\mathfrak{E}(t) = \begin{cases} \alpha' t^{n/2}, & t \le t_0, \\ \beta' t^{\nu/2}, & t \ge t_0, \end{cases}$$

which implies, by Theorem VIII.6.1, that

$$\|P_t\|_{1\to\infty} \le \text{const.} \begin{cases} t^{-n/2}, & t \le t_0, \\ t^{-\nu/2}, & t \ge t_0, \end{cases}$$

that is, we have an alternate proof of Theorem VIII.5.4 above.

Example VIII.6.2 (Example VI.3.2 continued.) Let M satisfy a geometric g-isoperimetric inequality, with $g(v)$ given by

$$g(v) = \begin{cases} \alpha v^{1-1/n}, & v \le v_0, \\ \beta, & v \ge v_0, \end{cases}$$

where $\alpha {v_0}^{1-1/n} = \beta$. Proposition VI.3.2 implies that M satisfies a eigenvalue Λ-isoperimetric inequality, where

$$\Lambda(v) = \begin{cases} \alpha v^{-2/n}/4, & v \le v_0, \\ \eta = \beta^2/4, & v \ge v_0, \end{cases}$$

Therefore, one has the existence of $t_0 > 0$ such that

$$\mathfrak{E}(t) = \begin{cases} \alpha' t^{n/2}, & t \le t_0, \\ \beta' e^{\eta t}, & t \ge t_0, \end{cases}$$

which implies, by Theorem VIII.6.1, that

$$\|P_t\|_{1\to\infty} \le \text{const.} \begin{cases} t^{-n/2}, & t \le t_0, \\ e^{-\eta t}, & t \ge t_0, \end{cases}$$

which corresponds qualitatively to hyperbolic space [note that η is not necessarily $\lambda(M)$ — which is the sharp result, by Proposition VIII.1.2].

Definition We say that a function $f : (0, \infty) \to (0, \infty)$ *has at most polynomial decay* if there exists $\alpha > 0$ such that for all $t > 0$, $\beta \in [1, 2]$ we have

$$f(\beta t) \ge \alpha f(t).$$

An example would be a function f satisfying $f'(t)/f(t) \ge -N/t$ for all t.

Theorem VIII.6.2 *Given M, consider the functions $\Lambda(v)$ and $\mathfrak{E}(t)$ related by* (VIII.6.1), *and assume that $\mathfrak{E}'(t)/\mathfrak{E}(t)$ has at most polynomial decay. If we are also given that*

$$\|P_t\|_{1\to\infty} \le \frac{1}{\mathfrak{E}(t)}$$

for all $t > 0$, then

$$\lambda_k(\Omega) \ge \text{const.}\Lambda(V(\Omega)/k)$$

for all $k = 1, 2, \ldots$, where $\lambda_k(\Omega)$ denotes the k th Dirichlet eigenvalue of the domain Ω.

Proof Fix Ω. From the Sturm–Liouville eigenvalue–eigenfunction expansion we have

$$q(x, y, t) = \sum_{j=1}^{\infty} e^{-\lambda_j t}\phi_j(x)\phi_j(y),$$

where q is the Dirichlet heat kernel of Ω, and $\{\phi_1, \phi_2, \ldots\}$ is a complete orthonormal basis of $L^2(\Omega)$ consisting of eigenfunctions of $-\Delta_{\mathrm{dir}}$ with the eigenvalue of ϕ_j equal to λ_j. This implies

$$ke^{-\lambda_k t} \le \sum_{j=1}^{\infty} e^{-\lambda_j t} = \int_\Omega q(x, x, t)\,dV(x) \le \int_\Omega p(x, x, t)\,dV(x) \le V(\Omega)/\mathfrak{E}(t),$$

which implies

$$t\lambda_k \ge \ln \frac{k\mathfrak{E}(t)}{V(\Omega)}$$

for *all* $t > 0$. Pick τ such that $\mathfrak{E}(\tau) = V(\Omega)/k$, and pick $t = 2\tau$. Then

$$\begin{aligned}
\lambda_k &\ge \frac{1}{2\tau} \ln \frac{k\mathfrak{E}(2\tau)}{V(\Omega)} \\
&= \frac{1}{2\tau} \ln \frac{\mathfrak{E}(2\tau)}{\mathfrak{E}(\tau)} \\
&= \frac{1}{2} \left\{ \frac{d}{dt} \ln \mathfrak{E}(t) \right\}_{t=\tau+\theta} \qquad \text{for some } \theta \in (0, 1) \\
&\ge \frac{\alpha}{2} \frac{\mathfrak{E}'(\tau)}{\mathfrak{E}(\tau)} \\
&= \frac{\alpha}{2} \Lambda(\mathfrak{E}(\tau)) \\
&= \frac{\alpha}{2} \Lambda(V(\Omega)/k),
\end{aligned}$$

which implies the theorem. ■

VIII.7 Bibliographic Notes

§VIII.1 Proposition VIII.1.1 is from Cheeger, Gromov, and Taylor (1982); also see Chavel (1984, §VIII.4). Theorem VIII.1.2 was first proved in Chavel and Karp (1991), and Corollary VIII.1.2 was communicated privately by P. Li. (See also Li 1986a.) Example VIII.1.1 is from Varopoulos (1989).

The arguments we use in subsequent sections all revolve around semigroup theory, in one way or another, namely, the isoperimetric inequalities lead to Sobolev inequalities, which give upper bounds on $\|P_t\|_{L^\ell \to L^k}$, for good choices of ℓ and k. The arguments can also yield off-diagonal corrections – see Davies (1987, 1989), Fabes and Stroock (1986), and Carlen, Kusuoka, and Stroock (1987). Davies's arguments (1987, 1989) emphasize log Sobolev inequalities, instead of the Sobolev inequalities of the type considered here.

A different method, using parabolic Harnack principles *for Riemannian* manifolds, also gives strong results, including lower bounds on the heat kernel. The early fundamental paper is Li and Yau (1986b). In fact, Proposition VIII.1.3 is from that paper, but our use of the result gave no hint of its ultimate power. More recent fundamental papers in parabolic Harnack inequalities are Grigor'yan (1991) and Saloff-Coste (1992), as well as the very recent Yau (1994, 1995) and Bakry and Qian (1999). For arguments not relying on global uniformity of Harnack inequalities, see Davies (1997).

Still another method for producing on-diagonal upper bounds is the use of the isoperimetric function of a manifold. It has been used to date on compact manifolds (see §V.4); one can find a heat kernel comparison theorem, based on comparison of isoperimetric functions, in Bérard and Gallot (1983). See also the discussion in Bérard (1986, Chapter V). The discussion there includes a Schwarz symmetrization argument for the heat kernel, in the spirit of the original Faber–Krahn argument. See the original version of this argument in Bandle (1980, Chapter IV). Finally, see an abstract version of the Schwarz symmetrization argument for the heat kernel, based on Kato's inequality, in Besson's appendix to Bérard (1986).

§VIII.2 Theorem VIII.2.1 is from Nash (1958), and Theorem VIII.2.2 from Carlen, Kusuoka, and Stroock (1987).

§VIII.3 Proposition VIII.3.1, the Riesz–Thorin interpolation theorem, can be found in Reed and Simon (1980, Vol. I); Proposition VIII.3.2, the maximal theorem, in Stein (1970). Theorem VIII.3.2 is from Varopoulos (1986).

§VIII.4 Theorem VIII.4.2 is from Coulhon (1990).

§VIII.5 Lemma VIII.5.1 is from Dodziuk (1984), and Theorem VIII.5.2 from Coulhon (1992).

Theorem VIII.5.4 was first proved in Chavel and Feldman (1991), although we preferred here to give the proof of Coulhon (1992). The estimate (VIII.5.5) gives the best rate of decay possible, for such generality. The first such result was proved by Varopoulos, in Varopoulos (1984), with exponent $-\frac{1}{2} + \epsilon$ for any $\epsilon > 0$. Our result confirms Varopoulos's conjecture, there, that one can sharpen the result to $\epsilon = 0$.

§VIII.6 This section is from Grigor'yan (1994c). See also Grigor'yan's article in the 1998 Edinburgh Lectures (Davies and Safarov, 1999). The methods of Grigor'yan have been developed to give very strong results on off-diagonal corrections – strictly as consequences of on-diagonal upper bounds alone (see Grigor'yan, 1994a).

All of our results on upper bounds were unobtainable as consequences of lower bounds on volume growth alone (recall Example VIII.1.1). More recently, Barlow, Coulhon, and Grigor'yan (2000) have shown that merely postulating $V(x;r) \geq \text{const.}r^k$ as $r \uparrow +\infty$ only implies a rate of decay given by $p \leq t^{-k/(k+1)}$ as $t \uparrow +\infty$. Furthermore, the result is sharp!

Bibliography

C. ADAMS & F. MORGAN (1999). Isoperimetric curves on hyperbolic surfaces. *Proc. Amer. Math. Soc.* **129**, 1347–1356.

R.A. ADAMS (1975). *Sobolev Spaces*. New York: Academic Press.

A.D. ALEXANDROV (1962). A characteristic property of spheres. *Ann. Math. Pura Appl.* **58**, 303–315.

F. ALMGREN (1986). Optimal isoperimetric inequalities. *Indiana Univ. Math. J.* **35**, 451–547.

F. ALMGREN & E.H. LIEB (1989). Continuity and discontinuity of the spherically decreasing rearrangement. *J. Amer. Math. Soc.* **2**, 683–773.

T. AREDE (1985). Manifolds for which the heat kernel is given interms of geodesic lengths. *Lett. Math. Phys.* **9**, 121–131.

T. AUBIN (1982). *Nonlinear Analysis on Manifolds. Monge–Ampère Equations*. New York: Springer-Verlag.

R. AZENCOTT (1974). Behaviour of diffusion semigroups at infinity. *Bull. Soc. Math. France* **102**, 193–240.

A. BAERNSTEIN (1995). A unified approach to symmetrization. *Symp. Mat.* **35**, 47–91.

D. BAKRY, T. COULHON, M. LEDOUX, & L. SALOFF-COSTE (1995). Sobolev inequalities in disguise. *Indiana Univ. Math. J.* **44**, 1033–1074.

D. BAKRY & Z. QIAN (1999). Harnack inequalities on a manifold with positive or negative Ricci curvature. *Rev. Mat. Iberoamer.* **15**, 143–179.

C. BANDLE (1980). *Isoperimetric Inequalities and Applications*. Boston: Pitman.

M. BARLOW, T. COULHON, & A.A. GRIGOR'YAN (2000). Manifolds and graphs with slow heat kernel decay. Preprint.

N. BENABDALLAH (1973). Noyau de diffusion sur les espaces homogènes compacts. *Bull. Soc. Math. France* **101**, 265–283.

I. BENJAMIMI & J. CAO (1996). A new isoperimetric theorem for surfaces of variable curvature. *Duke Math. J.* **85**, 359–396.

I. BENJAMINI, I. CHAVEL, & E.A. FELDMAN (1996). Heat kernel lower bounds on Riemannian manifolds using the old ideas of Nash. *Proc. London Math. Soc.* **72**, 215–240.

P.H. BÉRARD (1986). *Spectral Geometry: Direct and Inverse Problems*, Lecture Notes Math. **1207**. Berlin: Springer-Verlag.

P.H. BÉRARD & S. GALLOT (1983). Inégalités isopérimétrique pour l'equation de la chaleur et application a l'estimation de quelques invariants. *Séminaire Goulaouic–Meyer–Schwartz* **XV** (1983–1984).

M. BERGER (1987). *Geometry*. Berlin: Springer-Verlag. 2 volumes.

W. BLASCHKE (1956). *Kreis und Kugel*. Berlin: de Gruyter, 2nd edn.

R. BROOKS (1981). The fundamental group and the spectrum of the Laplacian. *Comment. Mat. Helv.* **56**, 581–598.

YU.D. BURAGO & V.A. ZALGALLER (1980). *Geometric Inequalities*. Berlin: Springer-Verlag, 1988. Original Russian edition: *Geometricheskie neravenstva*, Leningrad.

A. BURCHARD (1997). Steiner symmetrization is continuous in $W^{1,p}$. *Geom. & Fcnl. Anal.* **7**, 823–860.

P. BUSER (1982). A note on the isoperimetric constant. *Ann. Sci. Éc. Norm. Sup., Paris* **15**, 213–230.

C. CARATHÉODORY & E. STUDY (1909). Zwei Beweise des Satzes dass der Kreis unter alle Figuren gleichen Umgangs den grössten Inhalt hat. *Math. Ann.* **68**, 133–144.

T. CARLEMAN (1921). Zur theorie der Minimalflächen. *Math. Z.* **9**, 154–160.

E.A. CARLEN, S. KUSUOKA, & D.W. STROOCK (1987). Upper bounds for symmetric Markov transition densities. *Ann. Inst. H. Poincaré Prob. Stat.* **23**, 245–287.

G. CARRON (1996). Inégalités isopérimétriques de Faber–Krahn et consequences. In *Actes de la table ronde de géométrie dufférentielle en l'honneur de Marcel Berger*, Collection SMF Séminaires et Congrés **1**, 205–232.

I. CHAVEL (1978). On A. Hurwitz' method in isoperimetric inequalities. *Proc. Amer. Math. Soc.* **71**, 275–279.

I. CHAVEL (1984). *Eigenvalues in Riemannian Geometry*. New York: Academic Press.

I. CHAVEL (1994). *Riemannian Geometry: a modern introduction*. Cambridge: Cambridge Univ. Press.

I. CHAVEL & E.A. FELDMAN (1991). Modified isoperimetric constants, and large time heat diffusion in Riemannian manifolds. *Duke Math. J.* **64**, 473–499.

I. CHAVEL & L. KARP (1991). Large time behavior of the heat kernel: the parabolic λ-potential alternative. *Comment. Math. Helv.* **66**, 541–556.

J. CHEEGER (1970). A lower bound for the smallest eigenvalue of the Laplacian. In *Problems in Analysis*, R. Gunning ed., 195–199. Princeton Univ. Press.

J. CHEEGER, M. GROMOV, & M. TAYLOR (1982). Finite propagation speed, kernel estimates for functions of the Laplace operator, and the geometry of complete Riemannian manifolds. *J. Diff. Geom.* **17**, 15–54.

S.Y. CHENG & P. LI (1981). Heat kernel estimates and lower bounds of eigenvalues. *Comment. Mat. Helv.* **56**, 327–338.

K.S. CHOU & X.P. ZHU (2001). *The Curve Shortening Problem*. Chapman & Hall. To appear.

T. COULHON (1990). Dimension à l'infini d'un semigroup analytique. *Bull. Soc. Math. France* **114**, 485–500.

T. COULHON (1992). Noyau de chaleur et discretisation d'une variété riemanniene. *Israel J. Math.* **80**, 289–300.

T. COULHON & A.A. GRIGOR'YAN (1997). On-diagonal lower bounds for heat kernels on non-compact manifolds and Markov chains. *Duke Math. J.* **89**, 133–199.

T. COULHON & M. LEDOUX (1994). Isopérimétrie, decroisaance du noyau de la chaleur et transformations de Riesz: un contra-example. *Ark. Math.* **32**, 63–77.

T. COULHON & L. SALOFF-COSTE (1993). Isopérimétrie pour les groupes et les variétés. *Rev. Mat. Iberoamer.* **9**, 293–314.

T. COULHON & L. SALOFF-COSTE (1995). Variétés riemanniennes isoperimetriques à la l'infini. *Rev. Mat. Iberoamer.* **3**, 687–726.

R. COURANT & H. ROBBINS (1941). *What is Mathematics?* Oxford: Oxford Univ. Press.

C.B. CROKE (1980). Some isoperimetric inequalities and eigenvalue estimates. *Ann. Sci. Éc. Norm. Sup. Paris* **13**, 419–435.

E.B. DAVIES (1980). *One-Parameter Semigroups*. London: Academic Press.
E.B. DAVIES (1987). Explicit constants for Gaussian upper bounds on heat kernels. *Amer. J. Math.* **109**, 319–334.
E.B. DAVIES (1989). *Heat Kernels and Spectral Theory*. Cambridge: Cambridge Univ. Press.
E.B. DAVIES (1992). Heat kernel bounds, conservation of probability and the Feller property. *J. Analyse Math.* **58**, 99–119.
E.B. DAVIES (1995). *Spectral Theory and Differential Operators*. Cambridge: Cambridge Univ. Press.
E.B. DAVIES (1997). Non-gaussian aspects of heat kernels behaviour. *J. London Math. Soc.* **55**, 105–125.
E.B. DAVIES & N. MANDOUVALOS (1987). Heat kernel bounds on manifolds with cusps. *J. Fcnl. Anal.* **75**, 311–322.
E.B. DAVIES & Y. SAFAROV, eds. (1999). *Spectral Theory and Geometry*. Cambridge: Cambridge Univ. Press.
U. DIERKES, S. HILDEBRANDT, A. KÜSTER, & O. WOHLRAB (1992). *Minimal Surfaces*. Berlin: Springer-Verlag.
J. DODZIUK (1983). Maximum principle for parabolic inequalities and heat on open manifolds. *Indiana Univ. Math. J.* **32**, 703–716.
J. DODZIUK (1984). Difference equations, isoperimetric inequality, and transience of certain random walks. *Trans. Amer. Math. Soc.* **284**, 787–794.
C. FABER (1923). Beweiss, dass unter allen homogenen Membrane von gleicher Fläche und gleicher Spannung die kreisförmige die tiefsten Grundton gibt. *Sitzungber. Bayer Akad. Wiss., Math.-Phys.*, Munich, 169–172.
E.B. FABES & D.W. STROOCK (1986). A new proof of Moser's parabolic inequality via the old ideas of Nash. *Arch. Rat. Mech. Anal.* **96**, 327–338.
H. FEDERER (1969). *Geometric Measure Theory*. New York: Springer-Verlag.
H. FEDERER & W.H. FLEMING (1960). Normal integral currents. *Ann. Math.* **72**, 458–520.
H.D. FEGAN (1976). The heat equation on a compact Lie group. *Trans. Amer. Math. Soc.* **246**, 339–357.
H.D. FEGAN (1983). The fundamental solution of the heat equation on a compact Lie group. *J. Diff. Geom.* **18**, 659–668.
T. FIGIEL, J. LINDENSTRAUSS, & V.D. MILMAN (1977). The dimension of almost spherical sections of convex bodies. *Acta Math.* **139**, 53–94.
H.R. FISCHER, J.J. JUNGSTER, & F.L. WILLIAMS (1984). The heat kernel on the 2-sphere. *Adv. in Math.* **54**, 226–232.
M.P. GAFFNEY (1954). A special Stokes' theorem for complete Riemannian manifolds. *Ann. Math.* **60**, 140–145.
M.P. GAFFNEY (1959). The conservation property of the heat equation on complete Riemannian manifolds. *Comm. Pure Appl. Math.* **12**, 1–11.
M. GAGE (1984). Curve shortening makes convex curves circular. *Invent. Math.* **76**, 357–364.
M. GAGE (1991). References for geometric evolution equations. Dept. of Math., Univ. of Rochester.
D. GILBARG & N.S. TRUDINGER (1977). *Elliptic Partial Differential Equations of Second Order*. Berlin: Springer-Verlag.
E. GIUSTI (1984). *Minimal Surfaces and Functions of Bounded Variation*. Boston: Birkhäuser.
A.A. GRIGOR'YAN (1985). Isoperimetric inequalities for Riemannian products (in Russian). *Mat. Zametki* **38**. *Eng. Transl. Math. Notes* **38**, 849–854. **MR:** 87g53062.

A.A. GRIGOR'YAN (1986). On stochastically complete manifolds. *Soviet Math. Dokl.* **290**. *AMS Transl.* **34** (1987), 310–313.

A.A. GRIGOR'YAN (1991). The heat equation on noncompact Riemannian manifolds. *Mat. Sb.* **182**. *AMS Transl.* **72** (1992), 47–77.

A.A. GRIGOR'YAN (1994a). Gaussian upper bounds for the heat kernel and for its derivatives on a Riemannian manifold. In *Classical and Modern Potential Theory and Applications*, K. GOWRISANKARAN et al., eds., 237–252. Kluwer Academic.

A.A. GRIGOR'YAN (1994b). Heat kernel on a manifold with a local Harnack inequality. *Comm. Anal. Geom.* **2**, 111–138.

A.A. GRIGOR'YAN (1994c). Heat kernel upper bounds on a non-compact manifold. *Rev. Mat. Iberoamer.* **10**, 395–452.

A.A. GRIGOR'YAN (1999). Analytic and geometric background of recurrence and non-explosion of the Brownian motion on Riemannian manifolds. *Bull. Amer. Math. Soc.* **36**, 135–249.

A.A. GRIGOR'YAN & M. NOGUCHI (1998). The heat kernel on hyperbolic space. *J. London Math. Soc.* **30**, 643–650.

M. GROMOV (1986). Isoperimetric inequalities in Riemannian manifolds. In *Asymptotic Theory of Finite Dimensional Normed Spaces*, Lecture Notes Math. **1200**, Appendix I, 114–129. Berlin: Springer-Verlag.

H. HADWIGER (1957). *Vorlesungen über Inhalt, Oberfläche, und Isoperimetrie*. Berlin: Springer-Verlag.

R.Z. HASMINKII (1960). Ergodic properties of recurrent diffusion processes and stabilisation of the solution of the Cauchy problem for parabolic equations. *Theor. Prob. Anal.* **5**, 179–196.

E. HEBEY (1999). Nonlinear Analysis on Manifolds: Sobolev Spaces and Inequalities. *Inst. Math. Sci.* New York: Courant.

M.W. HIRSCH (1976). *Differential Topology*. New York: Springer-Verlag.

I. HOLOPAINEN (1994). Rough isometries and p-harmonic functions with finite Dirichlet integral. *Rev. Math. Iberoamer.* **217**, 459–477.

H. HOPF (1950). Über Flächen mit einer Relation zwischen den Hauptkrümmungen. *Math. Nachr.* **4**, (1950–1), 232–249.

H. HOPF & W. RINOW (1931). Über den Begriff der vollständigen differential-geometrischen Flächen. *Comment. Mat. Helv.* **3**, 209–225.

H. HOWARDS, M. HUTCHINGS, & F. MORGAN (1999). The isoperimetric problem on surfaces. *Amer. Math. Monthly* **106**, 430–439.

A. HURWITZ (1901). Sur le problème des isopérimétres. *C. R. Acad. Sci. Paris* **132**, 401–403.

K. ICHIHARA (1982). Curvature, geodesics, and the Brownian motion on a Riemannain manifold ii. *Nagoya Math. J.* **87**, 115–125.

K. ICHIHARA (1984). Explosion problems for symmetric diffusion processes. *Proc. Japan Acad.* **60**, 243–245.

K. ICHIHARA (1986). Explosion problems for symmetric diffusion processes. *Trans. Amer. Math. Soc.* **298**, 515–536.

M. KANAI (1985). Rough isometries, and combinatorial approximations of geometries of noncompact Riemannian manifolds. *J. Math. Soc. Japan* **37**, 391–413.

M. KANAI (1986a). Analytic inequalities, and rough isometries between noncompact Riemannian manifolds. In *Curvature and Topology of Riemannian Manifolds*, K. SHIOHAMA, T. SAKAI, & T. SUNADA, eds., Lecture Notes Math. **1201**, 122–137. Berlin: Springer-Verlag.

M. KANAI (1986b). Rough isometries and the parabolicity of Riemannian manifolds. *J. Math. Soc. Japan* **38**, 227–238.

L. KARP (1984). Noncompact manifolds with purely continuous spectrum. *Mich. Math. J.* **31**, 339–347.

L. KARP & P. LI. The heat equation on complete Riemannian manifolds. Preprint.

B. KAWOHL (1985). *Rearrangements and Convexity of Level Sets in PDE*, Lecture Notes Math. **1150**. Berlin: Springer-Verlag.

H. KNOTHE (1957). Contributions to the theory of convex bodies. *Mich. Math. J.* **4**, 39–52.

E. KRAHN (1925). Über eine von Rayleigh formulierte Minimaleigenschafte des Kreises. *Math. Ann.* **94**, 97–100.

P. LI (1980). On the Sobolev constant and the p-spectrum of a compact Riemannian manifold. *Ann. Sci. Éc. Norm. Sup. Paris* **13**, 419–435.

P. LI (1986). Large time behavior of the heat equation on complete manifolds with nonnegative Ricci curvature. *Ann. Math.* **124**, 1–21.

P. LI & S.T. YAU (1986). On the parabolic kernel of the Schrödinger operator. *Acta Math.* **156**, 153–201.

E.H. LIEB & M. LOSS (1996). *Analysis*. Providence, RI: American Math. Soc.

H. LIEBMANN (1900). Über die Verbiegung der glossenen Flächen positiver Krümmung. *Math. Ann.* **53**, 81–112.

V.G. MAZ'YA (1960). Classes of domains and embedding theorems for functional spaces (in Russian). *Dokl. Acad. Nauk. SSSR* **133**, 527–530. Engl. transl., *Soviet Math. Dokl.* **1** (1961), 882–885.

J. MILNOR (1968). A note on curvature and the fundamental group. *J. Diff. Geom.* **2**, 1–7.

F. MORGAN (1988). *Geometric Measure Theory: A Beginner's Guide*. Boston: Academic Press.

J. MOSER (1964). A Harnack inequality for parabolic differential equations. *Comm. Pure Appl. Math.* **17**, 101–134.

J. MOSSINO (1984). *Inégalités isopérimétriques et application en physique*. Paris: Herman.

R. NARASIMHAN (1968). *Analysis on Real and Complex Manifolds*. Amsterdam: North Holland.

J. NASH (1958). Continuity of solutions of parabolic and elliptic equations. *Amer. J. Math.* **80**, 931–954.

M.N. NDUMU (1986). An elementary formula for the Dirichlet heat kernel on Riemannian manifolds. In *From Local Times to Global Geometry, Control and Physics*, K.D. ELWORTHY, ed., π-Pitman Res. Notes Math. **150**, 320–328. Essex: Longman Scientific & Technical.

L. NIRENBERG (1959). On elliptic partial differential equations. *Ann. Scuola Norm. Sup. Pisa* **13**, 115–162.

J. OPREA (1997). *Differential Geometry and its Applications*. Upper Saddle River, NJ: Prentice Hall.

Y. OSHIMA (1992). On conservativeness and recurrence criteria of Markov processes. Preprint. *Potential Analysis* **1**, 115–131.

R. OSSERMAN (1978). The isoperimetric inequality. *Bull. Amer. Math. Soc.* **84**, 1182–1238.

R. OSSERMAN (1979). Bonnesen-style inequalities. *Amer. Math. Monthly* **86**, 1–29.

M. PANG (1988). L^1 properties of two classes of singular second order elliptic operators. *J. London Math. Soc.* **38**, 525–543.

M. PANG (1996). L^1 and L^2 spectral properties of a class of singular second order elliptic operators with measurable coefficients on $\mathbb{R}^n$. *J. Diff. Eq.* **129**, 1–17.

P. PANSU (1982). Une inegalité isopérimètrique sur le groupe de Heisenberg. *C. R. Acad. Sci. Paris* **295**, 127–130.

L.E. PAYNE (1967). Isoperimetric inequalities and their applications. *SIAM Rev.* **9**, 453–488.

G. PÓLYA & G. SZEGÖ (1951). *Isoperimetric Inequalities in Mathematical Physics*, Ann. Math. Studies **27**. Princeton, NJ: Princeton Univ. Press.

M.H. PROTTER & H.F. WEINBERGER (1984). *Maximum Principles in Differential Equations*. New York: Springer-Verlag, 2nd edn.

(Lord) J.W.S. RAYLEIGH (1877). *The Theory of Sound*. New York: Macmillan. Reprinted, New York: Dover (1945).

M. REED & B. SIMON (1975). *Functional Analysis* (*Methods of Mathematical Physics, Vol. II*). New York: Academic Press.

M. REED & B. SIMON (1980). *Functional Analysis* (*Methods of Mathematical Physics, Vol. I*). New York: Academic Press, 2nd edn.

R.C. REILLY (1977). Applications of the Hessian operator in a Riemannian manifold. *Indiana Univ. Math. J.* **26**, 459–472.

F. RIESZ & B. SZ. NAGY (1955). *Functional Analysis*. New York: Frederick Ungar.

A. ROS (1988). Compact surfaces with constant scalar curvature and a congruence theorem. *J. Diff. Geom.* **27**, 215–220.

W. RUDIN (1966). *Real and Complex Analysis*. New York: McGraw-Hill.

L. SALOFF-COSTE (1992). A note on Poincaré, Sobolev, and Harnack inequalities. *Duke Math. J.* **65**, International Mathematics Research Notices, 27–38.

E. SCHMIDT (1948, 1949). Der Brunn–Minkowskische Satz und sein Spiegel-theorem sowie die isoperimetrische Eigenschaft der Kugel in der euklidischen und nichteuklidischen Geometrie I, II. *Math. Nachr.* **1**, 81–157; **2**, 171–244.

L.M. SIMON (1984). *Lectures on Geometric Measure Theory*. Canberra, Australia: Centre Math. Anal., Australian Nat. Univ.

E.M. STEIN (1970). *Topics in Harmonic Analaysis*, Ann. Math. Studies **63**. Princeton, NJ: Princeton Univ. Press.

J. STEINER (1838). Einfache Beweise der isoperimetrische Hauptsätze. *J. Reine Angew. Math.* **18**, 281–296. Reprinted, *Gesammelte Werke*. Bronx, NY: Chelsea (1971) (reprint of 1881–1882 edn.), Vol. II, 75–91.

J. STEINER (1881). Über Maximum und Minimum. In *Gesammelte Werke*. Bronx, NY: Chelsea (1971) (reprint of 1881–1882 edn.).

D. SULLIVAN (1987). Related aspects of positivity in Riemannian geometry. *J. Diff. Geom.* **25**, 327–351.

M. TAKEDA (1989). On a martingale method for symmetric diffusion processes and its applications. *Osaka Math. J.* **26**, 605–623.

M. TAKEDA (1991). On the conservativeness of the Brownian motion on Riemannian manifolds. *Bull. London Math. Soc.* **23**, 86–88.

G. TALENTI (1993). The standard isoperimetric theorem. In *Handbook of Convex Geometry*, Vol. A, P.M. Gruber & J.M. Wills, eds., 73–123. Amsterdam: North Holland.

P. TOPPING (1997). The optimal constant in Wente's L^∞ estimate. *Comment. Mat. Helv.* **139**, 316–328.

P. TOPPING (1998). Mean curvature flow and geometric inequalities. *J. Reine Angew. Math.* **503**, 47–61.

P. TOPPING (1999). The isoperimetric inequality on a surface. *Manuscripta Math.* **100**, 23–33.

N.TH. VAROPOULOS (1983). Potential theory and diffusion on Riemannian manifolds. In *Conf. on Harmonic Analysis in Honor of A. Zygmund*, W. BECKNER et al., eds. Wadsworth.

N.TH. VAROPOULOS (1984). Brownian motion and random walks on manifolds. *Ann. Inst. Fourier Grenoble* **34**, 243–269.

N.TH. VAROPOULOS (1985). Isoperimetric inequalities and Markov chains. *J. Fcnl. Anal.* **63**, 240–260.

N.TH. VAROPOULOS (1986). Hardy-Littlewood theory for semigroups. *J. Fcnl. Anal.* **66**, 406–431.

N.TH. VAROPOULOS (1989). Small time Gaussian estimates of heat diffusion kernels. Part i: The semigroup technique. *Bull. Soc. Math. France* **113**, 253–277.

S.T. YAU (1975). Isoperimetric constants and the first eigenvalue of a compact manifold. *Ann. Sci. Éc. Norm. Sup. Paris* **8**, 487–507.

S.T. YAU (1976). Some function-theoretic properties of complete Riemannian manifolds, and their applications to geometry. *Indiana Univ. Math. J.* **25**, 659–670. Also cf. *Ibid.* **31** (1982), 307.

S.T. YAU (1978). On the heat kernel of a complete Riemannian manifold. *J. Math. Pures Appl.* **57**, 191–201.

S.T. YAU (1994). On the Harnack inequalities of partial differential equations. *Comm. Anal. and Geom.* **2**, 431–450.

S.T. YAU (1995). Harnack inequality for non-self-adjoint evolution equations. *Math. Res. Lett.* **2**, 387–399.

K. YOSIDA (1978). *Functional Analysis*. Berlin: Springer-Verlag, 5th edn.

W.P. ZIEMER (1989). *Weakly Differentiable Functions*. New York: Springer-Verlag.

Author Index

Subject Index

Printed in the United States
By Bookmasters